LONDON MATHEMATICAL SOCIETY STUDENT TEXTS

Managing Editor: Ian J. Leary,
Mathematical Sciences, University of Southampton, UK

London Mathematical Society Student Texts 106

Compact Matrix Quantum Groups and Their Combinatorics

AMAURY FRESLON
University of Paris-Saclay

CAMBRIDGE
UNIVERSITY PRESS

Shaftesbury Road, Cambridge CB2 8EA, United Kingdom

One Liberty Plaza, 20th Floor, New York, NY 10006, USA

477 Williamstown Road, Port Melbourne, VIC 3207, Australia

314–321, 3rd Floor, Plot 3, Splendor Forum, Jasola District Centre,
New Delhi – 110025, India

103 Penang Road, #05–06/07, Visioncrest Commercial, Singapore 238467

Cambridge University Press is part of Cambridge University Press & Assessment,
a department of the University of Cambridge.

We share the University's mission to contribute to society through the pursuit of
education, learning and research at the highest international levels of excellence.

www.cambridge.org
Information on this title: www.cambridge.org/9781009345736
DOI: 10.1017/9781009345705

First published 2023

A catalogue record for this publication is available from the British Library.

A Cataloging-in-Publication data record for this book is available from the Library of
Congress

ISBN 978-1-009-34573-6 Hardback
ISBN 978-1-009-34569-9 Paperback

Contents

Preface

The term 'quantum group' carries an air of mystery and physics, making it both frightening and fascinating. It is a difficult task to give an elementary explanation of what a quantum group is. In fact, the purpose of the first chapter of this book is to motivate and introduce as clearly as possible that notion. Nevertheless, the reader deserves a few explanations before trying to read it, and we will therefore try to give some elements of context concerning the theory of quantum groups.

The notion of quantum group is not an easy one to grasp. One of the main reasons for that is that there is, in a sense, no definition of 'quantum group' in general but rather a number of different families of objects which are given such a name. Another reason is that the definition of all these families of objects require tools beyond standard undergraduate mathematics. We will nevertheless try to give some intuition of what is going on, focusing, since this is our setting, on the compact case.

Let us start with a compact abelian group G and consider its *Pontryagin dual*, that is to say, the set $\widehat{G}$ of continuous group homomorphisms from G to the group $\mathbb{T}$ of complex numbers of modulus one. It is easy to see that $\widehat{G}$ is again an abelian group for pointwise multiplication. Moreover, the compact-open topology makes it a discrete topological space. Conversely, if we start with a discrete abelian group, then the Pontryagin dual will be a compact abelian group. The *Pontryagin duality theorem* then states that $\widehat{\widehat{G}} = G$.

To see why Pontryagin duality is interesting, we will look at the fundamental example $G = \mathbb{T}$. In that case, any group homomorphism $\varphi \colon \mathbb{T} \to \mathbb{T}$ is of the form $\varphi_n \colon z \mapsto z^n$ for some $n \in \mathbf{Z}$, so that $\widehat{\mathbb{T}} = \mathbf{Z}$. Consider now a finitely supported sequence $(a_n)_{n \in \mathbf{Z}}$. Seeing it as a function $a \colon \mathbf{Z} \to \mathbf{C}$, we can change it into a function $\widehat{a} \colon \mathbb{T} \to \mathbf{C}$ through the formula

$$\widehat{a} \colon z \mapsto \sum_{n \in \mathbf{Z}} a(n) \varphi_n(z) = \sum_{n \in \mathbf{Z}} a_n z^n.$$

In other words, finitely supported sequences and trigonometric polynomials can be put in one-to-one correspondence by Pontryagin duality. It is then natural to investigate what happens if we try to extend that correspondence to spaces of sequences in which the finitely supported ones are dense (for instance, square summable or continuous), and this is the field of Fourier analysis.

The upshot of the previous short discussion is that Pontryagin duality on $\mathbb{T}$ is an abstract, group-theoretic version of the theory of Fourier series. Thus, Pontryagin duality provides in full generality an analogue of the Fourier transform for any compact (or even locally compact, but this is another story) abelian group. It is then tempting to ask whether something similar is doable for non-abelian groups. Unfortunately, any group homomorphism $G \to \mathbb{T}$ must vanish on commutators since $\mathbb{T}$ is abelian, and so one has to find a different approach.

The theory of compact matrix quantum groups was initiated by S. L. Woronowicz during the 1980s [75], under the name *compact matrix pseudo-groups*, to provide an approach to Pontryagin duality which works for arbitrary compact groups. Not only was it successful in doing so, but it also gave rise to many examples which do not relate to any compact or discrete group, the prominent ones being obtained by 'deforming' (see Section C.2 for an illustration) classical Lie groups like $SU(N)$. This was the birth of the theory.

Around the same time, V. Drinfeld coined the term *quantum group* in [35] to denote constructions of algebraic entities called *Hopf algebras* (see Appendix C for more on that notion), generalising the deformations of Lie algebras introduced by M. Jimbo in [44]. The correspondence between Lie groups and Lie algebras naturally suggests that there should be strong connections between the algebraic approach involving, for instance, the Lie algebra $\mathfrak{su}_N$ and the deformations of $SU(N)$ constructed by S. L. Woronowicz. And indeed, it was clear by then that there was a kind of duality relationship between both constructions, so that the terminology evolved in the work of S. L. Woronowicz to *compact quantum groups*. But despite their similarities, the two theories grew in different directions over the years so that they nowadays form distinct research topics.

The algebraic theory expanded quickly, partly because there was no strict definition of a quantum group in that setting, and so the term was applied to more and more classes of Hopf algebras with interesting properties, often connected to problems in mathematical physics. In the analytical theory, there was a precise definition to satisfy, in which the new algebraic examples did not necessarily fit. Consequently, the search for examples was slower. One important step (in particular, as far as this text is concerned) was the definition by S. Wang in [70] and [72] of the *universal compact quantum groups* which are nowadays denoted by S_N^+, O_N^+ and U_N^+. These objects, about which many things remain mysterious, have been a subject of intense study for the last twenty years and have led to important advances in the field, as we will see. Their crucial feature is that they are not deformations of classical groups, but rather 'liberations' obtained by removing the commutation relations in a suitable presentation of a classical group algebra. (This idea will be illustrated in Chapter 7.)

The first step in the study of a compact quantum group is the computation of its representation theory. For the previously mentioned examples of S. Wang, the representation theory was computed by T. Banica in a series of seminal works [3], [4] and [6]. These works revealed important combinatorial structures that proved crucial in the proof of many properties of the representation theory. The nice feature of these combinatorial structures is that they are based on *diagrams* in the sense that one represents operators by some drawings and can do computations directly on the drawings by simply using intuitive pictorial properties. After a period of maturation, the connection between these combinatorial structures and compact quantum groups was formalised by T. Banica and R. Speicher in [18] in a way which is both natural and quite elementary. The fundamental idea is to express the relations defining the algebras of functions on the quantum groups in terms of intertwiners, themselves defined from partitions of finite sets, which naturally give back the diagrams. Although the difference may seem small, partitions offer great flexibility for computations and suggest the use of many results from combinatorics and free probability theory. This change of paradigm was a milestone in the study of compact quantum groups and has had many important consequences. Moreover, it continues to lead to unexpected connections with other fields of mathematics like quantum information theory (see, for instance, the beginning of Chapter 1, as well as Chapter 8).

At the date of this writing, the study of compact quantum groups defined through the combinatorics of partitions is a very active topic with a well-established theoretical basis. However, even though there exist books on the general theory of compact quantum groups, there is none detailing the combinatorial aspects and their connections to other subjects. The purpose of this book is therefore to give a comprehensive introduction to compact matrix quantum groups as well as to the combinatorial theory of partition quantum groups. It has been written and designed for students and should be particularly fit for a one-semester graduate course as well as for self-study by a motivated graduate (or even a highly motivated undergraduate) student. It is nevertheless our hope that it can also be of use to more advanced mathematicians who want to learn the subject or have a glimpse of its recent outcomes, but are afraid to get lost in the growing scientific literature.

Before outlining the contents of the text, let us comment on its differences with other books on the subject. As already mentioned, the theory of compact quantum groups was initially developed in the setting of operator algebras, and that theory is detailed in [60] and in [69]. It is known, however, that there is an equivalent algebraic approach to the subject through specific Hopf algebras called *CQG-algebras* (see Appendix C), which is explained in detail in [69]. These introductory books share the same caveat: the reader is supposed to have learned some prerequisites, operator algebras or Hopf algebras, before starting to study the main subject. Because we wanted to teach a course on compact quantum groups in a university where no course on operator algebras or Hopf algebras was available, we had to find another way.

It turns out that by focusing on compact *matrix* quantum groups (i.e. those having a faithful finite-dimensional representation), one can prove all the important theorems without resorting to any result from the theory of operator algebras or Hopf algebras. As a consequence, we are able to provide here a text which is completely self-contained for anyone having completed undegradu-ate studies in mathematics in any university. More precisely, the prerequisites for reading the book are only basic algebra, possibly (but not necessarily) including some elementary notions concerning the representation theory of finite-dimensional complex algebras. To make the book as self-contained as possible, we have included, where necessary, a short treatment of some more advanced algebraic notions like universal algebras or tensor products as well as an appendix containing statements and proofs of two results from represen-tation theory (Burnside's theorem and the double commutant theorem) which are needed in the course of the book.

Let us now briefly outline the contents and organisation of the text. An important point is that since it was designed for a course, it is intended to be read rather linearly, at least as far as the first two parts are concerned. For clarity, it is divided into three main parts, themselves split into chapters.

- Part I contains the basic theory of compact matrix quantum groups, going from the very definition of these objects in Chapter 1 to the complete description of their representation theory in Chapter 2. The definition is motivated by the example of quantum permutations and their connection with quantum information theory through the *graph isomorphism game*. Along the way, we briefly introduce universal algebras and tensor products (over the complex numbers) in case the reader is not familiar with these notions.

- Part II introduces the connection between partitions of finite sets and quantum groups through the notion of *partition quantum groups*. Chapter 3 introduces all the necessary material as well as a proof of the celebrated Tannaka–Krein–Woronowicz theorem, which is the cornerstone of the definition of partition quantum groups. In Chapter 4 we give a detailed treatment of the representation theory of partition quantum groups, which is then applied to the basic examples S_N^+ and O_N^+. We conclude this part with the construction of the analogue of the Haar measure. We take this occasion to make the bridge with the analytical approach to quantum groups, though without full proofs. We also give some applications to non-commutative probability theory.
- Part III contains extra topics of two different types. Chapters 6 and 7 introduce the coloured versions of partition quantum groups which enable one to deal with the unitary quantum groups U_N^+ and H_N^{s+}. This is also an occasion for discussing side topics like free complexification, free wreath products and quantum automorphism groups of graphs, which are not available in any book so far. Chapter 8, on the other hand, treats the link between quantum permutation groups and quantum information theory. This motivates questions of residual finite-dimensionality which are detailed in the final section. This shows the power of the combinatorial methods developed in the text.
- At the end of the book there are three appendices. Appendix A gives the statements and proofs of two elementary results on complex representations of matrix algebras which are needed in the text. Then, Appendix B shows how one can recover the usual description of the representation theory of classical compact matrix groups out of what has been done in the quantum setting. Finally, Appendix C is an invitation to general compact quantum groups, where we indicate how one can build a larger theory from what has been done in the main text and hint at some subtle issues that then arise.

The material for this book comes from two sources. The first one is a mini-course taught in the masterclass 'Subfactors and Quantum Groups' at Copenhagen in May 2019. The second is a graduate course entitled 'Introduction to Compact Matrix Quantum Groups and Their Combinatorics' taught at University Paris-Saclay in 2020 and 2021. It is a pleasure to thank both institutions, and the organisers of the masterclass, Rubén Martos and Ryszard Nest, for giving me opportunities to teach this subject. I am also deeply indebted to all those who have journeyed with me in the fascinating world of compact quantum groups, and in particular to Etienne Blanchard, Roland Vergnioux, Pierre Fima, Julien Bichon, Adam Skalski and Moritz Weber.

Part I

GETTING STARTED

1

Introducing Quantum Groups

The purpose of the first part of this text is to introduce objects called *compact quantum groups* and to deal in full detail with their algebraic aspects and in particular their representation theory. It turns out that many interesting examples of compact quantum groups fall into a specific subclass called *compact matrix quantum groups*. This subclass has the advantage of being more intuitive, as well as allowing for a simplified treatment of the whole theory. We will therefore restrict to it, and the connection with the more general setting of compact quantum groups will be briefly explained in Appendix C.

We believe that there is no better way of introducing a new concept than giving examples. We will therefore spend some time introducing one of the most important families of examples of compact matrix quantum groups, first defined by S. Wang in [72], called the *quantum permutation groups*.

1.1 The Graph Isomorphism Game

There are several ways of motivating the definition of quantum permutation groups, because these objects are related to several important notions like quantum isometry groups in the sense of non-commutative geometry (see, for instance, [22] or [7]) or quantum exchangeability in the sense of free probability (see, for instance, [50]). In this text, we will start from a recent connection, first made explicit in [53], between quantum permutation groups and quantum information theory. That connection appears through a *game* which we now describe.

As always in quantum information theory, the game is played by two players named *Alice* (denoted by A) and *Bob* (denoted by B). In this so-called *graph isomorphism game*, they cooperate to win against the *Referee* (denoted by R)

3

leading the game. The rules are given by two finite graphs,[1] X and Y, with vertex sets $V(X)$ and $V(Y)$ respectively having the same cardinality, which are known to A and B. At each round of the game, R sends a vertex $v_A \in V(X)$ to A and a vertex $v_B \in V(X)$ to B. Each of them answers with a vertex $w_A \in V(Y)$, $w_B \in V(Y)$ of the other graph, and they win the round if the following condition is matched.

Winning condition: 'The relation[2] between v_A and v_B is the same as the one between w_A and w_B.'[3]

The crucial point is that once the game starts, A and B **cannot communicate in any way**. The situation can be summarised by the following picture:

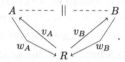

The question one asks is then, under which condition on the graphs X and Y can the players devise a strategy which wins whatever the given vertices are? It is not very difficult to see that the answer is the following (see Exercise 8.1 for a proof).

Proposition 1.1 *There exists a perfect classical strategy if and only if X and Y are isomorphic.*

This settles the problem in classical information theory, but in the quantum world, A and B can refine their strategy without communicating through the use of *entanglement*. This means that they can set up a quantum mechanical system and then split it into two parts, such that manipulating one part instantly modifies the other one. We will not go into the details right now, but it turns out that this gives more strategies, which are said to be *quantum*.[4] By using these

---

[1] The following discussion concerning graphs is only intended to motivate the introduction of quantum permutation groups, hence we do not give precise definitions. A rigorous treatment will be given in Chapter 8.

[2] Here, by 'relation' we mean either being equal, being adjacent or not being adjacent.

[3] This is not the most general version of the graph isomorphism game. We refer the reader to [2] for a more comprehensive exposition.

[4] The concept of quantum strategy turns out to be quite subtle, depending on the type of operators allowed. We here use the term in a purposely vague sense and refer the reader to the discussion at the beginning of Chapter 8 for more details.

quantum strategies, the previous proposition can be improved. Before giving a precise statement, let us fix some notations.

- Given a Hilbert space H, we denote by $\mathcal{B}(H)$ the algebra of bounded (i.e. continuous) linear maps from H to H;
- Given a graph X, we denote by A_X the adjacency matrix of A.

The following result is a combination of [2, theorem 5.8] and [53, theorem 4.4].

Theorem 1.2 (Atserias–Lupini–Mančinska–Roberson–Šamal–Severini–Varvitsiotis) *There is a perfect* quantum *strategy if and only if there exists a matrix* $P = (p_{ij})_{1 \leqslant i,j \leqslant N}$ *with coefficients in* $\mathcal{B}(H)$ *for some Hilbert space* H, *such that*

- p_{ij} *is an orthogonal projection for all* $1 \leqslant i, j \leqslant N$;
- $\displaystyle\sum_{k=1}^{N} p_{ik} = \mathrm{Id}_H = \sum_{k=1}^{N} p_{kj}$ *for all* $1 \leqslant i, j \leqslant N$;
- $A_X P = P A_Y$.

The proof of this result involves several tools coming from quantum information theory, graph theory and compact quantum group theory. For those reasons, we postpone it to Chapter 8.

Remark 1.3 From the perspective of quantum physics, this definition is at least reasonable. Indeed, a family of orthogonal projections summing up to one is a particular instance of a *Positive Operator Valued Measure* (see Definition 8.1). We are therefore considering a collection of such objects with compatibility conditions coming from the graphs.

Remark 1.4 It is not straightforward to produce a pair of graphs for which there is a perfect quantum strategy but no classical one. The first example, given in [2, section 6.2], has 24 vertices and is the smallest known at the time of this writing.

An intriguing point of Theorem 1.2 is the operator-valued matrices which appear in the statement. To understand them, let us consider the case $H = \mathbf{C}$. Then, the coefficients are scalars, and since they are projections, they all equal either 0 or 1. Moreover, the sum over any row is 1, hence there is exactly one non-zero coefficient on each row. The same being true for the columns, we have a permutation matrix! We should therefore think of the operator-valued matrices as quantum versions of permutations, and this leads to the following definition.

Definition 1.5 Let H be a Hilbert space. A *quantum permutation matrix* in H is a matrix $P = (p_{ij})_{1 \leqslant i,j \leqslant N}$ with coefficients in $\mathcal{B}(H)$ such that

- p_{ij} is an orthogonal projection for all $1 \leqslant i,j \leqslant N$;
- $\displaystyle\sum_{k=1}^{N} p_{ik} = \mathrm{Id}_H = \sum_{k=1}^{N} p_{kj}$ for all $1 \leqslant i,j \leqslant N$.

Moreover, with this point of view the last point of Theorem 1.2 has a nice interpretation. To explain it, let us first do a little computation.

Exercise 1.1 Let X, Y be graphs on N vertices and let $\sigma \in S_N$. Numbering the vertices from 1 to N, σ induces a bijection between the vertex sets of X and Y. Prove this is a graph isomorphism if and only if

$$A_X P_\sigma = P_\sigma A_Y.$$

Solution Denoting by $E(X)$ and $E(Y)$ the edge sets of X and Y respectively, the (i,j)-th coefficient of $A_X P_\sigma$ is

$$\sum_{k=1}^{N} (A_X)_{ik} (P_\sigma)_{kj} = \sum_{k=1}^{N} \delta_{(i,k) \in E(X)} \delta_{\sigma(k)j}$$

$$= \delta_{(i,\sigma^{-1}(j)) \in E(X)},$$

while the corresponding coefficient of $P_\sigma A_Y$ is

$$\sum_{k=1}^{N} (P_\sigma)_{ik} (A_X)_{kj} = \sum_{k=1}^{N} \delta_{(k,j) \in E(Y)} \delta_{\sigma(i)k}$$

$$= \delta_{(\sigma(i),j) \in E(Y)}.$$

These are equal if and only if

$$(i, \sigma^{-1}(j)) \in E(X) \Leftrightarrow (\sigma(i), j) \in E(Y).$$

Setting $k = \sigma^{-1}(j)$, the condition is equivalent to

$$(i, k) \in E(X) \Leftrightarrow (\sigma(i), \sigma(k)) \in E(Y),$$

which precisely means that σ induces a graph automorphism. □

In view of this, the last point of Theorem 1.2 can be interpreted as saying that the quantum permutation respects the edges of the graphs, so that one says that the graphs are *quantum isomorphic*.

1.2 The Quantum Permutation Algebra

1.2.1 Universal Definition

The brief discussion of Section 1.1 suggests that quantum permutation matrices are interesting objects which require further study. However, their definition lacks several important features of classical permutation matrices. In particular, there is no obvious way to 'compose' quantum permutation matrices, especially if they do not act on the same Hilbert space, so that one could recover an analogue of the group structure of permutations. To overcome this problem, it is quite natural from an (operator) algebraic point of view to introduce a universal object associated to quantum permutation matrices. Note that, in order to translate the fact that the operators p_{ij} are orthogonal projections, it is convenient to use the natural involution on $\mathcal{B}(H)$ given by taking adjoints. For this purpose, we will consider *-*algebras*, that is to say, complex algebras $\mathcal{A}$ endowed with an anti-linear and anti-multiplicative involution $x \mapsto x^*$.

Definition 1.6 Let $\mathcal{A}_s(N)$ be the universal *-algebra[5] generated by N^2 elements $(p_{ij})_{1 \leqslant i,j \leqslant N}$ such that

1. $p_{ij}^2 = p_{ij} = p_{ij}^*$;
2. For all $1 \leqslant i, j \leqslant N$, $\sum_{k=1}^{N} p_{ik} = 1 = \sum_{k=1}^{N} p_{kj}$;
3. For all $1 \leqslant i, j, k, \ell \leqslant N$, $p_{ij}p_{ik} = \delta_{jk}p_{ij}$ and $p_{ij}p_{\ell j} = \delta_{i\ell}p_{ij}$.

This will be called the *quantum permutation algebra* on N points.

Remark 1.7 The third condition in the definition may seem redundant since it is automatically satisfied for projections in a Hilbert space. However, a *-algebra may not have a faithful representation on a Hilbert space, hence Condition (3) does not necessarily follow from the two other ones.

Definition 1.6 refers to a so-called *universal object* and we will give a few details about it for the sake of completeness. This roughly means that we want the 'largest possible' algebra generated by elements that we call p_{ij} and such that the relations in the statement are satisfied. Proving that such an object exists and is well-behaved is not very difficult but requires a bit of abstraction. The intuition is to start with a full algebra of *non-commutative* polynomials and

---

[5] As the following relations show, we are in fact considering, here and throughout the text, universal *unital* algebras. For convenience we will drop the term 'unital' because we will never consider non-unital algebras.

then quotient by the desired relations. As for usual polynomials, it is easier to use a definition based on sequences.

Definition 1.8 Given a set I, we denote by $\mathcal{U}_I$ the complex vector space of all finite linear combinations of finite sequences of elements of I. It is endowed with the algebra structure induced by the concatenation of sequences, with the empty sequence acting as a unit.

If we denote by X_i the sequence (i), then the elements $(X_i)_{i \in I}$ generate $\mathcal{U}_I$ and any element can therefore be written as a linear combination of products of these generators, the latter products being called *monomials*. Note that this decomposition is unique up to the commutativity of addition. We therefore may, and should (and will) see $\mathcal{U}_I$ as the algebra of all non-commutative polynomials over the set I, and denote it by $\mathbf{C}\langle X_i \mid i \in I \rangle$. For our purpose, we will turn this into a $*$-algebra by setting $X_i^* = X_i$ for all $i \in I$.

Assuming now that we have a subset $\mathcal{R} \subset \mathbf{C}\langle X_i \mid i \in I \rangle$ called *relations*, here is how we can build our universal object.

Definition 1.9 The *universal $*$-algebra* generated by $(X_i)_{i \in I}$ with the relations $\mathcal{R}$ is the quotient of $\mathbf{C}\langle X_i \mid i \in I \rangle$ by the intersection of all the $*$-ideals containing $\mathcal{R}$. We will again denote its generators by $(X_i)_{i \in I}$.

That this is the correct definition is confirmed by the following *universal property*.

Exercise 1.2 Let $\mathcal{A}$ be a $*$-algebra generated by elements $(x_i)_{i \in I}$ and let $\mathcal{R} \subset \mathbf{C}\langle X_i \mid i \in I \rangle$. Prove that if $P(x_i) = 0$ for all $P \in \mathcal{R}$, then there exists a unique surjective $*$-homomorphism from the universal $*$-algebra generated by $(X_i)_{i \in I}$ with the relations $\mathcal{R}$ to $\mathcal{A}$ mapping X_i to x_i.

Solution We first construct a $*$-homomorphism from $\mathbf{C}\langle X_i \mid i \in I \rangle$. The requirements of the statements force $\pi(X_i) = x_i$, and the fact that π is a $*$-algebra homomorphism uniquely determines it on the whole of $\mathbf{C}\langle X_i \mid i \in I \rangle$, that is,

$$\pi(X_{i_1} \cdots X_{i_n}) = x_{i_1} \cdots x_{i_n}.$$

Note that this makes sense because, by definition, the monomials are a basis of $\mathbf{C}\langle X_i \mid i \in I \rangle$. Moreover, it is surjective because the x_i's are generators. By assumption, $\ker(\pi)$ is a $*$-ideal containing $\mathcal{R}$, hence it also contains the intersection J of all the $*$-ideals containing it. As a consequence, π factors through $\mathbf{C}\langle X_i \mid i \in I \rangle / J$, which is precisely the universal $*$-algebra. $\quad\square$

We now have a nice object to study, but the link to the classical permutation group is somewhat blurred. To clear it up, let us consider the functions $c_{ij} \colon S_N \to \mathbf{C}$ defined by

$$c_{ij}(\sigma) = \delta_{\sigma(i)j}.$$

This is nothing but the function sending the permutation matrix of σ to its (i, j)-th coefficient. In particular, c_{ij} always takes the value 0 or 1, hence

$$c_{ij}^* = c_{ij} = c_{ij}^2.$$

Similarly, it is straightforward to check that Conditions (2) and (3) of Definition 1.6 are satisfied. Hence, by the universal property of Exercise 1.2, there is a unique $*$-homomorphism

$$\pi_{\mathrm{ab}} \colon \begin{cases} \mathcal{A}_s(N) & \to & F(S_N) \\ p_{ij} & \mapsto & c_{ij}, \end{cases}$$

where $F(S_N)$ is the algebra of all functions from S_N to $\mathbf{C}$. Moreover, since the functions c_{ij} obviously generate the whole algebra $F(S_N)$, π_{ab} is onto. The subscript 'ab' is meant to indicate that π_{ab} is, in fact, the abelianisation map, that is to say, the quotient by the ideal generated by all commutators. In other words, we are claiming that $F(S_N)$ is the largest possible commutative $*$-algebra satisfying the defining relations of $\mathcal{A}_s(N)$. The proof of that fact is an easy exercise that we leave to the curious reader.

Exercise 1.3 Let $\mathcal{B}_N$ be the universal $*$-algebra generated by N^2 elements $(p_{ij})_{1 \leqslant i,j \leqslant N}$ satisfying Conditions (1), (2) and (3) as well as the relations

$$p_{ij}p_{k\ell} = p_{k\ell}p_{ij},$$

for all $1 \leqslant i, j, k, \ell \leqslant N$.

1. For a permutation $\sigma \in S_N$, we set

$$p_\sigma = \prod_{i=1}^N p_{i\sigma(i)}.$$

Prove that $(p_\sigma)_{\sigma \in S_N}$ spans $\mathcal{B}_N$.
2. Deduce that there is a $*$-isomorphism $\mathcal{B}_N \to F(S_N)$ sending p_{ij} to c_{ij}.

Solution 1. Let us first observe that $\mathcal{B}_N$ is by definition spanned by monomials in the generators. Moreover, we claim that in such a monomial $p = p_{i_1 j_1} \cdots p_{i_k j_k}$, we may assume that $i_\ell \neq i_{\ell'}$ and $j_\ell \neq j_{\ell'}$ for all $\ell \neq \ell'$. Indeed, otherwise we can assume by commutativity that $\ell = \ell + 1$

and, without loss of generality, that $i_\ell = i_{\ell+1}$. It then follows from the defining relations that either $j_\ell = j_{\ell+1}$, in which case we can remove one of these two terms since $p_{i_\ell j_\ell}^2 = p_{i_\ell j_\ell}$, or $j_\ell \neq j_{\ell+1}$, in which case $p = 0$. A straightforward consequence of this is that, by the pigeonhole principle, $\mathcal{B}_N$ is spanned by monomials of length at most N.

Let us set denote by E the span of the elements in the statement. We will prove by induction on k that any monomial of length $N - k$ is in E, for $0 \leqslant k \leqslant N$. The case $k = 0$ follows from the observations in the previous paragraphs: since $(i_1, \dots, i_N)$ and $(j_1, \dots, j_N)$ are tuples of pairwise distinct elements of $\{1, \dots, N\}$, there exists a permutation $\sigma \in S_N$ such that $j_\ell = \sigma(i_\ell)$ for all $1 \leqslant \ell \leqslant N$. Assume now that the result holds for some k and consider a monomial

$$p = p_{i_1 j_1} \cdots p_{i_{N-k-1} j_{N-k-1}}.$$

Let us choose an element $i_{N-k} \in \{1, \cdots, N\} \setminus \{i_1 \cdots i_{N-k-1}\}$. Then,

$$p = \sum_{j=1}^N p_{i_1 j_1} \cdots p_{i_{N-k-1} j_{N-k-1}} p_{i_{N-k} j}$$

and the proof is complete.

2. By universality, there is a surjective $*$-homomorphism $\mathcal{B}_N \to F(S_N)$ sending p_{ij} to c_{ij}. But from the first question we know that

$$\dim(\mathcal{B}_N) \leqslant N! = \dim(F(S_N)),$$

therefore the surjection must be injective. □

We will now use this link to investigate a possible 'group-like' structure on $\mathcal{A}_s(N)$. At the level of the coefficient functions, the group law of S_N satisfies the equation

$$c_{ij}(\sigma_1 \sigma_2) = \sum_{k=1}^N c_{ik}(\sigma_1) c_{kj}(\sigma_2).$$

The trouble here is that the right-hand side is an element of $F(S_N \times S_N)$, which has no analogue in terms of quantum permutations so far. It would be more helpful to express the product solely in terms of $F(S_N)$. It turns out that there is an algebraic construction which exactly does this: the *tensor product*.

1.2.2 The Tensor Product

Our problem is to build the algebra of functions on $S_N \times S_N$ using only algebraic constructions on $F(S_N)$. One may try to consider the direct product

$F(S_N) \times F(S_N)$, but it has dimension $2N!$ while $F(S_N \times S_N)$ has dimension $(N!)^2$, so that we need something else. Let us nevertheless focus on the direct product to get some insight. Given two functions P and Q on S_N, we can see PQ as a two-variable function. However, the set theoretic map

$$\Phi \colon (P, Q) \in F(S_N) \times F(S_N) \mapsto PQ \in F(S_N \times S_N)$$

fails to be linear. Indeed, we have the two following issues: first,

$$\begin{aligned}
\Phi((P, Q) + (P', Q')) &= \Phi(P + P', Q + Q') \\
&= (P + P')(Q + Q') \\
&\neq PQ + P'Q' \\
&= \Phi(P, Q) + \Phi(P', Q')
\end{aligned}$$

and second

$$\begin{aligned}
\Phi(\lambda(P, Q)) &= \Phi(\lambda P, \lambda, Q) \\
&= \lambda^2 PQ \\
&\neq \lambda \Phi(P, Q).
\end{aligned}$$

In order to remedy this, we can use a universal construction, as we already did to define $\mathcal{A}_s(N)$. In other words, we will start from the largest vector space on which the map Φ can be defined as a linear map.

Definition 1.10 Given two vector spaces V and W, the *free vector space on* $V \times W$ is the vector space $\mathcal{F}(V \times W)$ of all finite linear combinations of elements of $V \times W$.

One must be careful that the elements of $V \times W$ form a basis of $\mathcal{F}(V \times W)$, hence

$$(v, w) + (v', w') \neq (v + v', w + w')$$

in that space. The point of this construction is that the map Φ, defined on $F(S_N) \times F(S_N)$ by $\Phi(P, Q) = PQ$, has by definition a unique extension to a linear map

$$\widetilde{\Phi} \colon \mathcal{F}(F(S_N) \times F(S_N)) \to F(S_N \times S_N).$$

The problem is, of course, that this map is far from injective, and we have to identify its kernel. Here are three obvious ways of building vectors on which $\widetilde{\Phi}$ vanishes:

- $\widetilde{\Phi}((P,Q) + (P,Q')) = PQ + PQ' = P(Q + Q') = \widetilde{\Phi}(P, Q + Q')$,
- $\widetilde{\Phi}((P,Q) + (P',Q)) = PQ + P'Q = (P + P')Q = \widetilde{\Phi}(P + P', Q)$,
- $\widetilde{\Phi}(\lambda P, Q) = \lambda PQ = \widetilde{\Phi}(P, \lambda Q)$.

The main result of this section is that this is enough to generate the kernel. Before proving this, let us give a formal definition.

Definition 1.11 Given two vector spaces V and W, we denote by $\mathcal{I}(V, W)$ the linear subspace of $\mathcal{F}(V \times W)$ spanned by the vectors

- $(v, w) + (v, w') - (v, w + w')$,
- $(v, w) + (v', w) - (v + v', w)$,
- $(\lambda v, w) - (v, \lambda w)$,

for all $(v, w) \in V \times W$. Then, the *tensor product* of V and W is the quotient vector space

$$V \otimes W = \mathcal{F}(V \times W)/\mathcal{I}(V, W).$$

The image of (v, w) in this quotient will be denoted by $v \otimes w$.

This construction may seem weird at first sight, since we are quotienting a 'huge' vector space by a 'huge' vector subspace. However, it turns out that the result is very tractable and perfectly fits our requirements. Before proving this, let us elaborate a bit more on the general construction by identifying a basis.

Proposition 1.12 *Let* $(e_i)_{i \in I}$ *and* $(f_j)_{j \in J}$ *be bases of* V *and* W *respectively. Then,*

$$(e_i \otimes f_j)_{(i,j) \in I \times J}$$

is a basis of $V \otimes W$.

Proof Let $v \in V$ and $w \in W$. By assumption, they can be written as

$$v = \sum_{i \in I_v} \lambda_i e_i \text{ and } w = \sum_{j \in J_w} \mu_j f_j$$

for some finite subsets $I_v \subset I$ and $J_w \subset J$. Thus,

$$(v, w) - \sum_{(i,j) \in I_v \times J_w} \lambda_i \mu_j (e_i, f_j) \in \mathcal{I}(V, W)$$

by definition. In other words, we have in $V \otimes W$ the equality

$$v \otimes w = \sum_{(i,j) \in I_v \times J_w} \lambda_i \mu_j e_i \otimes f_j,$$

proving that the family is generating.

To show linear independence, let us consider for some fixed $(i, j) \in I \times J$ the unique linear map

$$\varphi_{ij} \colon \mathcal{F}(V \times W) \to \mathbf{C}$$

sending (v, w) to $e_i^*(v) \times e_j^*(w)$ and all other basis vectors to 0. By construction, the kernel of φ_{ij} contains $\mathcal{I}(V, W)$, hence it factors through the quotient map $\pi \colon \mathcal{F}(V \times W) \to V \otimes W$ to a linear map $\psi_{ij} \colon V \otimes W \to \mathbf{C}$. It then follows that

$$\psi_{ij}(e_{i'} \otimes f_{j'}) = \delta_{ii'}\delta_{jj'},$$

and this clearly implies that the family is linearly independent, concluding the proof. □

As a consequence, we can elucidate the tensor product construction for finite-dimensional vector spaces.

Corollary 1.13 *Let V and W be vector spaces of dimension n and m respectively. Then, $V \otimes W$ has dimension $n \times m$.*

In particular, the dimension issue with the direct product disappears when considering tensor products. Back to our problem now, we want to prove that $F(S_N \times S_N)$ is isomorphic to $F(S_N) \otimes F(S_N)$. We will do this in greater generality, since we may need similar results later on in slightly diffferent contexts. We will consider algebras of the form $\mathcal{O}(X) = \mathbf{C}[X_1, \cdots, X_N]/I$ for some ideal I.[6]

Proposition 1.14 *Let $I \subset \mathbf{C}[X_1, \cdots, X_n]$ and $J \subset \mathbf{C}[Y_1, \cdots, Y_m]$ be ideals. Then, the map*

$$\Phi \colon (a + I, b + J) \mapsto ab + (I + J)$$

factors through a linear isomorphism

$$\mathbf{C}[X_1, \cdots, X_n]/I \otimes \mathbf{C}[Y_1, \cdots, Y_m]/J \simeq \mathbf{C}[X_1, \cdots, X_n, Y_1, \cdots, Y_m]/(I+J).$$

Proof To lighten notations, let us denote by A_n the complex polynomial algebra on n indeterminates. If $a \in A_n$, $b \in A_m$, $x \in I$ and $y \in J$, then

$$(a + x)(b + y) = ab + ay + xb + xy$$

and $ay + xb + xy \in I + J$ so that there is a well-defined linear map

$$\widetilde{\Phi} \colon \mathcal{F}(A_n/I \times A_m/J) \to A_{n+m}/(I + J).$$

---

[6] The notation here is the standard one from commutative algebra, since we are now considering algebras of *commutative polynomials*.

One easily checks that $\mathcal{I}(A_n/I, A_m/J) \subset \ker(\widetilde{\Phi})$, hence there is a well-defined induced map

$$\Phi \colon (A_n/I) \otimes (A_m/J) \to A_{n+m}/(I+J).$$

Conversely, let us set for $1 \leqslant i \leqslant n$ and $1 \leqslant j \leqslant m$

$$\widetilde{X}_i = \overline{X}_i \otimes 1 \in (A_n/I) \otimes (A_m/J),$$
$$\widetilde{Y}_j = 1 \otimes \overline{Y}_j \in (A_n/I) \otimes (A_m/J),$$

where the bar denotes the image in the quotient. Because these elements commute, there exists by universality a unique $*$-homomorphism

$$\widetilde{\Psi} \colon A_{n+m} \to (A_n/I) \otimes (A_m/J)$$

such that $\widetilde{\Psi}(X_i) = \widetilde{X}_i$ and $\widetilde{\Psi}(Y_j) = \widetilde{Y}_j$ for all $1 \leqslant i \leqslant n$ and $1 \leqslant j \leqslant m$. Obviously, $\widetilde{\Psi}$ vanishes on I and J, hence on $I + J$, allowing us to factor it through a map

$$\Psi \colon A_{n+m}/(I+J) \to (A_n/I) \otimes (A_m/J).$$

Now, applying $\Phi \circ \Psi$ and $\Psi \circ \Phi$ to the basis vectors $\overline{X}_i^k \overline{Y}_j^\ell$ and $\overline{X}_i^k \otimes \overline{Y}_j^\ell$ respectively shows that both compositions are the identity, concluding the proof. $\qquad\qquad\qquad\qquad\qquad\qquad\qquad\qquad\qquad\qquad\qquad\qquad\square$

The result is quite satisfying, except that we do not want to deal with vector spaces but with algebras. The construction is, however, easy to generalise. First note that, if A and B are algebras, then there is an algebra structure on $A \otimes B$ defined by

$$(a \otimes b)(a' \otimes b') = aa' \otimes bb'.$$

The fact that this corresponds to a well-defined bilinear map can be checked, for instance on a basis using Proposition 1.12. If, moreover, A and B are $*$-algebras, then there is a $*$-algebra structure on $A \otimes B$ given by

$$(a \otimes b)^* = a^* \otimes b^*.$$

We can now state and prove our main result:

Theorem 1.15 *Let $I \subset A_n$ and $J \subset A_m$ be $*$-ideals. Then, the map*

$$(a + I, b + J) \mapsto ab + (I + J)$$

factors through an algebra $*$-isomorphism

$$(A_n/I) \otimes (A_m/J) \simeq A_{n+m}/(I+J).$$

Proof One simply has to check that the linear isomorphisms Φ and Ψ from Proposition 1.14 are algebra $*$-homomorphisms, which is straightforward. $\square$

Applying this to the case of the algebra of functions on S_N is a good way to understand what precisely is going on.

Corollary 1.16 *The map* $\imath \colon F(S_N) \otimes F(S_N) \to F(S_N \times S_N)$ *sending* $f \otimes g$ *to the map*

$$(\sigma, \tau) \mapsto f(\sigma)g(\tau)$$

extends to a $*$-algebra isomorphism.

Proof Let I be the ideal of $A = \mathbf{C}[X_{ij} \mid 1 \leqslant i, j \leqslant N]$ generated by the polynomials giving the relations of Definition 1.6, so that $A/I = F(S_N)$ by Exercise 1.3. Theorem 1.15 yields an isomorphism

$$F(S_N) \otimes F(S_N) \to \mathbf{C}[X_{ij}, Y_{ij} \mid 1 \leqslant i, j \leqslant N]/\widetilde{I},$$

where $\widetilde{I}$ is generated by the two copies of I and the image of $P \otimes Q$ is $P \times Q$. Any element of the right-hand side can be written as a linear combination of products $P \times Q$ and therefore defines a function on $S_N \times S_N$ so that the map induces a surjection onto $F(S_N \times S_N)$. By equality of the dimensions, such a surjection must be an isomorphism, concluding the proof. $\square$

As a conclusion, we can identify canonically $F(S_N \times S_N)$ with $F(S_N) \otimes F(S_N)$, so that we have an analogue of the algebra of functions on pairs of quantum permutation matrices, which is simply $\mathcal{A}_s(N) \otimes \mathcal{A}_s(N)$.

Now that we are talking about tensor products, let us take the occasion to define the corresponding construction on linear maps, so that it is ready for use in the next chapters.

Exercise 1.4 Let V_i, W_i be vector spaces for $i \in \{1, 2\}$ and let $T_i \colon V_i \to W_i$ be linear maps. Prove that there exists a unique linear map

$$T_1 \otimes T_2 \colon V_1 \otimes V_2 \to W_1 \otimes W_2$$

such that for any $(v_1, v_2) \in V_1 \otimes V_2$,

$$(T_1 \otimes T_2)(v_1 \otimes v_2) = T_1(v_1) \otimes T_2(v_2).$$

Solution We can define a map

$$T_1 \odot T_2 \colon \mathcal{F}(V_1 \times V_2) \to \mathcal{F}(W_1 \times W_2) \to W_1 \otimes W_2$$

by the formula

$$(T_1 \odot T_2)(v_1, v_2) = T_1(v_1) \otimes T_2(v_2).$$

Then, the linearity of T_1 and T_2 implies that $T_1 \odot T_2$ vanishes on $\mathcal{I}(V_1, V_2)$, hence it factors through $V_1 \otimes V_2$, yielding the result. $\qquad\square$

1.2.3 Coproduct

Back to our formula for the product, we can now write, making the isomorphism implicit,

$$c_{ij}(\sigma_1\sigma_2) = \sum_{k=1}^{N}(c_{ik} \otimes c_{kj})(\sigma_1, \sigma_2).$$

Considering the elements p_{ij} as 'coefficient functions', this suggests to encode a kind of 'group law' through the map

$$\Delta: p_{ij} \to \sum_{k=1}^{N} p_{ik} \otimes p_{kj}. \tag{1.1}$$

But for this to work, one must first prove that such a map Δ exists.

Proposition 1.17 *There exists a unique $*$-homomorphism*

$$\Delta: \mathcal{A}_s(N) \to \mathcal{A}_s(N) \otimes \mathcal{A}_s(N)$$

satisfying formula (1.1).

Proof Let us set, for $1 \leqslant i, j \leqslant N$,

$$q_{ij} = \sum_{k=1}^{N} p_{ik} \otimes p_{kj}.$$

We claim that the q_{ij}'s satisfy Conditions (1) to (3) of Definition 1.6. The existence of Δ then follows from the universal property. $\qquad\square$

Exercise 1.5 Prove the claim in the preceding proof.

Solution It is clear that $q_{ij}^* = q_{ij}$. Let us now compute the square

$$q_{ij}^2 = \sum_{k,\ell=1}^{N} p_{ik}p_{i\ell} \otimes p_{kj}p_{\ell j}$$

$$= \sum_{k,\ell=1}^{N} \delta_{k\ell}p_{ik} \otimes p_{kj}$$

$$= q_{ij}.$$

We have therefore checked Condition (1). Moreover,

$$
\begin{aligned}
\sum_{i=1}^{N} q_{ij} &= \sum_{k,i=1}^{N} p_{ik} \otimes p_{kj} \\
&= \sum_{k=1}^{N} \left(\sum_{i=1}^{N} p_{ik} \right) \otimes p_{kj} \\
&= \sum_{k=1}^{N} 1 \otimes p_{kj} \\
&= 1 \otimes 1,
\end{aligned}
$$

hence Condition (2) also is satisfied. Eventually, for $j \neq j'$,

$$
q_{ij} q_{ij'} = \sum_{k,\ell=1}^{N} p_{ik} p_{i\ell} \otimes p_{kj} p_{\ell j'}.
$$

The first tensor in the sum vanishes unless $k = \ell$, but in that case the second one vanishes and Condition (3) follows. The argument for $i \neq i'$ is similar. $\square$

The map Δ is called the *coproduct* and is a reasonable substitute for matrix multiplication (i.e. the group law of a matrix group). In particular, it satisfies an analogue of the associativity property of the group law, called *coassociativity*, which reads

$$
(\Delta \otimes \mathrm{id}) \circ \Delta = (\mathrm{id} \otimes \Delta) \circ \Delta. \tag{1.2}
$$

Exercise 1.6 Prove that the coproduct on $\mathcal{A}_s(N)$ is indeed coassociative. Check also that the corresponding equation on the coefficient functions in S_N is equivalent to the associativity of the composition of permutations.

Solution Because Δ is a $*$-algebra homomorphism, it is enough to check coassociativity on the generators,

$$
\begin{aligned}
(\Delta \otimes \mathrm{id}) \circ \Delta(p_{ij}) &= \sum_{k=1}^{N} \Delta(p_{ik}) \otimes p_{kj} \\
&= \sum_{k,\ell=1}^{N} p_{i\ell} \otimes p_{\ell k} \otimes p_{kj} \\
&= \sum_{\ell=1}^{N} p_{i\ell} \otimes \Delta(p_{\ell j}) \\
&= (\mathrm{id} \otimes \Delta) \circ \Delta(p_{ij}).
\end{aligned}
$$

As for the second assertion, we have already seen that

$$\Delta(c_{ij})(\sigma_1, \sigma_2) = c_{ij}(\sigma_1\sigma_2).$$

Thus,

$$(\Delta \otimes \mathrm{id}) \circ \Delta(c_{ij})(\sigma_1, \sigma_2, \sigma_3) = \Delta(c_{ij})(\sigma_1\sigma_2, \sigma_3) = c_{ij}((\sigma_1\sigma_2)\sigma_3)$$

while

$$(\mathrm{id} \otimes \Delta) \circ \Delta(c_{ij})(\sigma_1, \sigma_2, \sigma_3) = \Delta(c_{ij})(\sigma_1, \sigma_2\sigma_3) = c_{ij}(\sigma_1(\sigma_2\sigma_3))$$

so that coassociativity is equivalent to $f((\sigma_1\sigma_2)\sigma_3) = f(\sigma_1(\sigma_2\sigma_3))$ for all $f \in \mathcal{F}(S_N)$ and $\sigma_1, \sigma_2, \sigma_3 \in S_N$, which is, in turn, equivalent to the associativity of the group law. $\qquad\square$

The coproduct certainly indicates that we are on the right track to produce a group-like structure on the quantum permutation algebra. However, we still need a neutral element and an inverse. But instead of trying to translate each of them, we will take advantage of the fact that we are considering a matrix group. Indeed, for any permutation σ, the corresponding matrix is orthogonal, hence

$$\sum_{k=1}^{N} c_{ik}(\sigma)c_{jk}(\sigma) = \delta_{ij} = \sum_{k=1}^{N} c_{ki}(\sigma)c_{kj}(\sigma). \tag{1.3}$$

Since this holds for any σ, it can be written as an equality of functions in $F(S_N)$, and it turns out that the same equality holds in $\mathcal{A}_s(N)$.

Proposition 1.18 *For any* $1 \leqslant i, j \leqslant N$,

$$\sum_{k=1}^{N} p_{ik}p_{jk} = \delta_{ij} = \sum_{k=1}^{N} p_{ki}p_{kj}. \tag{1.4}$$

Proof This is a direct consequence of Conditions (1) to (3). $\qquad\square$

This means that the quantum permutation algebra is somehow 'made of orthogonal quantum matrices' (see the beginning of Subsection 1.3.3 for a more precise statement), and this property should contain all information about the unit and the inverse. Another way to state this is that the matrix $P = (p_{ij})_{1 \leqslant i,j \leqslant N} \in M_N(\mathcal{A}_s(N))$ is orthogonal in the sense that its inverse equals its transpose. As a conclusion, the algebra $\mathcal{A}_s(N)$ with its generators $(p_{ij})_{1 \leqslant i,j \leqslant N}$ seem to have all the properties one can expect for a group-like object. It therefore deserves the name of *quantum group* that we will define in the next section.

The fact that Condition (1.4) yields a full group-like structure can be encoded in the two following maps, whose existence follows from the universal property of $\mathcal{A}_s(N)$.

- The *antipode* $S\colon \mathcal{A}_s(N) \to \mathcal{A}_s(N)$, which is the unique $*$-antihomomorphism induced by

$$p_{ij} \mapsto p_{ji}.$$

Since the transpose of P is its inverse, this plays the role of the inverse map.
- The *counit* $\varepsilon\colon \mathcal{A}_s(N) \to \mathbf{C}$, which is the unique $*$-homomorphism induced by

$$p_{ij} \mapsto \delta_{ij}.$$

Since the matrix $(\delta_{ij})_{1 \leqslant i,j \leqslant N}$ is the identity, this plays the role of the neutral element.

Exercise 1.7 Prove the existence of the maps S and ε.

Solution 1. We start with the antipode S. The uniqueness is clear, and we have to prove existence. Let us denote by $\mathcal{A}$ the opposite algebra of $\mathcal{A}_s(N)$, that is to say, the algebra with the same underlying vector space but such that $a \times_{\mathcal{A}} b = ba$. Let us also consider the elements $q_{ij} = p_{ji}$ in $\mathcal{A}$. Then, the matrix $(q_{ij})_{1 \leqslant i,j \leqslant N}$ is a quantum permutation matrix and its coefficients generate $\mathcal{A}$, hence there is a surjective $*$-homomorphism

$$\widetilde{S}\colon \mathcal{A}_s(N) \to \mathcal{A}$$

such that $\widetilde{S}(p_{ij}) = q_{ij}$. Composing with the identity map seen as a linear isomorphism $I\colon \mathcal{A} \to \mathcal{A}_s(N)$ yields a map $S = I \circ \widetilde{S}\colon \mathcal{A}_s(N) \to \mathcal{A}_s(N)$. By construction, $S(p_{ij}) = p_{ji}$ and, moreover,

$$
\begin{aligned}
S(p_{ij}p_{k\ell}) &= I \circ \widetilde{S}(p_{ij}p_{k\ell}) \\
&= I(q_{ij} \times_{\mathcal{A}} q_{k\ell}) \\
&= I(q_{k\ell}q_{ij}) \\
&= p_{\ell k}p_{ji} \\
&= S(p_{k\ell})S(p_{ij})
\end{aligned}
$$

so that S is anti-multiplicative.
2. We now turn to the counit ε. Noticing that the identity matrix is a quantum permutation matrix, the universal property of $\mathcal{A}_s(N)$ directly yields a

$*$-homomorphism $\varepsilon\colon \mathcal{A}_s(N) \to \mathbf{C}$ sending p_{ij} to the corresponding coefficient of the identity matrix, which is δ_{ij}. $\square$

It is worth working out the analogues of these maps for the classical permutation group to be convinced that they encode the complete group structure.

Exercise 1.8 Write down the explicit form of the analogues of the counit ε and antipode S for $F(S_N)$ in terms of permutations.

Solution The functions corresponding to the p_{ij}'s are the functions $c_{ij}\colon \sigma \mapsto \delta_{\sigma(i)j}$. Thus,

$$S(c_{ij})(\sigma) = c_{ji}(\sigma) = \delta_{\sigma(j)i} = \delta_{\sigma^{-1}(i)j} = c_{ij}(\sigma^{-1})$$

so that S is induced by the inverse map on S_N. Similarly,

$$\varepsilon(c_{ij}) = \delta_{ij} = c_{ij}(\mathrm{id})$$

so that ε corresponds to the identity permutation. $\square$

With these maps, Equation (1.4) becomes

$$m \circ (\mathrm{id} \otimes S) \circ \Delta = \varepsilon = m \circ (S \otimes \mathrm{id}) \circ \Delta,$$

where $m\colon \mathcal{A}_s(N) \otimes \mathcal{A}_s(N) \to \mathcal{A}_s(N)$ is the multiplication map. Our focus in this text is on the matricial aspect of quantum groups, and we will therefore never use these maps.[7] Note, however, that $(\mathcal{A}_s(N), \Delta, \varepsilon, S)$ is what is called a *Hopf algebra*. The theory of Hopf algebras is vast and has many connections to other fields. The reader may, for instance, read [62] for a detailed introduction or [48] for more categorical aspects and important applications.

1.3 Compact Matrix Quantum Groups

Our study of the quantum permutation algebra has given us enough motivation to introduce a notion of compact quantum group. There is a nice and complete theory of these objects, which was developed by S. L Woronowicz in [77]. There are two published books explaining this theory in detail, [69] and [60] to which the reader may refer for alternative expositions emphasising other aspects.

---

[7] This is a lie. We will use them at some point to get a computationally tractable description of representations, but in a way which has no consequence for the remainder of the development of the theory.

1.3.1 A First Definition

The purpose of this text is to give some examples of the interaction between the combinatorics of partitions and the theory of compact quantum groups. The most striking examples involve compact quantum groups which belong to a specific class which is, in a sense, simpler to define and handle. It was introduced by S. L Woronowicz in [75] as a generalisation of compact groups of matrices and as a first attempt at a definition of compact quantum groups. We will therefore focus on this class, even though our definition differs from [75, definition 1.1] and is closer to [70, definition 2.1'].

Before giving the definition, we have to give an important warning. We will use throughout this text a specific assumption on all compact quantum groups which is somehow hidden in the definition. In plain terms, all the objects that we will consider will be of so-called *Kac type*. However, for simplicity and because this is a consequence of our axioms, we will never mention that specificity again. But the reader should be aware that our terminology does not exactly match the literature, because one should, any time the words 'compact quantum group' are written hereafter, add the words 'of Kac type' (for more comments on this, see the end of Appendix C).

Definition 1.19 An *orthogonal compact matrix quantum group* of size N is given by a $*$-algebra $\mathcal{A}$ generated by N^2 elements $(u_{ij})_{1 \leqslant i,j \leqslant N}$ such that

1. $u_{ij} = u_{ij}^*$ for all $1 \leqslant i,j \leqslant N$;
2. For all $1 \leqslant i,j \leqslant N$,

$$\sum_{k=1}^{N} u_{ik}u_{jk} = \delta_{ij} = \sum_{k=1}^{N} u_{ki}u_{kj};$$

3. There exists a $*$-homomorphism $\Delta \colon \mathcal{A} \to \mathcal{A} \otimes \mathcal{A}$ such that for all $1 \leqslant i,j \leqslant N$,

$$\Delta(u_{ij}) = \sum_{k=1}^{N} u_{ik} \otimes u_{kj}.$$

Denoting by $u \in M_N(\mathcal{A})$ the matrix with coefficients $(u_{ij})_{1 \leqslant i,j \leqslant N}$, we will denote the orthogonal compact matrix quantum group by $(\mathcal{A}, u)$.

By analogy with our reasoning on S_N, $\mathcal{A}$ will be thought of as the algebra of functions on a non-existent 'quantum space'. However, if we consider general 'compact quantum spaces', we cannot use all the functions like for S_N. Here our crucial intuition will be that compact groups of matrices are completely determined by their algebra of *regular functions*, that is to say, functions which are polynomial in the matrix coefficients (see the beginning of Section 5.1 for some details on this). The usual notation for this is $\mathcal{O}(G)$, whence the notation $\mathcal{A} = \mathcal{O}(\mathbb{G})$ if $\mathbb{G} = (\mathcal{A}, u)$ denotes the orthogonal compact matrix quantum group. We can now formalise the properties of the quantum permutation algebras established in Section 1.2.

Definition 1.20 For any integer N, the pair $(\mathcal{A}_s(N), P)$ is an orthogonal compact matrix quantum group, where $P = (p_{ij})_{1 \leqslant i, j \leqslant N}$. It is called the *quantum permutation group* on N points and is usually referred to using the notation S_N^+.

Consequently, we may from now on write $\mathcal{O}(S_N^+)$ instead of $\mathcal{A}_s(N)$. This quantum group was first defined by S. Wang in [72]. It is natural (and crucial for our purpose) to wonder whether this is really different from S_N.

Exercise 1.9 Prove that for $N = 1, 2, 3$, $S_N^+ = S_N$ in the sense that π_{ab} is injective. Prove moreover that for any $N \geqslant 4$, $\mathcal{O}(S_N^+)$ is non-commutative, hence not isomorphic to $F(S_N)$.

Solution For $N = 1$, $\mathcal{A}_s(1)$ is generated by one self-adjoint projection, hence is isomorphic to $\mathbf{C} = F(S_1)$. For $N = 2$, observe that the relations force

$$P = \begin{pmatrix} p_{11} & 1 - p_{11} \\ 1 - p_{11} & p_{11} \end{pmatrix}$$

making $\mathcal{A}_s(2)$ abelian, hence equal to $F(S_2)$.

For $N = 3$, we must again prove that $\mathcal{A}_s(3)$ is abelian. Here is a simple argument from [53]. It is enough to prove that p_{11} commutes with p_{22}, since any independent permutation of the rows and columns of P yields an automorphism of $\mathcal{A}_s(N)$ by the universal property. We start by observing that

$$p_{11}p_{22} = p_{11}p_{22}(p_{11} + p_{12} + p_{13})$$
$$= p_{11}p_{22}p_{11} + p_{11}p_{22}p_{13}.$$

But

$$p_{11}p_{22}p_{13} = p_{11}(1 - p_{21} - p_{23})p_{13}$$
$$= p_{11}p_{13} - p_{11}p_{21}p_{13} - p_{11}p_{23}p_{13}$$
$$= 0,$$

hence

$$\begin{aligned}
p_{11}p_{22} &= p_{11}p_{22}p_{11} \\
&= (p_{11}p_{22}p_{11})^* \\
&= (p_{11}p_{22})^* \\
&= p_{22}^*p_{11}^* \\
&= p_{22}p_{11}.
\end{aligned}$$

For $N \geqslant 4$, let p and q be the orthogonal projections onto the lines spanned by the vectors $(0, 1)$ and $(1, 1)$ respectively in $\mathbf{C}^2$, so that $pq \neq qp$. Then, consider the matrix

$$\begin{pmatrix}
p & 1-p & 0 & 0 \\
1-p & p & 0 & 0 \\
0 & 0 & q & 1-q \\
0 & 0 & 1-q & q
\end{pmatrix}$$

and complete it to an $N \times N$ matrix by putting it in the upper left corner, setting the other diagonal coefficients to 1 and all the other coefficients to 0. This yields a quantum permutation matrix, hence a $*$-homomorphism $\pi \colon \mathcal{O}(S_N^+) \to \mathcal{B}(H)$. Because $\pi(u_{11}) = p$ and $\pi(u_{33}) = q$ do not commute, we infer that $\mathcal{O}(S_N^+)$ is not commutative. $\qquad\square$

To get a better understanding of Definition 1.19, it is worth working out the link with the classical case. This requires the identification of the kernel of a tensor product of linear maps, that we give here as a lemma.

Lemma 1.21 *Let $T_1 \colon V_1 \to W_1$ and $T_2 \colon V_2 \to W_2$ be linear maps. Then,*

$$\ker(T_1 \otimes T_2) = \ker(T_1) \otimes V_2 + V_1 \otimes \ker(T_2).$$

Proof We may assume without loss of generality that the maps are surjective. Moreover, we have decompositions

$$V_i = \ker(T_i) \oplus V_i'$$

such that the maps restrict to isomorphisms on V_i'. Let us denote by $\widetilde{T}_i$ the projection onto V_i' parallel to $\ker(T_i)$. It is easy to see that we have a decomposition

$$V_1 \otimes V_2 = (\ker(T_1) \otimes \ker(T_2)) \oplus (\ker(T_1) \otimes V_2') \oplus (V_1' \oplus \ker(T_2)) \oplus (V_1' \otimes V_2').$$

By definition, $\widetilde{T}_1 \otimes \widetilde{T}_2$ vanishes on the first three summands and is the identity on the last one so that its kernel is

$$(\ker(T_1) \otimes \ker(T_2)) \oplus (\ker(T_1) \otimes V_2') \oplus (V_1' \oplus \ker(T_2)) = \ker(T_1) \otimes V_2 \\ + V_1 \otimes \ker(T_2).$$

The result now follows from the fact that

$$T_1 \otimes T_2 = (\mathrm{id} \oplus T_{1|V_1'}) \otimes (\mathrm{id} \oplus T_{2|V_2'}) \circ (\widetilde{T}_1 \otimes \widetilde{T}_2)$$

has the same kernel as $\widetilde{T}_1 \otimes \widetilde{T}_2$.   □

We would like now to describe all orthogonal compact matrix quantum groups $(\mathcal{A}, u)$ where $\mathcal{A}$ is a commutative $*$-algebra, and to prove that they come from compact groups of orthogonal matrices. However, there is a rather non-trivial step in such a proof: one needs to prove that for a given $*$-ideal $I \subset \mathbf{C}[X_{ij} \mid 1 \leqslant i, j \leqslant N]$, any polynomial vanishing on the intersection of all the zeros if elements of I is again in I. This is reminiscent of the famous Nüllstelensatz from algebraic geometry, and taking this path would lead us to showing that some $*$-algebra does not contain nilpotent elements, which is difficult. There is, nevertheless, another way around, using operator algebras. But this will only be possible once the connection between our algebraic framework and functional analysis is made in Chapter 5 (see, more precisely, Corollary 5.18). We will therefore restrict ourselves to partial results hereafter, hoping that they nontheless give enough motivation to the reader to keep reading this text.

Exercise 1.10 Let $(\mathcal{A}, u)$ be an orthogonal compact matrix quantum group such that $\mathcal{A}$ is commutative. We set

$$\mathcal{O}(M_N(\mathbf{C})) = \mathbf{C}[X_{ij} \mid 1 \leqslant i, j \leqslant N].$$

1. Show that there exists a surjective $*$-homomorphism $\pi \colon \mathcal{O}(M_N(\mathbf{C})) \to \mathcal{A}$. We set $I = \ker(\pi)$.
2. Let us set

$$G = \{M \in M_N(\mathbf{C}) \mid P(M) = 0 \text{ for all } P \in I\}.$$

   Prove that G is a closed subgroup of O_N. *Hint: a compact bisimplifiable semigroup is a group (see, for instance, [60, example 1.1.2] for a proof).*
3. We now set

$$J = \{P \in \mathcal{O}(M_N(\mathbf{C})) \mid P(M) = 0 \text{ for all } M \in G\},$$

   so that $\mathcal{O}(G) = \mathcal{O}(M_N(\mathbf{C}))/J$. Check that $I \subset J$.
4. We now assume that $\mathcal{A}$ is finite-dimensional.
   (a) Show that G is then finite.
   (b) Conclude that $\mathcal{A} = \mathcal{O}(G)$.
5. We assume instead that for any $x \in \mathcal{A} \backslash \{0\}$, there exists a $*$-homomorphism $f \colon \mathcal{A} \to \mathbf{C}$ such that $f(x) \neq 0$. Conclude again that $\mathcal{A} = \mathcal{O}(G)$.

Solution 1. This follows from the fact that $\mathcal{O}(M_N(\mathbf{C}))$ is the universal $*$-algebra generated by N^2 self-adjoint pairwise commuting variables, so that setting $\pi(X_{ij}) = u_{ij}$ works.

2. We first note that G is closed by definition and consists of orthogonal matrices because u is orthogonal. Let us therefore prove that G is stable under product, which is enough according to the hint. Let us start by observing that the elements

$$Y_{ij} = \sum_{k=1}^{N} X_{ik} \otimes X_{kj}$$

are self-adjoint and pairwise commute, hence there exists a unique $*$-homomorphism $\Delta \colon \mathcal{O}(M_N(\mathbf{C})) \to \mathcal{O}(M_N(\mathbf{C})) \otimes \mathcal{O}(M_N(\mathbf{C}))$ such that $\Delta(X_{ij}) = Y_{ij}$. If now $P \in I$, we have

$$(\pi \otimes \pi) \circ \Delta(P) = \Delta \circ \pi(P) = 0,$$

so that by Lemma 1.21 we can write

$$\Delta(P) = \sum_i P_i \otimes Q_i \in \mathcal{O}(M_N(\mathbf{C})) \otimes \mathcal{O}(M_N(\mathbf{C}))$$

such that, for all i, either P_i or Q_i belongs to I. Thus, for any $M_1, M_2 \in G$,

$$\begin{aligned} P(M_1 M_2) &= \Delta(P)(M_1, M_2) \\ &= \sum_i P_i(M_1) Q_i(M_2) \\ &= 0 \end{aligned}$$

and $M_1 M_2 \in G$.

3. By definition of G, $P(M) = 0$ for any $M \in G$ if $P \in I$, hence the inclusion.

4. (a) Because $\mathcal{O}(G)$ is a quotient of $\mathcal{A}$, it is also finite-dimensional and our strategy will be to prove that if G is infinite, then $\mathcal{O}(G)$ is infinite-dimensional. To do this, let us first define, for $M \in O_N$,

$$P_M(X_{ij}) = \sum_{1 \leqslant i,j \leqslant N} (X_{ij} - M_{ij})^* (X_{ij} - M_{ij}).$$

This is a polynomial and $P_M(M') = 0$ if and only if $M' = M$. Therefore, if $F \subset G$ is a finite set, we can define the polynomial

$$P_{M,F}(X_{ij}) = \frac{1}{\displaystyle\prod_{M' \in F \setminus \{M\}} P_{M'}(M)} \prod_{M' \in F \setminus \{M\}} P_{M'}(X_{ij})$$

for any $M \in F$. That polynomial evaluates to 1 on M and 0 on all other elements of F. As a consequence, the family $(\pi(P_{M,F}))_{M \in F}$ is linearly independent in $\mathcal{O}(G)$, proving that

$$\dim(\mathcal{O}(G)) \geqslant |F|.$$

If now G is infinite, it contains arbitrary large finite subsets, hence $\dim(\mathcal{O}(G)) = +\infty$.

(b) We have to prove that $I = J$. Observe that by the previous question the polynomials $(P_{M,G})_{M \in G}$ span a complement of J in $\mathcal{O}(M_N(\mathbf{C}))$. Therefore, any $P \in I$ can be written as

$$P = Q + \sum_{M \in G} \lambda_M P_M$$

for some $Q \in J$ and $\lambda_M \in \mathbf{C}$ for all $M \in G$. Then, for any $M \in G$ evaluating the previous expression at M yields

$$\sum_{N \in G \setminus \{M\}} \lambda_N = 0$$

and the only solution to this linear system is $\lambda_M = 0$ for all $M \in G$. Therefore, $I = J$ and the proof is complete.

5. If $f \colon \mathcal{A} \to \mathbf{C}$ is a $*$-homomorphism, then we claim that the matrix $\hat{f} = (f(u_{ij}))_{1 \leqslant i,j \leqslant N}$ is in G. Indeed, if $P \in I$, then

$$P(\hat{f}) = (f(P(u_{ij}))_{1 \leqslant i,j \leqslant N} = 0.$$

As a consequence, f is nothing but the evaluation map at $\hat{f}$. Therefore, if

$$\pi' \colon \mathcal{A} \to \mathcal{A}/\pi(J) = \mathcal{O}(G)$$

is the canonical surjection and $x \in \ker(\pi')$, then

$$f(x) = \pi'(x)(\hat{f}) = 0$$

and the condition in the question ensures that $\ker(\pi') = 0$, hence $\mathcal{A} = \mathcal{O}(G)$. $\qquad\square$

1.3.2 The Quantum Orthogonal Group

Before delving into the general theory of compact quantum groups, let us give another fundamental example which is also due to S. Wang, but earlier in [70]. After a look at Definition 1.19, it is natural to wonder about the largest possible orthogonal compact matrix quantum group. Its definition relies on the following simple fact.

Exercise 1.11 Let N be an integer and let $\mathcal{A}_o(N)$ be the universal $*$-algebra generated by N^2 elements $(U_{ij})_{1 \leqslant i,j \leqslant N}$ such that

- $U_{ij}^* = U_{ij}$ for all $1 \leqslant i, j \leqslant N$;
- $\displaystyle\sum_{k=1}^{N} U_{ik} U_{jk} = \delta_{ij} = \sum_{k=1}^{N} U_{ki} U_{kj}.$

Then, there exists a unique $*$-homomorphism

$$\Delta \colon \mathcal{A}_o(N) \to \mathcal{A}_o(N) \otimes \mathcal{A}_o(N)$$

such that for all $1 \leqslant i, j \leqslant N$,

$$\Delta(U_{ij}) = \sum_{k=1}^{N} U_{ik} \otimes U_{kj}.$$

Solution The proof is similar to that of Proposition 1.17. We set

$$V_{ij} = \sum_{k=1}^{N} U_{ik} \otimes U_{kj}$$

and have to check that the corresponding matrix V is orthogonal. Indeed,

$$
\begin{aligned}
\sum_{k=1}^{N} V_{ik} V_{jk} &= \sum_{k,\ell,m=1}^{N} U_{i\ell} U_{jm} \otimes U_{\ell k} U_{mk} \\
&= \sum_{\ell,m=1}^{N} U_{i\ell} U_{jm} \otimes \left(\sum_{k=1}^{N} U_{\ell k} U_{mk} \right) \\
&= \sum_{\ell,m=1}^{N} U_{i\ell} U_{jm} \otimes \delta_{\ell m} \\
&= \sum_{\ell=1}^{N} U_{i\ell} U_{j\ell} \otimes 1 \\
&= \delta_{ij} 1 \otimes 1.
\end{aligned}
$$

The other equality is proved similarly, and it then follows from universality that there exists a $*$-homomorphism sending U_{ij} to V_{ij}. $\qquad\square$

Denoting by U the matrix $(U_{ij})_{1 \leqslant i,j \leqslant N} \in M_N(\mathcal{A}_o(N))$, this motivates the following definition.

Definition 1.22 The pair $(\mathcal{A}_o(N), U)$ is an orthogonal compact matrix quantum group called the *quantum orthogonal group*. It is usually referred to using the notation O_N^+.

As the name suggests, O_N^+ is linked to the orthogonal group. Indeed, if $c_{ij} \colon O_N \to \mathbf{C}$ are the matrix coefficient functions, then there is a surjective *-homomorphism

$$\pi_{\mathrm{ab}} \colon \mathcal{O}(O_N^+) \to \mathcal{O}(O_N)$$

sending U_{ij} to c_{ij}. Thus, O_N^+ is a quantum version of O_N just in the same way as S_N^+ is a quantum version of S_N.

Remark 1.23 The preceding comments, as well as the notation π_{ab}, suggest that $\mathcal{O}(O_N)$ is the largest abelian quotient of $\mathcal{O}(O_N^+)$. This would be easy to prove if we knew that any orthogonal compact matrix quantum group with commutative *-algebra comes from a group. It will therefore be a consequence of Corollary 5.18.

Note that it follows from the universal property that there is a surjective *-homomorphism

$$\mathcal{O}(O_N^+) \to \mathcal{O}(S_N^+),$$

so that $\mathcal{O}(O_N^+)$ is not commutative as soon as $N \geqslant 4$. However, more is true in that case.

Proposition 1.24 *The *-algebra $\mathcal{O}(O_N^+)$ is non-commutative as soon as $N \geqslant 2$.*

Proof Let $r_1, r_2 \in \mathcal{B}(\mathbf{C}^2)$ be two reflections with axis generated by the vectors $(1, 0)$ and $(1, 1)$ respectively, so that they do not commute. Then, the diagonal matrix with first coefficient r_1 and all other coefficients equal to r_2 is orthogonal, hence there exists a *-homomorphism

$$\pi \colon \mathcal{O}(O_N^+) \to \mathcal{B}(\mathbf{C}^2)$$

sending U_{ij} to 0 if $i \neq j$, to r_1 if $i = 1 = j$ and to r_2 otherwise. In particular, $\pi(U_{11})$ and $\pi(U_{22})$ do not commute, so that $\mathcal{O}(O_N^+)$ is not commutative. $\square$

1.3.3 The Unitary Case

The intuition that orthogonal compact matrix quantum groups generalise subgroups of O_N can be made rigorous in the following way: by universality, for any orthogonal compact matrix quantum group $\mathbb{G} = (\mathcal{O}(\mathbb{G}), u)$, there is a surjective *-homomorphism

$$\pi \colon \mathcal{O}(O_N^+) \to \mathcal{O}(\mathbb{G})$$

sending U_{ij} to u_{ij} and therefore satisfying

$$\Delta \circ \pi(x) = (\pi \otimes \pi) \circ \Delta(x)$$

for all $x \in \mathcal{O}(\mathbb{G})$. Thus, orthogonal compact matrix quantum groups are 'quantum subgroups' of O_N^+.

One may wonder whether it is possible to consider analogues of closed subgroups of the unitary group U_N instead of the orthogonal one. This is possible, but we will not need it until the last part of this text. Moreover, this more general setting involves subtleties which make some arguments tricky. This can already be seen in the following definition.

Definition 1.25 A *unitary compact matrix quantum group* of size N is given by a $*$-algebra $\mathcal{A}$ generated by N^2 elements $(u_{ij})_{1 \leqslant i,j \leqslant N}$ such that

1. There exist a $*$-homomorphism $\Delta : \mathcal{A} \to \mathcal{A} \otimes \mathcal{A}$ such that for all $1 \leqslant i$, $j \leqslant N$,

$$\Delta(u_{ij}) = \sum_{k=1}^{N} u_{ik} \otimes u_{kj};$$

2. For all $1 \leqslant i, j \leqslant N$,

$$\sum_{k=1}^{N} u_{ik} u_{jk}^* = \delta_{ij} = \sum_{k=1}^{N} u_{ki}^* u_{kj}$$

and

$$\sum_{k=1}^{N} u_{ki} u_{kj}^* = \delta_{ij} = \sum_{k=1}^{N} u_{ik}^* u_{jk}.$$

Remark 1.26 The relations in the previous definition mean that both the matrix u and its conjugate $\bar{u}$ (the matrix where each coefficient is replaced with its adjoint) are unitary. The second one does not follow from the first one in general (see [70, section 4.1] for a counter-example), so that both need to be included in the definition.

Once again, there is an obvious example obtained by considering the largest possible such quantum group at a fixed size N.

Definition 1.27 Let $\mathcal{A}_u(N)$ be the universal $*$-algebra generated by N^2 elements $(V_{ij})_{1 \leqslant i,j \leqslant N}$ such that

$$\sum_{k=1}^{N} V_{ik} V_{jk}^* = \delta_{ij} = \sum_{k=1}^{N} V_{ki}^* V_{kj}$$

and

$$\sum_{k=1}^{N} V_{ik}^* V_{jk} = \delta_{ij} = \sum_{k=1}^{N} V_{ki} V_{kj}^*.$$

One can construct as before a compact quantum group structure on this, and by now the reader should be able to do this alone.

Exercise 1.12 1. Prove that there exists a unique $*$-homomorphism

$$\Delta \colon \mathcal{A}_u(N) \to \mathcal{A}_u(N) \otimes \mathcal{A}_u(N)$$

such that

$$\Delta(V_{ij}) = \sum_{k=1}^{N} V_{ik} \otimes V_{kj}.$$

2. Prove that $\mathcal{A}_u(N)$ is non-commutative for $N \geqslant 2$.
3. What is $\mathcal{A}_u(1)$?

Solution 1. Let us set

$$W_{ij} = \sum_{k=1}^{N} V_{ik} \otimes V_{kj}.$$

Then,

$$\sum_{k=1}^{N} W_{ik} W_{jk}^* = \sum_{k=1}^{N} \sum_{\ell,\ell'=1}^{N} V_{i\ell} V_{j\ell'}^* \otimes V_{\ell k} V_{\ell' k}^*$$

$$= \sum_{\ell,\ell'=1}^{N} V_{i\ell} V_{j\ell'}^* \otimes \left(\sum_{k=1}^{N} V_{\ell k} V_{\ell' k}^* \right)$$

$$= \sum_{\ell,\ell'=1}^{N} V_{i\ell} V_{j\ell'}^* \otimes (\delta_{\ell,\ell'})$$

$$= \sum_{\ell=1}^{N} V_{i\ell} V_{j\ell'}^*$$

$$= \delta_{ij}.$$

The other relations are proven in a similar way.

2. Note that there is a quotient map $\pi \colon \mathcal{A}_u(N) \to \mathcal{A}_o(N)$ given by the quotient by the relations $V_{ij} = V_{ij}^*$ for all $1 \leqslant i, j \leqslant N$. The result therefore follows from the fact that $\mathcal{A}_o(N)$ is non-commutative for $N \geqslant 2$.

3. Observe that $\mathcal{A}_u(1)$ is the quotient of $\mathbf{C}\langle X \rangle$ by the relations

$$XX^* = 1 = X^*X.$$

In particular, the relations make $\mathcal{A}_u(1)$ commutative, and we will show that it corresponds to the circle group

$$\mathbb{T} = \{z \in \mathbf{C} \mid |z| = 1\}.$$

Indeed, denoting by e_k the function $z \mapsto z^k$ for $k \in \mathbf{Z}$, we know that $(e_k)_{k \in \mathbf{Z}}$ is a basis of $\mathcal{O}(\mathbb{T})$. Therefore, there exists a linear map $\Phi \colon \mathcal{O}(\mathbb{T}) \to \mathcal{A}_u(1)$ sending e_k to X^k, and it is surjective by definition. Moreover, because $e_k e_\ell = e_{k+\ell}$ and $e_k^* = e_{-k}$, Φ is in fact a $*$-homomorphism. To conclude, simply observe that $\mathcal{O}(\mathbb{T})$ is generated by e_1, which satisfies $e_1^* e_1 = 1 = e_1 e_1^*$, so that by universality there is a surjective $*$-homomorphism $\Psi \colon \mathcal{A}_u(1) \to \mathcal{O}(\mathbb{T})$ sending X to e_1. It is clear that Ψ is inverse to Φ, thence $\mathcal{A}_u(1) = \mathcal{O}(\mathbb{T})$. $\square$

The pair $U_N^+ = (\mathcal{A}_u(N), U)$ is called a *quantum unitary group*. Once again, abelianisation provides a link with the classical unitary group, with the same caveat as in Remark 1.23. Even though concrete examples of unitary compact quantum groups are more difficult to deal with than orthogonal ones, most of the general theory is exactly the same (see Chapter 6). As a consequence, we will state and prove general results in the setting of unitary compact matrix quantum groups as soon as this does not entail any additional technicality in the proof. As for the other statements which require some adaptation, we will treat them in Chapter 6.

2

Representation Theory

Now that we have a definition of orthogonal compact matrix quantum groups, it is time to investigate their general structure. It turns out that compact groups are tractable objects because they have a very nice representation theory. It is therefore natural to start by looking for a suitable notion of representation for compact quantum groups. For the sake of generality, we will state and prove results in the general setting of unitary compact matrix quantum groups as soon as this does not entail any further technicalities.

2.1 Finite-Dimensional Representations

2.1.1 Two Definitions

Following our basic strategy, we will restate the notion of representation in terms of functions. Recall that for a group G, a representation on a vector space V is a specific map

$$\rho \colon G \to \mathcal{L}(V).$$

Assume, for instance, that V is finite-dimensional so that we can identify $\mathcal{L}(V)$ with $M_n(\mathbf{C})$ for some n. Composing ρ with the coefficient functions produces a bunch of functions $(\rho_{ij})_{1 \leqslant i,j \leqslant n}$ from G to $\mathbf{C}$. Given our setting, it is natural to consider those finite-dimensional representations for which the coefficient functions belong to $\mathcal{O}(G)$. In groups of matrices, this holds for all finite-dimensional representations (see Theorem B.5 in Appendix B for a proof). The fact that ρ is a representation translates into two properties:

1. The matrix $[\rho_{ij}(g)]$ is invertible for all $g \in G$.
2. For any $1 \leqslant i,j \leqslant n$, $\rho_{ij}(gh) = \displaystyle\sum_{k=1}^{n} \rho_{ik}(g)\rho_{kj}(h)$.

The second point is reminiscent of the discussion around the definition of the coproduct in Section 1.2.3, and we therefore know how to translate it. As for the first one, it means that ρ is an invertible element in $M_n(\mathcal{O}(G))$. As a conclusion, we may give the following definition.

Definition 2.1 Let $G = (\mathcal{O}(G), u)$ be a unitary compact matrix quantum group and let n be an integer. An n-*dimensional representation* of G is an element $v \in M_n(\mathcal{O}(G))$ such that:

1. v is invertible.
2. $\Delta(v_{ij}) = \sum_{k=1}^{n} v_{ik} \otimes v_{kj}$ for all $1 \leqslant i, j \leqslant N$.

If, moreover, v is unitary in the sense that $v^* v = \mathrm{Id}_{M_n(\mathcal{O}(G))} = vv^*$, then it is said to be a *unitary representation*.

Example 2.2 The first, extremely important example is u, which is a unitary representation. Since it defines the quantum group, it ought to determine all the representations. This idea will be made clear by the Tannaka–Krein reconstruction Theorem 3.7, and because of this peculiar role, u is called the *fundamental representation* of G.

Example 2.3 The second important example is the element

$$\varepsilon = 1 \in M_1(\mathcal{O}(G)) = \mathcal{O}(G),$$

which is also a representation called the *trivial representation*.

The standard operations on representations generalise to this setting and will be crucial in the sequel. For instance, if v and w are two finite-dimensional representations of dimension n and m respectively, then

$$v \oplus w = \begin{pmatrix} v & 0 \\ 0 & w \end{pmatrix} \in M_{n+m}(\mathcal{O}(G))$$

is their *direct sum*, while

$$v \otimes w = (v_{ij} w_{k\ell})_{1 \leqslant i,j \leqslant n | 1 \leqslant k,\ell \leqslant m} \in M_{nm}(\mathcal{O}(G))$$

is their *tensor product*. Here, our convention is that the rows of $v \otimes w$ are indexed by the pairs (i, k), and the columns by the pairs (j, ℓ).

Exercise 2.1 Prove that the direct sum and the tensor product of two representations is again a representation, and that unitarity is preserved under these operations.

Solution Let us start with the direct sum $z = v \oplus w$. First,

$$\begin{pmatrix} v^{-1} & 0 \\ 0 & w^{-1} \end{pmatrix} z = \begin{pmatrix} v^{-1}v & 0 \\ 0 & w^{-1}w \end{pmatrix}$$

$$= \begin{pmatrix} \text{Id} & 0 \\ 0 & \text{Id} \end{pmatrix}$$

$$= \begin{pmatrix} vv^{-1} & 0 \\ 0 & ww^{-1} \end{pmatrix}$$

$$= z \begin{pmatrix} v^{-1} & 0 \\ 0 & w^{-1} \end{pmatrix},$$

so that this is an invertible matrix as soon as v and w are. The same computation, moreover, shows that $z^* = z^{-1}$ if v and w are unitary. Second, the property of the coproduct is checked case by case depending on whether the indices are smaller or larger than $n = \dim(v)$. For instance, for $1 \leqslant i,j \leqslant n$,

$$\Delta(z_{ij}) = \Delta(v_{ij}) = \sum_{k=1}^{n} v_{ik} \otimes v_{kj} = \sum_{k=1}^{n} z_{ik} \otimes z_{kj} = \sum_{k=1}^{n+m} z_{ik} \otimes z_{kj},$$

and similarly for the three other cases.

We now turn to the tensor product $z = v \otimes w$. Invertibility is again easily checked, following our indexing conventions: setting $z^{-1} = w^{-1} \otimes v^{-1}$,

$$\sum_{a=1}^{n} \sum_{b=1}^{m} z_{(i,k),(a,b)}^{-1} z_{(i',k'),(a,b)} = \sum_{b=1}^{m} (w^{-1})_{kb} \left(\sum_{a=1}^{n} (v^{-1})_{ia} v_{i'a} \right) w_{k',b}$$

$$= \sum_{b=1}^{m} (w^{-1})_{kb} \delta_{ii'} w_{k',b}$$

$$= \delta_{(i,k),(i',k')},$$

and similarly the other way round. This then implies that $z^* = z^{-1}$ when v and w are unitary. The other property is a consequence of the multiplicativity of the coproduct:

$$\Delta\left(z_{(i,k),(j,\ell)}\right) = \Delta(v_{ij})\Delta(w_{k\ell})$$

$$= \sum_{a=1}^{n} \sum_{b=1}^{m} v_{ia} w_{kb} \otimes v_{aj} w_{b\ell}$$

$$= \sum_{a=1}^{n} \sum_{b=1}^{m} z_{(i,k),(a,b)} \otimes z_{(a,b),(k,\ell)}. \qquad \square$$

Our notion of a representation is sufficient to develop the whole theory. However, there are situations where it is easier to think of representations as a particular kind of action. Such a description is also available in the quantum setting, and we will now introduce it. A representation of G on V can be seen as a map

$$\alpha \colon G \times V \to V$$

satisfying the axioms of an action and the conditions making the maps $\alpha(g, \cdot)$ linear. To write such a map in terms of $\mathcal{O}(G)$, we will need the following result.

Lemma 2.4 *Let G be a compact group of matrices and let V be a vector space. Then, the map*

$$f \otimes x \mapsto (g \mapsto f(g)x)$$

is well defined and yields a linear isomorphism between $\mathcal{O}(G) \otimes V$ and the space $\mathcal{O}(G, V)$ of functions from G to V for which all coordinates are polynomial in the matrix coefficients.

Proof As usual, we first consider the obviously well-defined map

$$\widetilde{\Phi} \colon \mathcal{F}(\mathcal{O}(G) \times V) \to \mathcal{O}(G, V)$$

satisfying the same formula as in the statement. A straightforward computation shows that $\mathcal{I}(\mathcal{O}(G), V) \subset \ker(\widetilde{\Phi})$, and so we have a well-defined quotient map

$$\Phi \colon \mathcal{O}(G) \otimes V \to \mathcal{O}(G, V)$$

as in the statement. Fixing a basis $(e_i)_{1 \leqslant i \leqslant \dim(V)}$ of V, consider now an element

$$x = \sum_{i=1}^{\dim(V)} f_i \otimes e_i \in \ker(\Phi).$$

By linear independence (Proposition 1.12), it follows that, for all $1 \leqslant i \leqslant \dim(V)$ and $g \in G$, $f_i(g) = 0$. This implies that $f_i = 0$ as an element of $\mathcal{O}(G)$, hence $x = 0$ and Φ is injective. As for surjectivity, for $f \in \mathcal{O}(G, V)$ set $f_i = e_i^* \circ f \in \mathcal{O}(G)$. Then,

$$f = \sum_{i=1}^{\dim(V)} f_i e_i = \Phi \Big(\sum_{i=1}^{\dim(V)} f_i \otimes e_i \Big),$$

so that Φ is surjective, concluding the proof. $\qquad\square$

It is now easy to characterise representations as particular actions.

Proposition 2.5 *Let G be a compact group of matrices and let $\alpha\colon G \times V \to V$ be a map which is linear in the second variable. It is a linear action if and only if the corresponding map*

$$\rho\colon \begin{cases} V & \to & \mathcal{O}(G,V) = \mathcal{O}(G) \otimes V \\ x & \mapsto & (g \mapsto \alpha(g)(x)) \end{cases}$$

is injective and satisfies

$$(\mathrm{id} \otimes \rho) \circ \rho = (\Delta \otimes \mathrm{id}) \circ \rho. \tag{2.1}$$

Proof Assume first that α is an action. Then, the map $\eta\colon \mathcal{O}(G,V) \to V$ sending a function f to $f(e)$ is a left inverse for ρ, hence ρ is injective. Moreover, Equation (2.1) translates, for any $x \in V$ and $g, h \in G$, into

$$\alpha(h, \alpha(g, x)) = \alpha(gh, x),$$

which holds by definition of an action.

Conversely, if ρ satisfies Equation (2.1), then α is almost an action, the only thing to be checked being that $\alpha(e, \cdot) = \mathrm{id}_V$. Note that, because $\alpha(e, x) = \alpha(e, \alpha(e, x))$, the operator $\alpha(e, \cdot)$ is a projection. If it is not the identity, then, for any x in its kernel and any $g \in G$,

$$\alpha(g, x) = \alpha(g, \alpha(e, x)) = 0,$$

so that $\rho(x) = 0$, contradicting injectivity. $\qquad\square$

This leads us to an alternative definition, usually called a corepresentation of $\mathcal{O}(\mathbb{G})$. We will keep this distinction in order to avoid confusion with our previous definition.

Definition 2.6 Let $\mathbb{G}$ be a unitary compact matrix quantum group. A *corepresentation* of $\mathcal{O}(\mathbb{G})$ is a vector space V together with a linear map

$$\rho\colon V \to \mathcal{O}(G) \otimes V,$$

which is injective and satisfies Equation (2.1).

Given two corepresentations ρ_1 and ρ_2 on V_1 and V_2 respectively, their direct sum is simply defined by

$$\rho_1 \oplus \rho_2\colon (x_1, x_2) \mapsto \rho_1(x_1) + \rho_2(x_2) \in (\mathcal{O}(\mathbb{G}) \otimes (V_1 \oplus V_2))$$
$$\simeq (\mathcal{O}(\mathbb{G}) \otimes V_1) \oplus (\mathcal{O}(\mathbb{G}) \otimes V_2).$$

We used here a distributivity property of the tensor product, the proof of which we leave as an exercise.

Exercise 2.2 Prove that, given three vector spaces V_1, V_2 and V_3, there exists a canonical isomorphism

$$V_1 \otimes (V_2 \oplus V_3) \simeq (V_1 \otimes V_2) \oplus (V_1 \otimes V_3).$$

Solution Let us consider the map

$$\Phi\colon \mathcal{F}(V_1 \times (V_2 \times V_3)) \to \mathcal{F}(V_1 \times V_2) \oplus \mathcal{F}(V_1 \times V_3) \to (V_1 \otimes V_2) \oplus (V_1 \otimes V_3)$$

sending $(x_1, (x_2, x_3))$ to $((x_1 \otimes x_2), (x_1 \otimes x_3))$. Then, $\mathcal{I}(V_1, V_2 \oplus V_3) \subset \ker(\Phi)$, and so it factors as

$$\widetilde{\Phi}\colon V_1 \otimes (V_2 \oplus V_3) \to (V_1 \otimes V_2) \oplus (V_1 \otimes V_3).$$

In the same way, there exists a linear map

$$\widetilde{\Psi}\colon (V_1 \otimes V_2) \oplus (V_1 \otimes V_3) \to V_1 \otimes (V_2 \oplus V_3)$$

sending $((x_1 \otimes x_2), (x_1' \otimes x_3))$ to $x_1 \otimes (x_2, 0) + x_1' \otimes (0, x_3)$. To conclude, it suffices to check that these two maps are inverse to one another, which is straightforward. $\qquad\square$

The description of the tensor product is a bit more involved: if

$$\rho(x_1) = \sum_i a_1^i \otimes x_1^i \text{ and } \rho(x_2) = \sum_j a_2^j \otimes x_2^j,$$

then

$$\rho_1 \otimes \rho_2\colon x_1 \otimes x_2 \mapsto \sum_{i,j} a_1^i a_2^j \otimes x_1^i \otimes x_2^j.$$

The two points of view are, of course, equivalent. To see this, let us start with a representation v of dimension n and let $(e_i)_{1 \leqslant i \leqslant N}$ be the canonical basis of $\mathbf{C}^n$. We set

$$\rho_v\colon e_i \mapsto \sum_{j=1}^{n} v_{ij} \otimes e_j$$

and extend it to a linear map $\mathbf{C}^n \to \mathcal{O}(\mathbb{G}) \otimes \mathbf{C}^n$. Conversely, if ρ is a coaction on V and if a basis is fixed, then there exist elements $(v(\rho)_{ij})_{1 \leqslant i,j \leqslant \dim(V)}$ in $\mathcal{O}(\mathbb{G})$ such that

$$\rho(e_i) = \sum_{j=1}^{\dim(V)} v(\rho)_{ij} \otimes e_j.$$

Remark 2.7 There are canonical isomorphisms (whose existence the reader is invited to prove)

$$\mathcal{L}(V, \mathcal{O}(\mathbb{G}) \otimes V)) \simeq \mathcal{O}(G) \otimes V \otimes V^* \simeq \mathcal{O}(\mathbb{G}) \otimes \mathcal{L}(V)$$

providing a basis-free version of the preceding correspondence between representations and corepresentations.

As one may hope, the direct sum and tensor product constructions behave well with respect to the operations $v \to \rho_v$ and $\rho \to v(\rho)$.

Proposition 2.8 *With the preceding notations, ρ is a corepresentation if and only if $v(\rho)$ is a representation, and v is a representation if and only if ρ_v is a corepresentation. Moreover, the following identities hold:*

1. $\rho_{v(\rho)} = \rho$ and $v(\rho_v) = v$.
2. $\rho_{v \oplus w} = \rho_v \oplus \rho_w$.
3. $\rho_{v \otimes w} = \rho_v \otimes \rho_w$.
4. $v(\rho \oplus \delta) = v(\rho) \oplus v(\delta)$.
5. $v(\rho \otimes \delta) = v(\rho) \otimes v(\delta)$.

Proof We start by noticing that

$$(\mathrm{id} \otimes \rho) \circ \rho(e_i) = \sum_{j=1}^{n} \sum_{k=1}^{n} v(\rho)_{ij} \otimes v(\rho)_{jk} \otimes e_k$$

while

$$(\Delta \otimes \mathrm{id}) \circ \rho(e_i) = \sum_{k=1}^{n} \Delta(v(\rho)_{ik}) \otimes e_k.$$

By linear independence, this shows that Equation (2.1) is equivalent to the compatibility of $v(\rho)$ with the coproduct.

Let us now show that if ρ is injective, then $v(\rho)$ is invertible. For this we will need an analogue of the evaluation map at the identity, which is given by the counit ε: this is the unique $*$-homomorphism $\varepsilon \colon \mathcal{O}(\mathbb{G}) \to \mathbf{C}$ sending u_{ij} to δ_{ij}. (Here u denotes the fundamental representation of $\mathbb{G}$.)[1] It satisfies the equation

$$(\varepsilon \otimes \mathrm{id}) \circ \Delta = \mathrm{id}$$

(check this on the generators, where it is obvious), and so, setting $P = (\varepsilon \otimes \mathrm{id}) \circ \rho \in \mathcal{L}(V)$,

---

[1] See Proposition 2.9 for the existence of this map.

$$P = (\varepsilon \otimes \varepsilon \otimes \mathrm{id}) \circ (\Delta \otimes \mathrm{id}) \circ \rho$$
$$= (\varepsilon \otimes \varepsilon \otimes \mathrm{id}) \circ (\mathrm{id} \otimes \rho) \circ \rho$$
$$= (\varepsilon \otimes P) \circ (\mathrm{id} \otimes \mathrm{id}) \circ \rho$$
$$= P \circ (\varepsilon \otimes \mathrm{id}) \circ \rho$$
$$= P^2,$$

hence P is a projection. Since, moreover,

$$\rho \circ P = \rho \circ (\varepsilon \otimes \mathrm{id}) \circ \rho$$
$$= (\varepsilon \otimes \mathrm{id} \otimes \mathrm{id}) \circ (\mathrm{id} \otimes \rho) \circ \rho$$
$$= (\varepsilon \otimes \mathrm{id} \otimes \mathrm{id}) \circ (\Delta \otimes \mathrm{id}) \circ \rho$$
$$= \rho,$$

we see that, if $P \neq \mathrm{id}_V$, then ρ is not injective. We have therefore proven that

$$\sum_{j=1}^{\dim(V)} \varepsilon(v(\rho))_{ij} e_j = e_i$$

for all $1 \leqslant i \leqslant \dim(V)$, that is, $\varepsilon(v(\rho)_{ij}) = \delta_{ij}$. Let us consider now the unique $*$-antihomomorphism[2] S of $\mathcal{O}(\mathbb{G})^2$ such that $S(u_{ij}) = u_{ji}$. Then, one can check on the generators that

$$m \circ (\mathrm{id} \otimes S) \circ \Delta = \varepsilon = m \circ (S \otimes \mathrm{id}) \circ \Delta,$$

and applying this to $v(\rho)$ shows that $(S(v(\rho)_{ij}))_{1 \leqslant i,j \leqslant n}$ is an inverse for $v(\rho)$. Assume conversely that $v(\rho)$ is invertible and let w be its inverse. The matrix

$$M = (\varepsilon(v(\rho)_{ij}))_{1 \leqslant i,j \leqslant n}$$

is then invertible with inverse $(\varepsilon(w_{ij}))_{1 \leqslant i,j \leqslant n}$. Moreover,

$$M_{ij} = (\varepsilon \otimes \varepsilon) \circ \Delta(v(\rho)_{ij}) = \sum_{k=1}^{n} M_{ik} M_{kj} = (M^2)_{ij},$$

so that M is an invertible projection, hence the identity. In other words, $\varepsilon(v(\rho)_{ij}) = \delta_{ij}$, which yields

$$(\varepsilon \otimes \mathrm{id}) \circ \rho = \mathrm{id}_V,$$

that is, the injectivity of ρ.

---

[2] We are slightly cheating here; see Corollary 2.18 and the comments thereafter. Note, however, that in concrete situations this map can often be constructed by hand, like in Exercise 1.7.

The equalities $v(\rho_v) = v$ and $\rho_{v(\rho)} = \rho$ directly follow from the definitions. Moreover, applying the definition, we see that, for $x \in V$,

$$(\rho_v \oplus \rho_w)(x) = \rho_v(x) = \rho_{v \oplus w}(x),$$

and similarly for $x \in W$, yielding the first equality. The second equality in turn follows from

$$(\rho_v \otimes \rho_w)(e_i \otimes e_k) = \sum_{j=1}^{n} \sum_{\ell=1}^{m} v_{ij} w_{k\ell} \otimes e_j \otimes e_\ell$$

$$= (v \otimes w)_{(i,k),(j,\ell)} \otimes e_j \otimes e_\ell.$$

The remainder is a direct consequence of this. □

We insist that this result is only meant to make computations simpler in the sequel, but that one could write all the proofs using only representations. Therefore, the existence of the maps S and ε is not necessary to prove any of the subsequent results on representation theory. In any case, the existence of ε is easy to prove, using what we already know.

Proposition 2.9 *Let* $\mathbb{G} = (\mathcal{O}(\mathbb{G}), u)$ *be an orthogonal compact matrix quantum group. Then, there exists a unique $*$-homomorphism*

$$\varepsilon \colon \mathcal{O}(\mathbb{G}) \to \mathbf{C}$$

such that, for all $1 \leqslant i, j \leqslant N$,

$$\varepsilon(u_{ij}) = \delta_{ij}.$$

Proof Let us denote by I the $*$-ideal in $\mathcal{O}(\mathbb{G})$ generated by the following elements:

- $u_{ij} - u_{ij}^2$, for all $1 \leqslant i, j \leqslant N$;
- $\sum_{k=1}^{N} u_{ik} - 1$ and $\sum_{k=1}^{N} u_{kj} - 1$, for all $1 \leqslant i, j \leqslant N$;
- $u_{ij} u_{k\ell} - u_{k\ell} u_{ij}$ for all $1 \leqslant i, j, k, \ell \leqslant N$.

The quotient $\mathcal{A} = \mathcal{O}(\mathbb{G})/I$, together with the image of u, is an orthogonal compact matrix quantum group. Moreover, by Exercise 1.3, it is a quotient of $\mathcal{O}(S_N)$, hence finite-dimensional. Applying the result of Question (4) in Exercise 1.10, we conclude that $\mathcal{A} = \mathcal{O}(G)$ for some finite group G. If π denotes the quotient map and $\mathrm{ev}_{\mathrm{Id}}$ denotes the evaluation map at the identity element in G, then

$$\varepsilon = \mathrm{ev}_{\mathrm{Id}} \circ \pi$$

satisfies the desired property. Uniqueness follows from the fact that the image of all the generators is prescribed.   □

2.1.2 Intertwiners

The heart of representation theory is understanding the link between various representations, and this link is given by the notion of intertwiner, which we now introduce. In the following definitions, we see scalar matrices as matrices with coefficients in $\mathcal{O}(\mathbb{G})$ through the embedding $\mathbf{C}.1_{\mathcal{O}(\mathbb{G})} \subset \mathcal{O}(\mathbb{G})$.

Definition 2.10 Let $\mathbb{G}$ be a unitary compact matrix quantum group and let v, w be representations of $\mathbb{G}$ of dimension n and m respectively.

- An *intertwiner* between v and w is a linear map $T \colon \mathbf{C}^n \to \mathbf{C}^m$ such that

$$Tv = wT$$

  as matrices in $M_{n,m}(\mathcal{O}(\mathbb{G}))$.
- The representations v and w are said to be *equivalent* if there exists an invertible intertwiner between them. If this intertwiner, moreover, is unitary, then they are said to be *unitarily equivalent*.
- The representation w is said to be a *subrepresentation* of v if there exists an injective intertwiner between v and w.
- A representation is said to be *irreducible* if it has no subrepresentation except for itself.

It turns out that concrete computations with intertwiners are often easier to do using corepresentations. As one may expect, the concept is easily translated.

Exercise 2.3 Let v and w be representations of dimension n and m respectively. Prove that $T \colon \mathbf{C}^n \to \mathbf{C}^m$ intertwines v and w if and only if

$$(\mathrm{id} \otimes T) \circ \rho_v = \rho_w \circ T. \tag{2.2}$$

Solution Let us denote by $(e_i)_{1 \leqslant i \leqslant n}$ and $(f_i)_{1 \leqslant i \leqslant m}$ respectively the canonical bases of $\mathbf{C}^n$ and $\mathbf{C}^m$. To match our definition of $\rho(v)$, we will use the (unconventional) notation

$$T(e_i) = \sum_{j=1}^{m} T_{ij} f_j.$$

Applying both sides of Equation (2.2) to e_i yields

$$\sum_{j=1}^{n}\sum_{k=1}^{m} v_{ij} \otimes T_{jk}f_k = \sum_{j=1}^{m}\sum_{k=1}^{m} w_{jk} \otimes T_{ij}f_k.$$

This equality holds if and only if it holds for each k separately, yielding

$$\sum_{j=1}^{n} T_{jk}v_{ij} = \sum_{j=1}^{m} w_{jk}T_{ij}$$

for all $1 \leqslant i \leqslant n$ and $1 \leqslant k \leqslant m$. With our unusual convention, this is equivalent to $Tv = wT$.   □

This correspondence allows for another description of subrepresentations, using the following notion of invariant subspace.

Definition 2.11 Let ρ be a corepresentation of $\mathcal{O}(\mathbb{G})$ on a vector space V. A vector subspace W of V is said to be *invariant* under ρ if

$$\rho(W) \subset \mathcal{O}(\mathbb{G}) \otimes W.$$

Proposition 2.12 *Let v be a* unitary *representation of dimension n, let $\{0\} \neq W \subset \mathbf{C}^n$ be a vector subspace and let P_W be the orthogonal projection onto W. The following are equivalent:*

- *P_W intertwines v with itself;*
- *Both W and $W^{\perp}$ are invariant subspaces of ρ_v.*

Moreover, in that case, the matrix $w = P_W v P_W$, seen as an element of

$$\mathcal{O}(\mathbb{G}) \otimes M_{\dim(W)}(\mathbf{C}) \simeq M_{\dim(W)}(\mathcal{O}(\mathbb{G})),$$

is a unitary representation, and the inclusion map $i_W \colon W \hookrightarrow \mathbf{C}^n$ is an intertwiner between w and v.

Proof By Exercise 2.3, P_W intertwines v with itself if and only if

$$\rho_v \circ P_W = (\mathrm{id} \otimes P_W) \circ \rho_v. \tag{2.3}$$

If this holds, then for all $x \in W$, $\rho_v(x) = (\mathrm{id} \otimes P_W) \circ \rho_v(x)$, that is, $\rho_v(x) \in \mathcal{O}(\mathbb{G}) \otimes W$. Moreover, because $P_{W^{\perp}} = \mathrm{Id} - P_W$, the former also intertwines v with itself so that we also have $\rho_v(y) \in \mathcal{O}(\mathbb{G}) \otimes W^{\perp}$ for all $y \in W^{\perp}$. Conversely, let $z \in \mathbf{C}^N$ and decompose it as $z = x + y$ with $x \in W$ and $y \in W^{\perp}$. Then,

$$(\mathrm{id} \otimes P_W) \circ \rho_v(z) = (\mathrm{id} \otimes P_W) \circ \rho_v(x) + (\mathrm{id} \otimes P_W) \circ \rho_v(y)$$
$$= (\mathrm{id} \otimes P_W) \circ \rho_v(x)$$
$$= \rho_v(x)$$
$$= \rho_v \circ P_W(z).$$

Let us choose an orthonormal basis $(e_i)_{1 \leqslant i \leqslant n}$ such that, if $d = \dim(W)$, then $e_i \in W$ for all $1 \leqslant i \leqslant d$ and $e_i \in W^\perp$ for $d+1 \leqslant i \leqslant n$. Then, $w_{ij} = v_{ij}$ for all $1 \leqslant i, j \leqslant d$. Moreover, Equation (2.3) implies that $v_{ij} = 0$ if $i \leqslant d$ and $j > d$, hence

$$\Delta(w_{ij}) = \sum_{k=1}^{n} v_{ik} \otimes v_{kj} = \sum_{k=1}^{d} v_{ik} \otimes v_{kj} = \sum_{k=1}^{d} w_{ik} \otimes w_{kj}.$$

Moreover, we have $\rho_v \circ i_W = (\mathrm{id} \otimes i_W) \circ \rho_w$ so that i_W is an intertwiner. All that is left to prove is that w is unitary. This follows from the fact that $i_W^* v i_W = w$, which yields

$$w^* w = i_W^* v^* i_W i_W^* v i_W$$
$$= i_W^* v^* P_W v i_W$$
$$= i_W^* P_W v^* v i_W$$
$$= i_W^* P_W i_W$$
$$= \mathrm{id}_W,$$

and similarly for ww^*.   □

Note in particular that v is irreducible if and only if ρ_v has no invariant subspace apart from $\{0\}$ and $\mathbf{C}^n$.

Definition 2.13 With the previous notations, the representation w is called the *restriction of v to W*.

Let us illustrate this with some simple examples.

Exercise 2.4 Prove that the trivial representation is a subrepresentation of the fundamental representation P of S_N^+.

Solution Consider the fundamental representation P of S_N^+, let $(e_i)_{1 \leqslant i \leqslant N}$ be the canonical orthonormal basis of $\mathbf{C}^N$ and set

$$\xi = \sum_{i=1}^{N} e_i.$$

It is a straightforward consequence of Condition (2) in Definition 1.6 that

$$\rho_P(\xi) = 1 \otimes \xi.$$

In other words, ρ_P has a fixed vector, which is equivalent to P having a trivial subrepresentation given by $\mathbf{C}.(1 \otimes \xi) \in \mathcal{O}(S_N^+) \otimes \mathbf{C}^n$. □

This phenomenon is analogous to a well-known fact for permutation groups: the permutation representation ρ of S_N on $\mathbf{C}^N$ decomposes as the direct sum of the trivial representation and an irreducible representation.[3] Let us show that the same holds for S_N^+ by exploiting the idea that S_N is a 'subgroup' of S_N^+.

Exercise 2.5 We set $V = \xi^\perp \subset \mathbf{C}^N$.

1. Prove that the restriction of P to V is a subrepresentation.
2. Let $W \subset V$ be a subspace which is a subrepresentation of P. Prove that W is stable under the action of S_N by permutation of the canonical basis.
3. Conclude that V is irreducible.

Solution 1. Note that because

$$(\mathrm{id} \otimes \xi^*)\rho_P(e_i - e_j) = \sum_{k=1}^{N} p_{ik}\langle \xi, e_k \rangle - \sum_{\ell=1}^{N} p_{j\ell}\langle \xi, e_k \rangle$$
$$= \sum_{k=1}^{N} p_{ik} - \sum_{\ell=1}^{N} p_{j\ell}$$
$$= 0,$$

$\rho_P(V) \subset \mathcal{O}(\mathbb{G}) \otimes V$. Therefore, V is a subrepresentation.
2. Let ρ be the permutation representation of S_N and let

$$\widetilde{\rho}\colon V \to \mathcal{O}(S_N) \otimes V \simeq \mathcal{O}(S_N, V)$$

be the map sending x to $g \mapsto \rho(g)(x)$. We have the equality

$$(\pi_{\mathrm{ab}} \otimes \mathrm{id}_V) \circ \rho_P = \widetilde{\rho},$$

from which it follows that W is invariant under $\widetilde{\rho}$.
3. We know that V is irreducible for the representation ρ of S_N. Thus, $W = \{0\}$ or $W = V$, and so V is irreducible for P. □

We can use the same strategy for the fundamental representation of O_N^+.

---

[3] Simply notice that $\xi^\perp$ contains the vector $e_1 - e_2$ and that by letting S_N act on it, one can obtain all the vectors $e_i - e_j$ for any $i \neq j$. Note also that these vectors generate $\xi^\perp$.

Exercise 2.6 Let $N \geqslant 2$ be an integer and consider the fundamental representation U of O_N^+.

1. Show that U is irreducible.
2. Let $(e_i)_{1 \leqslant i \leqslant N}$ be the canonical basis of $\mathbf{C}^N$. Show that the vector

$$\xi = \sum_{i=1}^{N} e_i \otimes e_i$$

is fixed for $\rho_{U \otimes U}$.

Solution The strategy is the same as for S_N^+, so let us denote by ρ the defining representation of O_N on $V = \mathbf{C}^N$ and by $\tilde{\rho}$ the map $x \mapsto (g \mapsto \rho(g)(x))$.

1. Because $(\pi_{ab} \otimes \mathrm{id}) \circ \rho_U = \tilde{\rho}$ and the right-hand side is irreducible, we infer that ρ_U, hence also U, is irreducible.
2. We compute

$$\rho_{U \otimes U}(\xi) = \sum_{i=1}^{N} \sum_{k,\ell=1}^{N} U_{ik} U_{i\ell} \otimes e_k \otimes e_\ell$$

$$= \sum_{k,\ell=1}^{N} \left(\sum_{i=1}^{N} U_{ik} U_{i\ell} \right) \otimes e_k \otimes e_\ell$$

$$= \sum_{k,\ell=1}^{N} \delta_{k\ell} \otimes e_k \otimes e_\ell$$

$$= 1 \otimes \xi. \qquad \square$$

As one sees from these examples, pushing the study further by considering higher tensor powers of U or P, one will have to deal with the full representation theory of S_N and O_N. We will see in Chapter 3 that there is another way of investigating the representation theory of these quantum groups which completely avoids the use of classical groups. This is fortunate because the representation theory of O_N^+, for instance, turns out to be much simpler than that of O_N (see, for instance, Exercise 2.8).

2.1.3 Structure of the Representation Theory

In order to go further, we need a better understanding of the representation theory of unitary compact matrix quantum groups. Observe that $\mathcal{O}(\mathbb{G})$ is spanned by products of coefficients of u, which are nothing but the coefficients of tensor powers of u. In other words, the whole unitary compact matrix quantum group can be recovered from its finite-dimensional representations. The main theorem of this section, which is fundamental, turns this observation into a tractable tool for the study of unitary compact matrix quantum groups. Before getting to this, let us warm up by observing that Schur's celebrated lemma still holds in our setting.

Lemma 2.14 (Schur's Lemma) *Let $\mathbb{G}$ be a unitary compact matrix quantum group and let v and w be* irreducible *representations of dimension n and m respectively. If T is an intertwiner between v and w, then either $T = 0$ or T is invertible (hence v is equivalent to w). Moreover, in the latter case the space of intertwiners between v and w is one-dimensional.*

Proof Consider the subspace $Z = \ker(T) \subset \mathbf{C}^n$. Then, for any $x \in Z$,

$$(\mathrm{id} \otimes T) \circ \rho_v(x) = \rho_w \circ T(x) = 0,$$

so that

$$\rho_v(Z) \subset \ker(\mathrm{id} \otimes T) = \mathcal{O}(\mathbb{G}) \otimes Z$$

(the latter equality follows from Lemma 1.21). Hence, Z is stable, meaning that the restriction of v to it is a subrepresentation. By irreducibility, we conclude that either $Z = \mathbf{C}^n$, in which case $T = 0$, or $Z = \{0\}$. In the second case, consider the subspace $Z' = \mathrm{ran}(T)$. For $y \in Z'$, there exists $x \in \mathbf{C}^n$ such that $y = T(x)$, hence

$$\rho_w(y) = \rho_w \circ T(x) = (\mathrm{id} \otimes T) \circ \rho_v(x) \in \mathcal{O}(\mathbb{G}) \otimes Z'.$$

This means that Z' is invariant, and by irreducibility of w it is either $\{0\}$ or $\mathbf{C}^m$. Since the first case is excluded by the injectivity of T, $Z' = \mathbf{C}^m$ and T is surjective.

If T is invertible, let S be another intertwiner from v to W and observe that $T^{-1}S$ intertwines v with itself. If $\lambda \in \mathbf{C}$ is an eigenvalue of $T^{-1}S$, then $T^{-1}S - \lambda.\,\mathrm{id}$ is an intertwiner between v and itself and is not invertible. Thus $T^{-1}S = \lambda.\,\mathrm{id}$ by the first part of the statement, that is, $S = \lambda T$. $\qquad\square$

We can now turn to the statement and proof of the main theorem on the structure theory of unitary compact matrix quantum groups. It was first proven (in a slightly different version) by S. L Woronowicz in [75, proposition 4.6 and lemma 4.8] and it is the cornerstone of the study of compact quantum groups.

Theorem 2.15 (Woronowicz) *Let $\mathbb{G} = (\mathcal{O}(\mathbb{G}), u)$ be a unitary compact matrix quantum group. Then,*

1. *Any finite-dimensional unitary representation splits as a direct sum of irreducible unitary representations;*
2. *Coefficients of inequivalent irreducible finite-dimensional representations are linearly independent;*
3. *Any finite-dimensional representation is equivalent to a direct sum of irreducible unitary ones, hence equivalent to a unitary one.*

Proof 1. Let v be a finite-dimensional unitary representation of dimension n on a Hilbert space V and assume that it is not irreducible. This means that there exists another representation w on a vector space W together with an injective intertwiner $i_W : W \to V$. Equipping W with the pull-back of the inner product on $i_W(W)$ induced by the inner product of V, we may assume that i_W is an isometry. Then, the orthogonal projection $P_W = i_W i_W^*$ on W intertwines v with itself. By Proposition 2.12, W is an therefore an invariant subspace for ρ_v. Picking an orthonormal basis of W and completing it into an orthonormal basis of V, we get a unitary matrix $B \in M_n(\mathbf{C})$ such that $z = B^* v B$ is block upper triangular, that is, has the form

$$z = \begin{pmatrix} A_{11} & A_{12} \\ 0 & A_{22} \end{pmatrix}.$$

Let us check that this implies $A_{12} = 0$. We compute

$$z^* z = \begin{pmatrix} A_{11}^* A_{11} & A_{11}^* A_{12} \\ A_{12}^* A_{11} & A_{12}^* A_{12} + A_{22}^* A_{22} \end{pmatrix} = \begin{pmatrix} \mathrm{Id} & 0 \\ 0 & \mathrm{Id} \end{pmatrix}$$

and observe that A_{11} is conjugate to $w = P_W v P_W$. In particular, A_{11} is invertible as an element of $M_{\dim(W)}(\mathcal{O}(\mathbb{G}))$ by Proposition 2.12, and the upper left corner of $z^* z$ shows that its inverse is A_{11}^*. But then, looking at the upper right corner yields

$$A_{12} = A_{11} A_{11}^* A_{12} = A_{11} 0 = 0.$$

As a consequence,

$$w = \begin{pmatrix} A_{11} & 0 \\ 0 & A_{22} \end{pmatrix}$$

is a direct sum of unitary subrepresentations. The result then follows by induction.

2. Let $\{v^{(1)}, \cdots, v^{(n)}\}$ be pairwise inequivalent irreducible representations with $v^{(i)}$ acting on a finite-dimensional vector space $V^{(i)}$. For a linear form $f \in \mathcal{O}(\mathbb{G})^*$, we denote by $\widehat{f}(v)$ the matrix with coefficients $(f(v_{ij}))_{1 \leqslant i, j \leqslant \dim(v)}$ and we set

$$B = \left\{ \sum_{i=1}^{n} \widehat{f}\left(v^{(i)}\right) \mid f \in \mathcal{O}(\mathbb{G})^* \right\} \subset \bigoplus_{i=1}^{n} \mathcal{L}\left(V^{(i)}\right) = \mathcal{B}.$$

We claim that this inclusion is an equality. The proof goes through several steps:

1. By definition, B is a vector space. Moreover, setting

$$f * g = (f \otimes g) \circ \Delta \in \mathcal{O}(\mathbb{G}),$$

we have

$$\left(\widehat{f}(v)\widehat{g}(v)\right)_{ij} = \sum_{k=1}^{\dim(v)} f(v_{ik})g(v_{kj}) = \left((\widehat{f * g})(v)\right)_{ij}$$

so that B is an algebra.

2. Let p_i be the minimal central projection in $\mathcal{B}$ corresponding to the ith summand and consider

$$\pi_i \colon x \in B \mapsto p_i x p_i \in \mathcal{L}\left(V^{(i)}\right).$$

This is an algebra representation which is, moreover, irreducible. Indeed, if $T \colon V^{(i)} \to V^{(i)}$, then the equality $T \circ \pi_i = \pi_i \circ T$ means that for all $f \in \mathcal{O}(\mathbb{G})^*$,

$$T\widehat{f}\left(v^{(i)}\right) = \widehat{f}\left(v^{(i)}\right)T.$$

By linearity, this is equivalent to

$$\widehat{f}\left(Tv^{(i)} - v^{(i)}T\right) = 0$$

which can in turn be written

$$f\left(\sum_{k=1}^{n} T_{jk}v_{k\ell}^{(i)} - v_{jk}^{(i)}T_{k\ell}\right) = 0 \text{ for all } 1 \leqslant j, \ell \leqslant \dim(v^{(i)}).$$

Since linear forms separate points, this implies that

$$\sum_{k=1}^{n} T_{jk}v_{k\ell}^{(i)} - v_{jk}^{(i)}T_{k\ell} = 0 \text{ for all } 1 \leqslant j, \ell \leqslant \dim(v^{(i)}),$$

which is equivalent to $Tv^{(i)} = v^{(i)}T$. Because $v^{(i)}$ is irreducible, Lemma 2.14 forces T to be a multiple of the identity, that is, π_i is irreducible. Thus, by Burnside's Theorem (see Theorem A.1 in the Appendix A),

$$\pi_i(B) = \mathcal{L}\left(V^{(i)}\right).$$

3. The proof of the irreducibility of π_i adapts easily to show that any inter-winer between π_i and π_j must be an intertwiner between $v^{(i)}$ and $v^{(j)}$. Because these are assumed to be irreducible and inequivalent, Lemma 2.14 forces $T = 0$. In other words, the representations are pairwise inequivalent.

4. Noticing that $\oplus_i \pi_i = \mathrm{id}_B$, we conclude that[4]

$$B = \mathrm{End}_B(B)$$

$$= \mathrm{End}_B\left(\bigoplus_{i=1}^{n} \pi_i(B)\right)$$

$$= \bigoplus_{i=1}^{n} \mathrm{End}_B\left(\pi_i(B)\right)$$

$$= \bigoplus_{i=1}^{n} \mathrm{End}_{\mathcal{L}(V^{(i)})}\left(\mathcal{L}\left(V^{(i)}\right)\right)$$

$$= \bigoplus_{i=1}^{n} \mathcal{L}\left(V^{(i)}\right)$$

$$= \mathcal{B}.$$

Now let, for each $1 \leqslant i \leqslant n$, $\left(\lambda_{k,\ell}^{(i)}\right)_{1 \leqslant k,\ell \leqslant \dim(v^{(i)})}$ be complex numbers and set

$$x = \sum_{i=1}^{n} \sum_{k,\ell=1}^{\dim(v^{(i)})} \lambda_{k,\ell}^{(i)} v_{k,\ell}^{(i)}.$$

If we denote by Λ_i the matrix with coefficients $\overline{\lambda}_{k,\ell}^{(i)}$, then

$$\Lambda = (\Lambda_1, \cdots, \Lambda_n) \in \mathcal{B} = B,$$

thus there exists $f \in \mathcal{O}(\mathbb{G})$ such that for all i, $\widehat{f}(v^{(i)}) = \Lambda_i$. As a consequence,

$$f(x) = \sum_{i=1}^{n} \sum_{k,\ell=1}^{\dim(v^{(i)})} |\lambda_{k,\ell}|^2$$

and x therefore vanishes if and only if all the coefficients vanish, proving linear independence.

---

[4] For the first and fifth lines, observe that for any unital algebra B, the map $b \mapsto (x \mapsto x.b)$ gives an isomorphism $B \mapsto \mathrm{End}_B(B)$ with inverse $T \mapsto T(1)$.

5. Observe that $\mathcal{O}(\mathbb{G})$ is spanned by the products of coefficients of u, that is to say coefficients of tensor powers of u which, by point 1, are linear combinations of coefficients of irreducible unitary representations. As a consequence, if v is a finite-dimensional representation of dimension d (without any unitarity assumption!), then there exists $\{v^{(1)}, \cdots, v^{(k)}\}$ unitary irreducible pairwise inequivalent representations such that all the coefficients of v are in the linear span of the coefficients of the $v^{(i)}$'s. This means, in particular, that if $f \in \mathcal{O}(\mathbb{G})^*$, then the value of f on a coefficient of v is completely determined by its value on the coefficients of all the $v^{(i)}$'s. This, together with the proof of the previous point, enables one to define an algebra homomorphism (where $n_i = \dim(v^{(i)})$)

$$\Phi \colon \bigoplus_i M_{n_i}(\mathbf{C}) \to M_d(\mathbf{C})$$

through the formula

$$\Phi\left(\widehat{f}(v^{(1)}) \oplus \cdots \oplus \widehat{f}(v^{(n)})\right) = \widehat{f}(v).$$

Now, $\ker(\Phi)$ being an ideal, it is equal to $\bigoplus_{i \notin I} M_{n_i}(\mathbf{C})$ for some subset $I \subset \{1, \cdots, k\}$. Thus, it factors through an injective algebra homomorphism

$$\widetilde{\Phi} \colon \bigoplus_{i \in I} M_{n_i}(\mathbf{C}) \simeq M_d(\mathbf{C}).$$

We furthermore claim that even though $\widetilde{\Phi}$ need not be surjective, the span of the ranges of elements in its image is $\mathbf{C}^d$. Indeed, let ξ be a vector orthogonal to the ranges of all the matrices $\widehat{f}(v)$. We have, for all $f \in \mathcal{O}(\mathbb{G})^*$ and all $1 \leqslant i \leqslant d$,

$$0 = \langle \xi, \widehat{f}(v)(e_i) \rangle = \sum_{j=1}^d \xi_j f(v_{ji}) = f\left(\sum_{j=1}^d v_{ji}\xi_j\right).$$

Because linear maps separate the points, this implies

$$\sum_{j=1}^d v_{ji}\xi_j = 0.$$

If now w is the inverse of v, multiplying by w_{ik} and summing over i yields

$$0 = \sum_{j=1}^d \sum_{k=1}^d v_{ji}w_{ik}\xi_j = \sum_{j=1}^d \delta_{jk}\xi_j = \xi_k,$$

and since this holds for all $1 \leqslant k \leqslant d$ we conclude that $\xi = 0$.

We can now finish the proof. Choosing a basis of the linear span of the ranges of elements in each matrix algebra yields an invertible matrix $Q \in M_d(\mathbf{C})$ such that

$$\widetilde{\Phi}\left(\bigoplus_{i \in I} \widehat{f}(v^{(i)})\right) = Q\left(\bigoplus_{i \in I} \widehat{f}(v^{(i)})\right) Q^{-1} = \widehat{f}\left(Q\left(\bigoplus_{i \in I} v^{(i)}\right) Q^{-1}\right).$$

Since this holds for any $f \in \mathcal{O}(\mathbb{G})^*$ and the latter separate the points, we conclude that

$$v = Q\left(\bigoplus_{i \in I} v^{(i)}\right) Q^{-1}.$$

$\square$

We have proved along the way a useful result which we restate here for later reference:

Lemma 2.16 *Let v and w be representations of a unitary compact matrix quantum group $\mathbb{G}$ of dimension n and m respectively. Then, $T \in \mathcal{L}(V,W)$ is an intertwiner if and only if, for all $f \in \mathcal{O}(\mathbb{G})^*$,*

$$T\widehat{f}(v) = \widehat{f}(w)T.$$

Proof The fact that T is an intertwiner reads, for any $1 \leqslant i \leqslant m$ and $1 \leqslant j \leqslant n$,

$$\sum_{k=1}^{n} T_{ik}v_{kj} = \sum_{k=1}^{m} T_{kj}w_{ik}.$$

Applying f to both sides yields the only if condition. The converse follows because linear maps separate the points. $\square$

Let us conclude this section by outlining some important consequences of Theorem 2.15.

Corollary 2.17 *Let $\mathbb{G} = (\mathcal{O}(\mathbb{G}), u)$ be an* orthogonal *compact matrix quantum group. Then, any irreducible representation is equivalent to a subrepresentation of $u^{\otimes k}$.*

Proof Because $\mathcal{O}(\mathbb{G})$ is generated by the coefficients of u, it is the linear span of coefficients of irreducible subrepresentations of tensor powers of u. Thus, any finite-dimensional representation is equivalent to one of these by point 2 of Theorem 2.15. $\square$

Corollary 2.18 *Let* $\mathbb{G} = (\mathcal{O}(\mathbb{G}), u)$ *be an orthogonal compact matrix quantum group. Then, there exists a unique $*$-antihomomorphism* $S \colon \mathcal{O}(\mathbb{G}) \to \mathcal{O}(\mathbb{G})$ *such that for all* $1 \leqslant i, j \leqslant N$,

$$S(u_{ij}) = u_{ji}.$$

Proof The uniqueness is clear; we only have to prove the existence. Let $\mathrm{Irr}(\mathbb{G})$ be a complete set of inequivalent irreducible unitary representations of $\mathbb{G}$ and let, for $\alpha \in \mathrm{Irr}(\mathbb{G})$, u^{α} be a representative of α. Then, the coefficients $(u_{ij}^{\alpha})_{\alpha \in \mathrm{Irr}(\mathbb{G}), 1 \leqslant i, j \leqslant \dim(\alpha)}$ form a basis of $\mathcal{O}(\mathbb{G})$, and so we can define S by

$$S(u_{ij}^{\alpha}) = \left(u_{ji}^{\alpha}\right)^{*}.$$

Splitting an arbitrary unitary representation into a sum of irreducible ones, we see that this formula holds for any unitary representation. Now, noticing that $u_{ij}^{\alpha} u_{k\ell}^{\beta} = u_{(i,k),(j,\ell)}^{\alpha \otimes \beta}$, we see that

$$S(u_{ij}^{\alpha} u_{k\ell}^{\beta}) = \left(u_{(\ell,j),(k,i)}^{\alpha \otimes \beta}\right)^{*} = \left(u_{\ell,k}^{\beta}\right)^{*} \left(u_{ji}^{\alpha}\right)^{*} = S(u_{k\ell}^{\beta})S(u_{ij}^{\alpha}). \qquad \square$$

Let us stress that even though the existence of S was used in Proposition 2.8 to prove that the matrix associated to a corepresentation is invertible, we have not used this afterwards, so that the proof of Theorem 2.15 does not rely upon this fact.

2.2 Interlude: Invariant Theory

Now that we have a nice abstract theory of representations of unitary compact matrix quantum groups, it should be high time we try to compute, say, all irreducible representations up to equivalence for some quantum groups like O_N^+. This is, however, not an easy task and requires the introduction of some combinatorial tools. To motivate this and to introduce the main object which will be needed, we start by revisiting the work of R. Brauer who, in [25], studied the orthogonal group O_N using partitions of finite sets.

2.2.1 The Orthogonal Group: Pair Partitions

Let us consider the group O_N. The defining representation ρ of O_N as matrices acting on $V = \mathbf{C}^N$ should by definition 'contain everything'. The precise meaning of this last expression is that, by Corollary 2.17, given any finite-dimensional representation π of O_N, there exists an integer k such that π is unitarily equivalent to a direct sum of subrepresentations of $\rho^{\otimes k}$.

This means that we can focus on subrepresentations of tensor powers of ρ. Going back to the definition, we see that subrepresentations can be read on the intertwiner spaces. Indeed, if T is an isometric intertwiner between π and $\rho^{\otimes k}$, then TT^* is an orthogonal projection intertwining $\rho^{\otimes k}$ with itself. In other words, it should in principle be possible to recover the whole representation theory of O_N from the algebra structure of the spaces

$$\mathrm{Mor}_{O_N}(\rho^{\otimes k}, \rho^{\otimes k}) = \{T \colon V^{\otimes k} \to V^{\otimes k} \mid T\rho^{\otimes k}(g)$$
$$= \rho^{\otimes k}(g)T \text{ for all } g \in O_N\}.$$

The study of such algebras is usually known under the name of *Schur–Weyl duality*.

Remark 2.19 The notation $\mathrm{Mor}_{O_N}(\rho^{\otimes k}, \rho^{\otimes k})$ emphasises the fact that intertwiners can be thought of as morphisms in a specific category whose objects are the (finite-dimensional) representations of O_N. This is why they are sometimes called *morphisms of representations* (see Section 3.3 for more details on the categorical aspects).

In our setting, orthogonality allows us to further reduce the problem thanks to the following elementary result.

Proposition 2.20 *For any integer k, there exists a canonical linear isomorphism*

$$\Phi_k \colon \mathrm{Mor}_{O_N}\left(\rho^{\otimes k}, \rho^{\otimes k}\right) \simeq \mathrm{Mor}_{O_N}\left(\rho^{\otimes 2k}, \varepsilon\right),$$

where ε denotes the trivial representation of O_N.

Proof Since the elements of O_N are unitary, they leave the inner product on V invariant. However, this inner product cannot be expressed as a linear form on $V \otimes V$ since it is not bilinear on $V \times V$ but only sesquilinear. It can, however, be made linear using the *duality map* $D_V \colon V \otimes V \to \mathbf{C}$ defined by

$$D_V(x \otimes y) = \langle x, \overline{y} \rangle,$$

where $\overline{y}$ is the image of y in the *conjugate Hilbert space*[5] $\overline{V}$. The key fact is that because the coefficients of matrices in O_N are real-valued, the map D_V is invariant under the representation $\rho^{\otimes 2}$. Now, any map

$$T \colon V^{\otimes k} \to V^{\otimes k}$$

---

[5] This is just the same abelian group as V but with scalar multiplication given by $\lambda.\overline{x} = \overline{\lambda}.\overline{x}$.

can be turned into a map

$$\tilde{T} \colon V^{\otimes 2k} \simeq V^{\otimes k} \otimes V^{\otimes k} \to \mathbf{C}$$

via the formula

$$\tilde{T} \colon x_1 \otimes \cdots \otimes x_k \otimes y_1 \otimes \cdots \otimes y_k \to D_{V^{\otimes k}} \left(T(x_1 \otimes \cdots \otimes x_k) \otimes y_k \otimes \cdots \otimes y_1 \right).$$

One easily checks that T is an intertwiner if and only if $\tilde{T}$ is, hence the result. $\qquad\square$

As a consequence, we are now looking for the *invariants* of the orthogonal group, that is to say the polynomial maps to $\mathbf{C}$ which are invariant under a given representation. Let us practice a little by computing $\mathrm{Mor}_{O_N}\left(\rho^{\otimes 2}, \varepsilon\right)$. We know that D_V yields a non-trivial element of this space. Moreover, because ρ is irreducible, it follows from Schur's Lemma 2.14 that

$$\dim\left(\mathrm{Mor}_{O_N}\left(\rho^{\otimes 2}, \varepsilon\right)\right) = \dim\left(\mathrm{Mor}_{O_N}\left(\rho, \rho\right)\right) = 1,$$

so that $\mathrm{Mor}_{O_N}\left(\rho^{\otimes 2}, \varepsilon\right) = \mathbf{C}.D_{\mathbf{C}^N}$.

We can extend this idea to build non-trivial elements of $\mathrm{Mor}_{O_N}\left(\rho^{\otimes 2k}, \varepsilon\right)$ for any $k \geqslant 1$ by pairing tensors using the map $D_{\mathbf{C}^N}$. To do this, we just need a partition p of $\{1, \cdots, 2k\}$ into subsets of size 2. Such a partition is called a *pair partition* and the set of pair partitions of $\{1, \cdots, 2k\}$ is denoted by $P_2(2k)$. Given such a pair partition p, we set

$$f_p \colon x_1 \otimes \cdots \otimes x_{2k} \mapsto \prod_{\{a,b\} \in p} D_{\mathbf{C}^N}(x_a \otimes x_b). \tag{2.4}$$

We can then produce many intertwiners by taking linear combinations, and one of the main results of R. Brauer's work [25] is that we indeed get everything.

Theorem 2.21 (Brauer) *For any integer k, we have*

$$\mathrm{Mor}_{O_N}\left(\rho^{\otimes 2k+1}, \varepsilon\right) = \{0\}$$
$$\mathrm{Mor}_{O_N}\left(\rho^{\otimes 2k}, \varepsilon\right) = \mathrm{Vect}\{f_p \mid p \in P_2(2k)\}.$$

2.2.2 The Quantum Orthogonal Group: Non-crossing Partitions

Building on the previous discussion, we now want to use the same strategy to investigate the spaces $\mathrm{Mor}_{O_N^+}\left(U^{\otimes 2k}, \varepsilon\right)$. One easily checks that the map $D_{\mathbf{C}^N}$ is still an invariant of $U^{\otimes 2}$.

$$(\mathrm{id} \otimes D_{\mathbf{C}^N}) \circ \rho_{U^{\otimes 2}}(e_i \otimes e_j) = (\mathrm{id} \otimes D_{\mathbf{C}^N}) \Big(\sum_{k,\ell} U_{ik} U_{j\ell} \otimes e_k \otimes e_\ell \Big)$$

$$= \sum_{k=1}^{N} U_{ik} U_{jk}$$

$$= \delta_{ij}$$

$$= \rho_\varepsilon (D_{\mathbf{C}^N}(e_i \otimes e_j)).$$

So what will be the difference between O_N and O_N^+ ? Let us look at $U^{\otimes 4}$ and the O_N-intertwiner of $\rho^{\otimes 4}$ given by the partition

$$p_{\mathrm{cross}} = \{\{1,3\}, \{2,4\}\}.$$

Does this yield an interwiner for O_N^+ ? Let us compute.

Exercise 2.7 Prove that for any orthogonal compact matrix quantum group $\mathbb{G} = (\mathcal{O}(\mathbb{G}), u)$,

$$(\mathrm{id} \otimes f_{p_{\mathrm{cross}}}) \circ \rho_{u^{\otimes 4}}(e_{i_1} \otimes e_{i_2} \otimes e_{i_3} \otimes e_{i_4}) = \sum_{k,\ell=1}^{N} u_{i_1 k} u_{i_2 \ell} u_{i_3 k} u_{i_4 \ell} \quad (2.5)$$

and

$$\rho_\varepsilon \circ f_{p_{\mathrm{cross}}}(e_{i_1} \otimes e_{i_2} \otimes e_{i_3} \otimes e_{i_4}) = \delta_{i_1 i_3} \delta_{i_2 i_4}. \quad (2.6)$$

Solution This is an elementary computation,

$$(\mathrm{id} \otimes f_{p_{\mathrm{cross}}}) \circ \rho_{u^{\otimes 4}}(e_{i_1} \otimes e_{i_2} \otimes e_{i_3} \otimes e_{i_4})$$

$$= \sum_{j_1, \cdots, j_4 = 1}^{N} u_{i_1 j_1} \cdots u_{i_4 j_4} f_p(e_{j_1} \otimes e_{j_2} \otimes e_{j_3} \otimes e_{j_4})$$

$$= \sum_{j_1, \cdots, j_4 = 1}^{N} u_{i_1 j_1} \cdots u_{i_4 j_4} \delta_{j_1 j_3} \delta_{j_2 j_4}$$

$$= \sum_{j_1, j_2 = 1}^{N} u_{i_1 j_1} u_{i_2 j_2} u_{i_3 j_1} u_{i_4 j_2},$$

and Equation (2.5) follows from the changes of indices $k = j_1$, $\ell = j_2$. As for Equation (2.6), this is the definition of $f_{p_{\mathrm{cross}}}$. $\square$

Understanding the meaning of this relation is the key to the world of partition quantum groups. We will therefore state it as a proposition.

Proposition 2.22 *Let* $\mathbb{G} = (\mathcal{O}(\mathbb{G}), u)$ *be an orthogonal compact matrix quantum group and let* $p_{cross} = \{\{1, 3\}, \{2, 4\}\}$. *Then,* $f_{p_{cross}} \in \mathrm{Mor}_{\mathbb{G}} \left(u^{\otimes 4}, \varepsilon \right)$ *if and only if* $\mathcal{O}(\mathbb{G})$ *is commutative.*

Proof We will play around with the equality (2.5) = (2.6) to show that the coefficients of u pairwise commute. More precisely, multiplying each side by $u_{i_4 j_2} u_{i_3 j_1}$, for two arbitrary indices $1 \leqslant j_1, j_2 \leqslant N$, and summing over i_4 and i_3 yields

$$\sum_{k, \ell, i_3, i_4 = 1}^{N} u_{i_1 k} u_{i_2 \ell} u_{i_3 k} u_{i_4 \ell} u_{i_4 j_2} u_{i_3 j_1} = \sum_{i_3, i_4 = 1}^{N} \delta_{i_1 i_3} \delta_{i_2 i_4} u_{i_4 j_2} u_{i_3 j_1}$$

$$= u_{i_2 j_2} u_{i_1 j_1}.$$

The left-hand side here can be simplified using the fact that u is orthogonal. Indeed,

$$\sum_{i_4 = 1}^{N} u_{i_4 \ell} u_{i_4 j_2} = \delta_{\ell j_2}$$

and similarly for i_3 so that

$$\sum_{k, \ell, i_3, i_4 = 1}^{N} u_{i_1 k} u_{i_2 \ell} u_{i_3 k} u_{i_4 \ell} u_{i_4 j_2} u_{i_3 j_1} = \sum_{k, \ell, i_3 = 1}^{N} u_{i_1 k} u_{i_2 \ell} u_{i_3 k} \delta_{\ell j_2} u_{i_3 j_1}$$

$$= \sum_{k, \ell = 1}^{N} u_{i_1 k} u_{i_2 \ell} \delta_{k j_1} \delta_{\ell j_2}$$

$$= u_{i_1 j_1} u_{i_2 j_2}.$$

Thus, we have proven that if f_p is an intertwiner, then the coefficients of u pairwise commute. Moreover, it is clear that the converse holds. $\quad\square$

Remark 2.23 Anticipating Corollary 5.18, we can restate the previous criterion as follows: $f_{p_{cross}} \in \mathrm{Mor}_{\mathbb{G}} \left(u^{\otimes 4}, \varepsilon \right)$ is an intertwiner if and only if $\mathbb{G}$ is a classical group.

As a consequence of Proposition 1.24, f_p is not an intertwiner of O_N^+ and in the construction of R. Brauer we have used, without noticing it, the commutativity of $\mathcal{O}(O_N)$. So what is really the smallest space of intertwiners that one can build from $D_{\mathbb{C}^N}$ using pair partitions? The answer relies on the following definition, where we call *block* any of the subsets defining the partition.

Definition 2.24 A partition is said to be *crossing* if there exists $k_1 < k_2 < k_3 < k_4$ such that

- k_1 and k_3 are in the same block,
- k_2 and k_4 are in the same block,
- the four elements are not in the same block.

Otherwise, it is said to be *non-crossing*.

To illustrate this notion we now introduce a very useful pictorial description of partitions: we draw k points in a row and then connect two points if and only if they belong to the same subset of the partition. It is clear that, for instance,

$$p_{\text{cross}} = \{\{1, 3\}, \{2, 4\}\} =$$

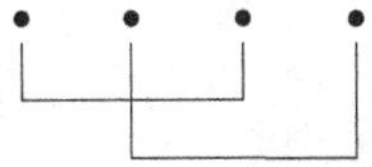

cannot be drawn without letting the lines cross. It is therefore a crossing partition.

It is certainly not clear at the moment that non-crossing partitions have to do with the quantum orthogonal groups O_N^+ and there is indeed still some work needed to see the link. As a motivation, let us state the quantum analogue of R. Brauer's result, which was proven by T. Banica in [3]. We will denote the set of non-crossing pair partitions on $2k$ points by $NC_2(2k)$.

Theorem 2.25 (Banica) *For any integer k, we have*

$$\begin{aligned} \mathrm{Mor}_{O_N^+}\left(\rho^{\otimes 2k+1}, \varepsilon\right) &= \{0\} \\ \mathrm{Mor}_{O_N^+}\left(\rho^{\otimes 2k}, \varepsilon\right) &= \mathrm{Vect}\{f_p \mid p \in NC_2(2k)\}. \end{aligned}$$

The proof will be a consequence of the general theory that we will develop in the sequel. As an application, we can completely decompose $U \otimes U$ into irreducible subrepresentations.

Exercise 2.8 Let U be the fundamental representation of O_N^+. Give the decomposition of $U \otimes U$ into irreducible subrepresentations.

Solution By Exercise 2.6 we already know that $U \otimes U$ contains a copy of the trivial representation and that $V = \xi^{\perp}$ is also a subrepresentation. Moreover,

$$\mathrm{Mor}_{O_N^+}(U^{\otimes 2}, U^{\otimes 2}) \cong \mathrm{Mor}_{O_N^+}\left(U^{\otimes 4}, \varepsilon\right)$$

has dimension at most 2, because there are only two non-crossing pairings on four points, namely $\{\{1, 4\}, \{2, 3\}\}$ and $\{\{1, 2\}, \{3, 4\}\}$. As a consequence, $U^{\otimes 2}$ has at most two irreducible subrepresentations. Therefore, ξ and V are irreducible and we are done. □

By way of comparison, the representation $\rho \otimes \rho$ of O_N splits into a direct sum of three irreducible representations: the trivial one, the orthogonal complement of the trivial one in the space of symmetric tensors, and the space of antisymmetric tensors. This can, for instance, be deduced from Theorem 2.21.

Part II

PARTITIONS ENTER THE PICTURE

3
Partition Quantum Groups

3.1 Linear Maps Associated to Partitions

As we have seen in Section 2.2, the idea to use partitions of finite sets to study the representation theory of compact groups dates back, at least, to the work of R. Brauer [25]. That connection was used several times and at several degrees of generality to study some specific families of compact quantum groups; but only with the breakthrough article of T. Banica and R. Speicher [18] did the systematic formalisation and study of the relationship between partitions and compact quantum groups start to spread as a research subject of its own.

Teodor Banica and R. Speicher were motivated by the question of *liberation*: how can one find a good presentation of the algebra of regular functions on a compact matrix group so that removing the commutation relations yields an interesting compact matrix quantum group? We will follow a different path based on the following summary of Section 2.2: the representation theory of O_N is determined by the pair partitions P_2 while the representation theory of O_N^+ is determined by the non-crossing pair partitions NC_2. This raises a natural question, namely: what other (quantum) groups have their representation theory determined by partitions?

The answer requires enlarging our setting. First, we will from now on consider arbitrary partitions of finite sets, not only those in pairs. To define a linear form associated to such a general partition p, we need to extend the definition of the maps f_p. This will be done by first defining a scalar function on tuples of integers. So let p be an arbitrary partition of the set $\{1, \ldots, k\}$ and consider integers $1 \leqslant i_1, \ldots, i_k \leqslant N$. Let us write, for $a, b \in \{1, \ldots, k\}$, $a \sim_p b$ if a and b belong to the same block of p. Then, we set

$$\delta_p(i_1, \ldots, i_k) = 1$$

61

if $i_a = i_b$ as soon as $a \sim_p b$. Otherwise, we set

$$\delta_p(i_1, \ldots, i_k) = 0.$$

The pictorial description of partitions enables one to easily compute such functions. This is done by 'placing' these indices on the points of p from left to right. Then, if whenever two indices are connected they are equal, we set $\delta_p(i_1, \ldots, i_k) = 1$ and otherwise $\delta_p(i_1, \ldots, i_k) = 0$. For instance, with the partition

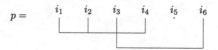

we get

$$\delta_p(i_1, i_2, i_3, i_4, i_5, i_6) = \delta_{i_1 i_2 i_4} \delta_{i_3 i_6}.$$

From now on, we will denote by $P(k)$ the set of all partitions of the set $\{1, \ldots, k\}$.

Definition 3.1 Let $p \in P(k)$ and let N be an integer. We define a map

$$f_p \colon \left(\mathbf{C}^N\right)^{\otimes k} \to \mathbf{C}$$

by the formula

$$f_p(e_{i_1} \otimes \cdots \otimes e_{i_k}) = \delta_p(i_1, \ldots, i_k).$$

Remark 3.2 Let us note the obvious but important fact that $f_p = f_q$ if and only if $p = q$.

As an example, let us consider the pair partition $\sqcup = \{\{1, 2\}\} \in P(2)$. It follows from a straightforward calculation using the canonical basis that $f_\sqcup$ is nothing but the duality map introduced in the proof of Proposition 2.20, that is to say

$$f_\sqcup = D_{\mathbf{C}^N} \colon x \otimes y \mapsto \langle x, \overline{y} \rangle.$$

This enables us to connect Definition 3.1 with the definition given by Equation (2.4).

Exercise 3.1 Check that, for pair partitions, this coincides with the previous definition.

Solution Simply observe that, for a pair partition,

$$\delta_p(i_1, \ldots, i_{2k}) = \prod_{\{a,b\} \subset p} \delta_\sqcup(i_a, i_b) = \prod_{\{a,b\} \subset p} D_{\mathbf{C}^N}\left(e_{i_a} \otimes e_{i_b}\right). \qquad \square$$

If k and ℓ are integers, then we claim that the space $\mathrm{Mor}_{\mathbb{G}}\left(V^{\otimes k}, V^{\otimes \ell}\right)$ should be recoverable from the invariants of $\mathbb{G}$. Let us explain how this follows from extending the construction of the maps Φ_k introduced in Proposition 2.20. Indeed, using the duality map D_V one can build an isomorphism

$$\Psi_{k,\ell} \colon \mathcal{L}\left(V^{\otimes k}, V^{\otimes \ell}\right) \to \mathcal{L}\left(V^{\otimes (k+1)}, V^{\otimes (\ell-1)}\right)$$

through the formula

$$\Psi_{k,\ell}(T)(x_1 \otimes \cdots \otimes x_k \otimes x_{k+1}) = \left(\mathrm{id}_V^{\otimes(\ell-1)} \otimes D_V\right)\left(T(x_1 \otimes \cdots \otimes x_k) \otimes x_{k+1}\right).$$

This can be pictorially seen as 'turning one point' of T as follows:

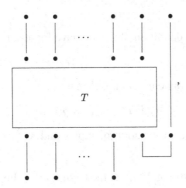

and iterating the construction gives isomorphisms

$$\Phi_{k,\ell} = \Psi_{k+\ell-1,1} \circ \cdots \circ \Psi_{k,\ell} \colon \mathcal{L}\left(V^{\otimes k}, V^{\otimes \ell}\right) \to \mathcal{L}\left(V^{\otimes(k+\ell)}, \mathbf{C}\right)$$

corresponding to 'rotating T on one line'. In particular, if p is a partition of $\{1, \ldots, k + \ell\}$ then we can define an operator

$$T_p = \Phi_{k,\ell}^{-1}(f_p)$$

which is an intertwiner as soon as f_p is.

The drawback of that construction is that we have lost the pictorial description of the operator. To recover it, notice that we can also draw a partition $p \in P(k + \ell)$ on two rows instead of one, drawing for instance k points corresponding to $\{1, \ldots, k\}$ in the upper row and ℓ points corresponding to $\{k + 1, \ldots, k + \ell\}$ in the lower row. With this point of view, the operator T_p admits an explicit description which generalises that of f_p. Let $(e_i)_{1 \leqslant i \leqslant N}$ be the canonical basis of $V = \mathbf{C}^N$. For a partition $p \in P(k + \ell)$ drawn with k upper points and ℓ lower points, we extend the definition of the function δ_p, taking now as arguments a k-tuple $i = (i_1, \ldots, i_k)$ and an ℓ-tuple $j = (j_1, \ldots, j_\ell)$, in the following way:

- We draw the indices of i on the upper points of the partition (from left to right) and the indices of j on the lower points of p (from left to right);
- If whenever two points are connected, then their indices are equal, we set $\delta_p(i, j) = 1$;
- Otherwise, we set $\delta_p(i, j) = 0$.

With this in hand, we have

Proposition 3.3 *For any tuples i and j,*

$$T_p(e_{i_1} \otimes \cdots \otimes e_{i_k}) = \sum_{j_1, \cdots, j_\ell = 1}^{N} \delta_p(i, j) e_{j_1} \otimes \cdots \otimes e_{j_\ell}. \qquad (3.1)$$

Proof Unwinding the definition of T_p, we have for any $j_1, \ldots, j_\ell$,

$$\langle T_p(e_{i_1} \otimes \cdots \otimes e_{i_k}), e_{j_1} \otimes \cdots \otimes e_{j_\ell} \rangle = f_p(e_{i_1} \otimes \cdots \otimes e_{i_k} \otimes e_{j_\ell} \otimes \cdots \otimes e_{j_1}).$$

Moreover, it follows from our definition[1] that

$$\delta_p(i, j) = \delta_p(i \otimes j^{-1}),$$

where $\otimes$ denotes the concatenation of tuples and j^{-1} is the reversed tuple, hence the result. □

For instance, the partition drawn at the beginning of this chapter can also be seen as

$$p = \quad \begin{array}{ccc} i_1 & i_2 & i_3 \\ & & \\ j_1 & j_2 & j_3 \end{array}$$

yielding

$$T_p(e_{i_1} \otimes e_{i_2} \otimes e_{i_3}) = \delta_{i_1, i_2} \sum_{j_2 = 1}^{N} e_{i_3} \otimes e_{j_2} \otimes e_{i_1}.$$

3.2 Operations on Partitions

Let us generalise the notation introduced at the end of Chapter 2 and write $\mathrm{Mor}_{\mathbb{G}}(v, w)$ for the set of all intertwiners between two finite-dimensional representations u and v of a compact matrix quantum group $\mathbb{G}$. Assume that we have a collection $\mathcal{C}$ of partitions made of subsets $\mathcal{C}(k, \ell) \subset P(k, \ell)$ for

---

[1] Note that we have two definitions of the symbol δ_p, which should not be confused: one has two arguments corresponding to the choice of an upper and a lower row in p, while the original one has only one argument.

all integers k and ℓ and that we want to find an orthogonal compact matrix (quantum) group $\mathbb{G}$ such that, for all $k, \ell \in \mathbf{N}$,

$$\mathrm{Mor}_{\mathbb{G}} \left(u^{\otimes k}, u^{\otimes \ell} \right) = \mathrm{Vect} \left\{ T_p \mid p \in \mathcal{C}(k, \ell) \right\}.$$

Obviously, the set $\mathcal{C}$ must satisfy some stability conditions in order for the spaces above to be intertwiner spaces. For instance, when two intertwiners can be composed, their composition must again be an intertwiner so that we need to ensure that $T_q \circ T_p$ is a linear combination of maps of the form T_r. This condition is linked to the following operation on partitions: given two partitions $p \in P(k, \ell)$ and $q \in P(\ell, m)$, we can perform their *vertical concatenation* qp (or $q \circ p$ if ambiguity needs to be avoided) by placing q below p and connecting the lower points of p to the corresponding ones in the upper row of q. This process may nevertheless produce 'loops'.

Definition 3.4 For two partitions $p \in P(k, \ell)$ and $q \in P(\ell, m)$, consider the set L of elements in $\{1, \dots, \ell\}$ which are not connected to an upper point of p nor to a lower point of q. The lower row of p and the upper row of q both induce a partition of the set L. For $x, y \in L$, let us set $x \sim y$ if x and y belong either to the same block of the partition induced by p or to the one induced by q. The transitive closure of $\sim$ is an equivalence relation on L, and the corresponding partition is called the *loop partition* of L; its blocks are called *loops*.

When composing two partitions, we erase the loops and only remember their number, denoted by $\mathrm{rl}(q, p)$. Here is an example with $\mathrm{rl}(q, p) = 1$:

At the level of the operators T_p, this translates into the formula

$$T_q \circ T_p = \dim(V)^{\mathrm{rl}(q,p)} T_{qp}. \tag{3.2}$$

Exercise 3.2 Prove Equation (3.2).

Solution Assume that $p \in P(k, \ell)$ and $q \in P(\ell, m)$. For a tuple $\boldsymbol{i} = (i_1, \dots, i_k)$, we have

$$T_q \circ T_p(e_{i_1} \otimes \cdots \otimes e_{i_k}) = \sum_{j_1, \cdots j_\ell = 1}^{N} \delta_p(\boldsymbol{i}, \boldsymbol{j}) T_q(e_{j_1} \otimes \cdots \otimes e_{j_\ell})$$

$$= \sum_{j_1, \cdots j_\ell = 1}^{N} \sum_{s_1, \cdots s_m = 1}^{N} \delta_p(\boldsymbol{i}, \boldsymbol{j}) \delta_q(\boldsymbol{j}, \boldsymbol{s}) (e_{s_1} \otimes \cdots \otimes e_{s_m}).$$

It follows from the definition that if there is a j such that $\delta_p(i, j) \neq 0$ and $\delta_q(j, s) \neq 0$, then $\delta_{qp}(i, s) \neq 0$. Conversely, for given tuples i and s, any tuple j which coincides with i on the lower row of p and with s on the upper row of q will do. The points of such a tuple which are connected either to an upper point of p or to a lower point of q are already determined, so that the only freedom we have concerns points which are only connected to the lower row in p and to the upper row of q. The value of j must coincide on such blocks if they get connected in the composition; thus the number of compatible tuples j is the number of possible indexing of the loops which get removed in the composition $q \circ p$. This gives the formula in the statement. $\qquad\square$

Stability under vertical concatenation is not sufficient to produce commutants of (quantum) group representations. For instance, if T_p and T_q are intertwiners, then $T_p \otimes T_q$ also is and must therefore come from partitions. This corresponds to the *horizontal concatenation*: if p and q are two partitions of $\{1, \ldots, 2k\}$ and $\{1, \ldots, 2\ell\}$ respectively, then we build a partition $p \otimes q$ of $\{1, \ldots, 2(k + \ell)\}$ by simply drawing q on the right of p. For instance,

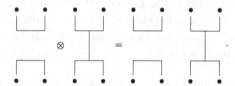

As for vertical concatenation, this operation has a nice interpretation at the level of operators, namely

$$T_p \otimes T_q = T_{p \otimes q}. \tag{3.3}$$

Exercise 3.3 Prove Equation (3.3).

Solution Assume that $p \in P(k, \ell)$ and $q \in P(m, s)$. For a tuple $i = (i_1, \ldots, i_{k+m})$, the image of $e_{i_1} \otimes \cdots \otimes e_{i_{k+m}}$ under $T_p \otimes T_q$ is

$$X = \left(\sum_{j_1, \ldots, j_\ell = 1}^{N} \delta_p \left((i_1, \ldots, i_k), (j_1, \ldots, j_\ell) \right) e_{j_1} \otimes \cdots \otimes e_{j_\ell} \right)$$

$$\otimes \left(\sum_{j_{\ell+1}, \ldots, j_{\ell+s} = 1}^{N} \delta_p \left((i_{k+1}, \ldots, i_{k+m}), (j_{\ell+1}, \ldots, j_{\ell+s}) \right) e_{j_{\ell+1}} \otimes \cdots \otimes e_{j_{\ell+s}} \right)$$

$$= \sum_{j_1, \ldots, j_{\ell+s} = 1}^{N} \delta_{p \otimes q} (i, j) \, e_{j_1} \otimes \cdots \otimes e_{j_{\ell+s}}$$

$$= T_{p \otimes q}(e_{i_1} \otimes \cdots \otimes e_{i_{k+m}}). \qquad\square$$

There is a third, very elementary operation which is needed. Because both $u^{\otimes k}$ and $u^{\otimes \ell}$ are unitary representations, if T is an intertwiner then so is its adjoint T^*. To translate this into an operation on partitions, let us denote by p^* the partition obtained by reflecting p with respect to an horizontal axis between the two rows. Here is an example:

Exercise 3.4 Prove that for any partition p,

$$T_p^* = T_{p^*}.$$

Solution The result follows from the following computation:

$$
\delta_p(\boldsymbol{i},\boldsymbol{t}) = \left\langle \sum_{j_1,\dots,j_\ell=1}^{N} \delta_p(\boldsymbol{i},\boldsymbol{j}) e_{j_1} \otimes \cdots \otimes e_{j_\ell}, e_{t_1} \otimes \cdots \otimes e_{t_\ell} \right\rangle
$$
$$
= \left\langle e_{i_1} \otimes \cdots \otimes e_{i_k}, \sum_{j_1,\dots,j_k=1}^{N} \delta_p(\boldsymbol{j},\boldsymbol{t}) e_{j_1} \otimes \cdots \otimes e_{j_k} \right\rangle. \qquad \square
$$

Do we now have everything needed to produce an orthogonal compact matrix quantum group out of $\mathcal{C}$? The answer is yes, but the proof requires a detour through a more abstract structure.

3.3 Tannaka–Krein Reconstruction

Reconstructing a group from its representations is the subject of *Tannaka–Krein duality*. The main point of this theory is that the representations of the group and their intertwiners can be assembled into one rich algebraic structure called a *tensor category*. The reader may, for instance, refer to the book [37] for a comprehensive introduction to tensor categories. For our purpose, we can restrict to a specific type of such categories for which we will use the following more 'down-to-earth' definition.

Definition 3.5 Let V be a finite-dimensional Hilbert space. A *concrete rigid orthogonal C*-tensor category*[2] $\mathfrak{C}$ is a collection of spaces $\mathrm{Mor}_{\mathfrak{C}}(k, \ell) \subset \mathcal{L}(V^{\otimes k}, V^{\otimes \ell})$ for all $k, \ell \in \mathbf{N}$ such that

1. If $T \in \mathrm{Mor}_{\mathfrak{C}}(k, \ell)$ and $T' \in \mathrm{Mor}_{\mathfrak{C}}(k', \ell')$, then $T \otimes T' \in \mathrm{Mor}_{\mathfrak{C}}(k + k', \ell + \ell')$;
2. If $T \in \mathrm{Mor}_{\mathfrak{C}}(k, \ell)$ and $T' \in \mathrm{Mor}_{\mathfrak{C}}(\ell, r)$, then $T' \circ T \in \mathrm{Mor}_{\mathfrak{C}}(k, r)$;
3. If $T \in \mathrm{Mor}_{\mathfrak{C}}(k, \ell)$, then $T^* \in \mathrm{Mor}(\ell, k)$;
4. $\mathrm{id} \colon x \mapsto x \in \mathrm{Mor}_{\mathfrak{C}}(1, 1)$;
5. $D_V \colon x \otimes y \mapsto \langle x, \overline{y} \rangle \in \mathrm{Mor}_{\mathfrak{C}}(2, 0)$.

If, moreover, $\sigma \colon x \otimes y \mapsto y \otimes x \in \mathrm{Mor}(2, 2)$, then $\mathfrak{C}$ is said to be *symmetric*.

The fundamental example is, of course, given by (quantum) groups:

Example 3.6 Let $\mathbb{G}$ be an orthogonal compact matrix quantum group with a fundamental representation u of size N. Set $V = \mathbf{C}^N$ and

$$\mathrm{Mor}_{\mathfrak{C}}(k, \ell) = \mathrm{Mor}_{\mathbb{G}}\left(u^{\otimes k}, u^{\otimes \ell}\right).$$

This defines a concrete rigid orthogonal C*-tensor category called the *representation category* of $\mathbb{G}$ and denoted by $\mathcal{R}(\mathbb{G})$. If, moreover, $\mathbb{G}$ is classical, then $\mathcal{R}(\mathbb{G})$ is symmetric.

That this example is, in fact, fully general is the content of the quantum Tannaka–Krein theorem proved by S. L Woronowicz in [76].

Theorem 3.7 (Woronowicz) *Let $\mathfrak{C}$ be a concrete rigid orthogonal C*-tensor category associated to a Hilbert space V. Then, there exists an orthogonal compact matrix quantum group $\mathbb{G}$ with a fundamental representation u of dimension $\dim(V)$ such that for all $k, \ell \in \mathbf{N}$,*

$$\mathrm{Mor}_{\mathbb{G}}\left(u^{\otimes k}, u^{\otimes \ell}\right) = \mathrm{Mor}_{\mathfrak{C}}(k, \ell).$$

Moreover, the quantum group $\mathbb{G}$ is unique up to isomorphism and is a classical group if and only if $\mathfrak{C}$ is symmetric.

---

[2] We are slightly abusing terminology here, since our definition is not the usual one for a concrete rigid C*-tensor category (that is, a C*-tensor category with duals equipped with a fibre functor; see for instance [60, chapter 2]). However, the two notions turn out to play equivalent roles in the theory, so this is not an issue. The adjective 'orthogonal' is not standard but is meant to emphasise the presence of D_V as a morphism, which is equivalent to the orthogonality assumption in the definition of compact matrix quantum groups.

There are several proofs of this result using a different amount of categorical machinery; see for instance [76] or [60, theorem 2.3.2]. However, due to our restricted definition of a concrete rigid orthogonal C*-tensor category, we can give an elementary proof which is close to that of [55].

Proof The idea is to build $\mathbb{G} = (\mathcal{O}(\mathbb{G}), u)$ as a quotient of the space of all 'non-commutative matrices' by all the relations making the maps T intertwiners of tensor powers of its fundamental representation. So let N be the dimension of V and consider the universal *-algebra[3] $\mathcal{A}$ generated by N^2 elements X_{ij} with the involution given by $X_{ij}^* = X_{ij}$. Fixing an orthonormal basis $(e_i)_{1 \leqslant i \leqslant N}$ of V once and for all, we introduce some shorthand notations: if $\boldsymbol{i} = (i_1, \ldots, i_k), \boldsymbol{j} = (j_1, \ldots, j_\ell)$ and $T \in \mathcal{L}(V^{\otimes k}, V^{\otimes \ell})$, then we set

$$e_{\boldsymbol{i}} = e_{i_1} \otimes \cdots \otimes e_{i_k}$$
$$T_{\boldsymbol{j}\boldsymbol{i}} = \langle T(e_{\boldsymbol{i}}), e_{\boldsymbol{j}} \rangle$$
$$X_{\boldsymbol{i}\boldsymbol{j}} = X_{i_1 j_1} \cdots X_{i_k j_k}.$$

Consider now the following non-commutative polynomial

$$P_{T, \boldsymbol{i}, \boldsymbol{j}} = \sum_{\boldsymbol{s}} X_{\boldsymbol{i}\boldsymbol{s}} T_{\boldsymbol{s}\boldsymbol{j}} - \sum_{\boldsymbol{r}} T_{\boldsymbol{i}\boldsymbol{r}} X_{\boldsymbol{r}\boldsymbol{j}}.$$

More explicitely, given an orthogonal compact matrix quantum group $(\mathcal{O}(\mathbb{G}), u)$, there is a corresponding surjective *-homomorphism $\pi_{\mathbb{G}} \colon \mathcal{A} \to \mathcal{O}(\mathbb{G})$ sending $(X_{ij})_{1 \leqslant i, j \leqslant N}$ to u coefficient-wise. We can then set

$$P_{T, \boldsymbol{i}, \boldsymbol{j}}(u) = \pi_{\mathbb{G}}(P_{T, \boldsymbol{i}, \boldsymbol{j}})$$

and therefore think of the left-hand side as a 'polynomial' applied to the coefficients of u. Then, T intertwines $u^{\otimes k}$ with $u^{\otimes \ell}$ if and only if

$$P_{T, \boldsymbol{i}, \boldsymbol{j}}(u) = 0$$

for all tuples $\boldsymbol{i}$ and $\boldsymbol{j}$. Let us therefore consider the sets

$$\mathcal{I}_{k, \ell} = \operatorname{Vect} \{ P_{T, \boldsymbol{i}, \boldsymbol{j}} \mid T \in \operatorname{Mor}_{\mathfrak{C}}(k, \ell), \boldsymbol{i} = (i_1, \ldots, i_k), \boldsymbol{j} = (j_1, \ldots, j_\ell) \},$$
$$\mathcal{I}_k = \bigoplus_{i,j=0}^{k} \mathcal{I}_{i,j},$$
$$\mathcal{I} = \bigcup_{k \in \mathbf{N}} \mathcal{I}_k.$$

According to our basic idea, let us set $\widetilde{\mathcal{A}} = \mathcal{A}/\mathcal{I}$ (this makes sense at least as a quotient of vector spaces) and let u_{ij} be the image of X_{ij} in this quotient.

---

[3] The abelianisation of $\mathcal{A}$ is nothing but the algebra of regular functions on $M_N(\mathbf{R})$. The algebra $\mathcal{A}$ can therefore be thought of as the regular functions on all 'non-commutative real matrices'.

If $(\tilde{\mathcal{A}}, u)$ is an orthogonal compact matrix quantum group, then it certainly is the one we are looking for. But for this to hold, $\mathcal{I}$ must satisfy some extra conditions; namely we have to prove that

1. $\mathcal{I}$ is an ideal, so that $\tilde{\mathcal{A}}$ is an algebra;
2. $\mathcal{I} = \mathcal{I}^*$, so that $\tilde{\mathcal{A}}$ is a $*$-algebra;
3. $\Delta(\mathcal{I}) \subset \mathcal{I} \otimes \mathcal{A} + \mathcal{A} \otimes \mathcal{I}$, so that the coproduct is well defined on $\tilde{\mathcal{A}}$. Indeed, let $\Delta \colon \mathcal{A} \to \mathcal{A} \otimes \mathcal{A}$ be the unique $*$-homomorphism such that $\Delta(X_{ij}) = \sum_k X_{ik} \otimes X_{kj}$ (whose existence follows from the universal property of $\mathcal{A}$). Then, if $\pi_{\mathcal{I}}$ denotes the quotient map, for any $x \in \mathcal{O}(O_N^+)$ and $y \in \mathcal{I}$ we have

$$(\pi_{\mathcal{I}} \otimes \pi_{\mathcal{I}}) \circ \Delta(x + y) = (\pi_{\mathcal{I}} \otimes \pi_{\mathcal{I}}) \circ \Delta(x)$$

so that the coproduct factors through the quotient.

As one may expect, these properties follow from the axioms of concrete rigid orthogonal C*-tensor categories, and here is how:

1. Observe that, if i', j' are m-tuples, then

$$X_{i'j'} P_{T,i,j} = P_{\mathrm{id}^{\otimes m} \otimes T, i' \otimes i, j' \otimes j},$$

where $\otimes$ denotes the concatenation of tuples. Thus, $\mathcal{I}$ absorbs monomials, hence also polynomials, on the left. The same property on the right follows from a similar argument. As a consequence, $\mathcal{I}$ is an ideal.
2. Denoting by $i^{-1} = (i_k, \dots, i_1)$ the reversed tuple, we have

$$
\begin{aligned}
P_{T,i,j}^* &= \sum_s X_{i^{-1}s^{-1}} \overline{T_{sj}} - \sum_r X_{r^{-1}j^{-1}} \overline{T_{ir}} \\
&= \sum_s X_{i^{-1}s^{-1}} T_{js}^* - \sum_r X_{r^{-1}j^{-1}} T_{ri}^* \\
&= \sum_s X_{i^{-1}s^{-1}} R(T^*)_{s^{-1}j^{-1}} - \sum_r X_{r^{-1}j^{-1}} R(T^*)_{i^{-1}r^{-1}} \\
&= P_{R(T^*), i^{-1}, j^{-1}},
\end{aligned}
$$

where $R(T)$ is defined to be the operator with coefficients

$$R(T)_{ij} = T_{j^{-1}i^{-1}}.$$

To see that $R(T)$ is in $\mathrm{Mor}_{\mathcal{C}}(\ell, k)$ if $T \in \mathrm{Mor}_{\mathcal{C}}(k, \ell)$, let us define inductively operators

$$D_k \colon V^{\otimes 2k} \to \mathbf{C}$$

by $D_1 = D_V$ and the formula

$$D_{k+1} = D_k \circ (\mathrm{id}_{V^{\otimes k}} \otimes D_V \otimes \mathrm{id}_{V^{\otimes k}}).$$

These are in fact the maps T_{p_k} corresponding to the partition

$$p_k = \begin{array}{c} \bullet \;\; \cdots \;\; \bullet \;\; \bullet \;\;\;\; \bullet \;\; \bullet \;\; \cdots \;\; \bullet \end{array}$$

obtained by nesting k pairings. An easy induction shows that

$$D_k(e_i \otimes e_j) = \delta_{i,j^{-1}}.$$

Using the map D_k we define, for a linear map T,

$$\widetilde{T} = (\mathrm{id}_{V^{\otimes k}} \otimes D_\ell)(\mathrm{id}_{V^{\otimes k}} \otimes T \otimes \mathrm{id}_{V^{\otimes \ell}})(D_k^* \otimes \mathrm{id}_{V^{\otimes \ell}}).$$

This can be seen pictorially as a 'reversed version' of T:

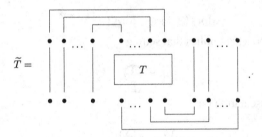

Now,

$$\widetilde{T}(e_i) = (\mathrm{id}_{V^{\otimes k}} \otimes D_\ell)(\mathrm{id}_{V^{\otimes k}} \otimes T \otimes \mathrm{id}_{V^{\otimes \ell}}) \Big(\sum_s e_s \otimes e_{s^{-1}} \otimes e_i \Big)$$

$$= \sum_s \sum_t (\mathrm{id}_{V^{\otimes k}} \otimes D_\ell)(e_s \otimes T_{ts^{-1}} e_t \otimes e_i)$$

$$= \sum_s T_{i^{-1}s^{-1}} e_s$$

$$= \sum_s R(T)_{si} e_s$$

$$= R(T)(e_i)$$

so that $R(T)$ is a morphism in the category $\mathfrak{C}$ as soon as T is. As a consequence, $P^*_{T,\mathbf{i},\mathbf{j}} \in \mathcal{I}$ hence $\mathcal{I}^* = \mathcal{I}$.

3. This is a straightforward calculation:

$$\Delta\left(P_{T,i,j}(X)\right) = \sum_{t,s} X_{it} \otimes X_{ts}T_{sj} - \sum_{m,r} T_{ir}X_{rm} \otimes X_{mj}$$

$$= \sum_{t} X_{it} \otimes P_{T,t,j} + \sum_{t,n} X_{it} \otimes T_{tn}X_{nj}$$

$$+ \sum_{m} P_{T,i,m} \otimes X_{mj} - \sum_{m,\ell} X_{i\ell}T_{\ell m} \otimes X_{mj}$$

$$= \sum_{t} X_{it} \otimes P_{T,t,j}(X) + \sum_{m} P_{T,i,m}(X) \otimes X_{mj}$$

$$+ \sum_{t,n} X_{it} \otimes T_{tn}X_{nj} - \sum_{t,n} X_{it}T_{tn} \otimes X_{nj}$$

$$= \sum_{t} X_{it} \otimes P_{T,t,j}(X) + \sum_{m} P_{T,i,m}(X) \otimes X_{mj}.$$

It follows from all this that setting $u_{ij} = \pi_{\mathcal{I}}(X_{ij})$, $\mathbb{G} = (\tilde{\mathcal{A}}, u)$ is an orthogonal compact matrix quantum group (observe that the relations $P_{D_V,i,j}(u) = 0$ are precisely the orthogonality condition in Definition 1.19). Moreover, by construction,

$$\mathrm{Mor}_{\mathcal{C}}(k,\ell) \subset \mathrm{Mor}_{\mathbb{G}}\left(u^{\otimes k}, u^{\otimes \ell}\right)$$

for all $k, \ell \in \mathbf{N}$. This implies that setting

$$u_k = \bigoplus_{i=0}^{k} u^{\otimes i},$$

seen as a representation on

$$V_k = \bigoplus_{i=0}^{k} V^{\otimes i},$$

we have

$$\mathcal{B}_k = \bigoplus_{i,j=0}^{k} \mathrm{Mor}_{\mathcal{C}}(i,j) \subset \mathrm{Mor}_{\mathbb{G}}\left(u_k, u_k\right).$$

Here, $\mathcal{B}_k$ is naturally seen as a matrix algebra consisting in block matrices with blocks given by the decomposition of V_k as a direct sum of tensor powers of V. To show the converse inclusion, let us first note that by universality, all the monomials in $\mathcal{A}$ are linearly independent. Hence denoting by $X^{(k)}$ the matrix with coefficients X_{ij} with indices running over all tuples of length k and setting [4]

$$X_k = \bigoplus_{i=0}^{k} X^{(i)},$$

---

[4] We are seeing here X_k as a block matrix in the same way as the elements of $\mathcal{B}_k$.

there exists for any operator $T \in \mathcal{L}(V_k)$ a linear map $f_T \in \mathcal{A}^*$ such that

$$\widehat{f_T}(X_k) = (f(X_{ij}))_{i,j}$$

is the matrix of T in the canonical basis. In particular, any element of the commutant $\mathcal{B}'_k$ of $\mathcal{B}_k$ is of the form $\widehat{f}(X_k)$ for some f. Setting

$$\mathcal{D}_k = \{ \widehat{f}(X_k) \mid f \in \mathcal{I}_k^\perp \},$$

we have

$$\widehat{f}(X_k) \in \mathcal{B}'_k \Rightarrow \omega\big([T, \widehat{f}(X_k)]\big) = 0, \ \forall\, T \in \mathcal{B}_k, \ \forall\, \omega \in \mathcal{L}(V_k)^*$$
$$\Rightarrow (\omega \otimes f)\big([T, X_k]\big) = 0, \ \forall\, T \in \mathcal{B}_k, \ \forall\, \omega \in \mathcal{L}(V_k)^*$$
$$\Rightarrow f\big((\omega \otimes \mathrm{id})\,([T, X_k])\big) = 0, \ \forall\, T \in \mathcal{B}_k, \ \forall\, \omega \in \mathcal{L}(V_k)^*.$$

Assume for the moment that $T \in \mathrm{Mor}_{\mathcal{C}}(i, j)$ and that ω is the linear form sending an operator to its (i, j)-th coefficient. Then, $(\omega \otimes \mathrm{id})\,([T, X_k])$ is nothing but $P_{T,\mathrm{i,j}}$. Now, any $T \in \mathcal{L}(V_k)$ is a sum of operators from one tensor power of V to another (simply cut T using the corresponding orthogonal projections) and any linear form is a linear combination of coefficient functions. Thus, the elements $(\omega \otimes \mathrm{id})\,([T, X_k])$ are exactly $\mathcal{I}_k$. We have therefore proven that

$$\widehat{f}(X_k) \in \mathcal{B}'_k \Rightarrow f \in \mathcal{I}_k^\perp.$$

As a consequence, $\mathcal{B}'_k \subset \mathcal{D}_k$ so that by the Double Commutant Theorem (see Theorem A.3 in Appendix A),

$$\mathcal{D}'_k \subset \mathcal{B}''_k = \mathcal{B}_k.$$

By Lemma 2.16, any $T \in \mathrm{Mor}_{\mathbb{G}}(u_k, u_k)$ commutes with $\widehat{f}(u_k)$ for all $f \in \mathcal{O}(\mathbb{G})^*$. Equivalently, T commutes with $\widehat{f}(X_k)$ for all $f \in \mathcal{I}_k^\perp$, that is to say with $\mathcal{D}_k$. In other words, $T \in \mathcal{D}'_k \subset \mathcal{B}_k$ and the equality is proved.

Let us turn to uniqueness. Consider an orthogonal compact matrix quantum group $\mathbb{H} = (\mathcal{O}(\mathbb{H}), v)$ such that

$$\mathrm{Mor}_{\mathcal{C}}(k, \ell) = \mathrm{Mor}_{\mathbb{H}}\big(v^{\otimes k}, v^{\otimes \ell}\big)$$

for all $k, \ell \in \mathbf{N}$. By definition there exists an ideal $\mathcal{J} \in \mathcal{A}$ such that $\mathcal{O}(\mathbb{H}) = \mathcal{A}/\mathcal{J}$ and we can, up to isomorphism, identify the copies of $\mathcal{A}$ used for $\mathbb{G}$ and for $\mathbb{H}$. Moreover, with the previous notations, $\mathcal{I} \subset \mathcal{J}$. If now $\mathcal{J}_k$ denotes the intersection of $\mathcal{J}$ with the span of coefficients of X_k, then the preceding computations show that

$$\left\{ \widehat{f}(X_k) \mid f \in \mathcal{J}_k \right\}' = \mathrm{Mor}_{\mathbb{H}}\left(v_k, v_k\right)$$
$$= \mathcal{D}'_k$$
$$= \mathcal{B}_k$$
$$= \left\{ \widehat{f}(X_k) \mid f \in \mathcal{I}_k \right\}'$$

so that $\mathcal{J}_k = \mathcal{I}_k$. Thus, $\mathcal{J} = \cup \mathcal{J}_k = \cup \mathcal{I}_k = \mathcal{I}$ and $\mathbb{H} \simeq \mathbb{G}$.

The last point concerns symmetry. But we have already seen that $\sigma \in \mathrm{Mor}_{\mathcal{C}}(2, 2)$ is equivalent to having $f_{p_{\text{cross}}} \in \mathrm{Mor}_{\mathcal{C}}(4, 0)$, and we proved in Proposition 2.22 that this is equivalent to $\mathcal{O}(\mathbb{G})$ being commutative. Anticipating Corollary 5.18, this is equivalent to $\mathbb{G}$ being classical.   □

Recall that we want to apply Tannaka–Krein reconstruction to build a compact quantum group out of a set of partitions. Considering the axioms in Definition 3.5, we see that Axioms (1) to (3) correspond to the three operations defined in Section 3.2. As for the last two axioms, they follow from the following elementary computations:

- $T_| = \mathrm{id}$,
- $T_\sqcup = D_{\mathbf{C}^N}$.

To simplify later statements, let us gather all the necessary properties in a definition.

Definition 3.8 A *category of partitions* $\mathcal{C}$ is a collection of sets of partitions $\mathcal{C}(k, \ell) \subset P(k, \ell)$ for all integers k and ℓ such that

1. If $p \in \mathcal{C}(k, \ell)$ and $q \in \mathcal{C}(k', \ell')$, then $p \otimes q \in \mathcal{C}(k + k', \ell + \ell')$;
2. If $p \in \mathcal{C}(k, \ell)$ and $q \in \mathcal{C}(\ell, r)$, then $q \circ p \in \mathcal{C}(k, r)$;
3. If $p \in \mathcal{C}(k, \ell)$, then $p^* \in \mathcal{C}(\ell, k)$;
4. $| \in \mathcal{C}(1, 1)$;
5. $\sqcup \in \mathcal{C}(2, 0)$.

If, moreover, $p_{\text{cross}} \in \mathcal{C}(2, 2)$, then $\mathcal{C}$ is said to be *symmetric*.

We conclude this section by a theorem which is rather a summary and rewriting of the whole construction that we have done.

Theorem 3.9 (Banica–Speicher) *Let N be an integer and let $\mathcal{C}$ be a category of partitions. Then, there exists a unique orthogonal compact matrix quantum group $\mathbb{G} = (\mathcal{O}(\mathbb{G}), u)$, where u has dimension N, such that for any $k, \ell \in \mathbf{N}$,*

$$\mathrm{Mor}_{\mathbb{G}}(u^{\otimes k}, u^{\otimes \ell}) = \mathrm{Vect}\,\{T_p \mid p \in \mathcal{C}(k, \ell)\}.$$

Moreover, $\mathbb{G}$ is a classical group if and only if $\mathcal{C}$ is symmetric. The compact quantum group $\mathbb{G}$ will be denoted by $\mathbb{G}_N(\mathcal{C})$ and called the partition quantum group *associated to N and $\mathcal{C}$.*

Remark 3.10 The quantum groups associated to a category of partitions as we defined them here were first introduced by T. Banica and R. Speicher in [18] under the name of *easy quantum groups*. It was already clear at the time that this setting could be extended to unitary compact matrix quantum groups by adding black and white colours to the points of the partition. The formal description of this extended formalism was made in [68] and will be detailed in Chapter 6. However, it had by then become apparent that one can even allow arbitrary colourings of the partitions, and that idea led to the general setting of *partition quantum groups* introduced in [39].

Remark 3.11 One of the first questions raised by the definition of partition quantum groups is their classification, based on the categories of partitions. For easy quantum groups, there is a complete classification, as a result of many works including [18], [73] and [64] (part of it is detailed in the beginning of Section 6.1). The classification of the unitary versions of easy quantum groups (see Chapter 6 for details) is more subtle, but many results are now known, starting with [68] and followed by several other works like [57, 58]. As for the general setting of partition quantum groups, the only known results beyond the previous ones are restricted to non-crossing partitions with two colours; see [40].

3.4 Examples of Partition Quantum Groups

We conclude this chapter with some examples. Here is the basic strategy to compute them: to determine $\mathbb{G}_N(\mathcal{C})$, one only has to consider *generators* of $\mathcal{C}$, that is, a subset F such that $\mathcal{C}$ is the smallest category of partitions containing F. In such a circumstance, we will write $\mathcal{C} = \langle F \rangle$. Then, $\mathcal{O}(\mathbb{G}_N(\mathcal{C}))$ is the quotient of $\mathcal{O}(O_N^+)$ by the relations given by the fact that T_p is an intertwiner for $p \in F$. This can be proven by observing that the ideal $\mathcal{I}$ in the proof of Theorem 3.7 is then generated by the elements $P_{T_p, i, j}$ for $p \in F$.

In order to apply that strategy, we will have to identify the generators of some specific categories of partitions. To do this, the following elementary lemma will prove useful.

Lemma 3.12 *Let p be a non-crossing partition lying on one line. Then, p contains an* interval, *that is to say a block consisting of consecutive points.*

Proof The proof is done by induction on the number of points of p. If it has at most two points, then the result is clear. Assume that the result holds for all partitions on at most n points and consider a partition p on $n + 1$ points. We will focus on its leftmost point, and there are two possibilities:

- If it belongs to an interval, then we are done;
- Otherwise, there is $i > 2$ such that it is connected to the ith point from the left and to none of the $\{2, \dots, i - 1\}$-th points. Because p is non-crossing, a point between the first and ith one can only be connected to another point between the first and the ith one. Here is a picture of the situation. Thus,

$$p = \begin{array}{c} 1 \\ \end{array} \boxed{\quad q \quad} \begin{array}{c} i \\ \end{array} \quad \dots$$

restricting to the points $\{2, \dots, i-1\}$ yields a subpartition q of p on at most $n - 1$ points. We then conclude by induction. $\qquad\qquad\qquad\qquad\Box$

What about partitions not lying on one line? The result still holds, once one sees that in fact we can always reduce to one line. This is done through a *rotation operation* which we now introduce. Let p be a partition and consider the leftmost point of its upper row. Let us rotate it to the left of the lower row, without changing the strings connecting it to other points. We thus obtain a new partition p', called a *rotated version* of p. The key fact is that categories of partitions are invariant under this operation.

Lemma 3.13 *If $\mathcal{C}$ is a category of partitions and if $p \in \mathcal{C}$, then $p' \in \mathcal{C}$. The same holds for rotations of points on the right.*

Proof This simply follows from the fact that if p has k upper points, then

$$p' = (\mid \otimes\, p) \circ (\sqcap \otimes \mid^{k-1}) \in \mathcal{C}.$$

Here is how this translates pictorially.

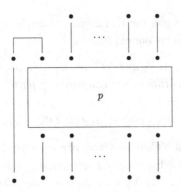

The proof for rotations on the right is analogous. Note that this is the partition version of the operation used to define the maps T_p from the maps f_p at the beginning of this chapter. □

The last thing that we need is to check a fact which, even though it looks obvious and is fundamental in the theory, needs a proof.

Lemma 3.14 *The set NC of all non-crossing partitions is a category of partitions.*

Proof The stability under horizontal concatenation and reflection is clear, and the only problem is vertical concatenation. If $p \in NC(k, \ell)$ and $q \in NC(\ell, m)$, then $q \circ p$ is a rotation of $(q \otimes |^{\otimes k}) \circ \widetilde{p}$, where $\widetilde{p}$ is obtained by rotating all the upper points of p to the right of the lower row. We may therefore assume without loss of generality that $p \in NC(0, \ell)$. Let us prove by induction on ℓ that $q \circ p$ is non-crossing.

If $\ell = 1$, then the composition $q \circ p$ amounts to erasing the unique point in the upper row of q. As a consequence, $q \circ p$ is the lower row of q. But a subpartition of a non-crossing partition is itself non-crossing, hence the result.

If the result holds for a given $\ell - 1 \geqslant 1$, assume by contradiction that there is a crossing in $q \circ p$, that is to say four points $k_1 < k_2 < k_3 < k_4$ such that k_1 is in the same block as k_3 while k_2 is in the same block as k_4, but the four points are not in the same block. By Lemma 3.12, there exists an interval b in p and it cannot contain any of the four previous points by definition. As a consequence, removing a point from that block and removing the corresponding point of q does not change the fact that there is a crossing. In other words, we can build $p' \in NC(0, \ell - 1)$ and $q' \in NC(\ell - 1, m)$ such that $q' \circ p'$ has a crossing, a contradiction. □

We are now ready to give our first explicit example of a partition quantum group, which is nothing but our old friend O_N^+.

Proposition 3.15 *The sets P_2 and NC_2 of respectively all pair partitions and all non-crossing pair partitions are categories of partitions. Moreover, there are isomorphisms*

$$\mathbb{G}_N(NC_2) \simeq O_N^+ \text{ and } \mathbb{G}_N(P_2) \simeq O_N.$$

Proof Let us prove that NC_2 is a category of partitions. As in Lemma 3.14, the only thing to prove is stability under vertical concatenation and we will proceed by induction in a similar way. We may assume without loss of generality p to lie on the lower row, and if it consists of two points the result is clear. If now $p \in NC_2(0, 2\ell)$ and $q \in NC_2(2\ell, 2m)$ for some $\ell > 1$, consider an interval in p. Because it has size two, it consists in two neighbouring points, say i and $i + 1$.

- If these two points are also connected in q, they form a block there and removing these two points in both partitions does not change the vertical concatenation;
- Otherwise, i is connected to some j and $i + 1$ is connected to some j' in q. But then, removing i and $i + 1$ in both partitions and connecting j and j' does not change the vertical concatenation.

In both cases, $q \circ p$ can be obtained by composing two non-crossing pair partitions of smaller size (note that j and j' were connected to distinct upper points, hence we have removed two blocks of size 2 and created a new one), hence is again a non-crossing pair partition. Note that this also proves that P_2 is a category of partitions.

The rest of the proof relies on the fact that NC_2 is the smallest category of partitions, in the sense that it is generated by ⊔. This is proven by induction on the number of points of a partition in NC_2 as follows:

- If the partition has two points, then it is ⊔.
- If the result holds for all partitions in NC_2 on at most $2n$ points, let $p \in NC_2$ be a partition on $2(n + 1)$ points. Up to rotating, we may assume that p lies on one line and that the first two points belong to an interval. Because we are considering pair partitions, this means that $p = ⊔ \otimes p'$. But then, $p' \in NC_2$ and has $2n$ points so that by induction it is in the category of partitions generated by ⊔, hence also p, concluding the proof.

Because any non-crossing pair partition can be obtained from ⊔ using the category operations, $\mathcal{O}(\mathbb{G}_N(NC_2))$ is obtained from the universal $*$-algebra $\mathcal{A}$

by adding the relations coming from the fact that $T_\sqcup = D_{\mathbb{C}^N}$ is an intertwiner. But these are exactly the relations making the fundamental representation orthogonal. This proves the first isomorphism.

As for the second one, observe[5] that $\mathcal{O}(O_N)$ is the quotient of $\mathcal{O}(O_N^+)$ by the relations making all the generators commute. As shown in Proposition 2.22, the relations corresponding to the fact that the crossing partition yields an intertwiner are exactly the commutation relations between all the coefficients of the fundamental representation. Thus, the category of partitions $\mathcal{C}$ of O_N is the one generated by $\sqcup$ and p_{cross} and we have to prove that this equals P_2. To do this, notice that because $\sqcup, p_{\mathrm{cross}} \in P_2$, they generate a subcategory of partitions so that we have an inclusion $\mathcal{C} \subset P_2$. On the other hand, we have explained in Section 2.2 how any pair partition yields an intertwiner between some tensor power of the fundamental representation of O_N and the trivial one. This means that $P_2 \subset \mathcal{C}$ so that we have proven equality (note that this proves in particular that P_2 is a category of partitions). $\qquad\qquad\square$

We can now recover R. Brauer's Theorem 2.21, as well as T. Banica's Theorem 2.25, as immediate corollaries of Theorem 3.9. As one may expect, the quantum permutation groups, of course, also fall into this class, but this requires a bit of work.

Exercise 3.5 Let $\mathbb{G} = (\mathcal{O}(\mathbb{G}), u)$ be an orthogonal compact matrix quantum group with u of size N. Prove that $\mathcal{O}(\mathbb{G})$ satisfies the defining relations of $\mathcal{A}_s(N)$ in Definition 1.6 if and only if it has the following intertwiners:

- Conditions (1) and (3): $T_{p_3} \in \mathrm{Mor}(u^{\otimes 2}, u)$ where $p_3 = \{\{1,2,3\}\}$ is the partition with one block in $P(2,1)$;
- Condition (2): $T_s \in \mathrm{Mor}(u, \varepsilon)$ where s is the singleton partition in $P(1,0)$.

Solution We compute on the one hand

$$(\mathrm{id} \otimes T_{p_3}) \circ \rho_{u^{\otimes 2}} (e_{i_1} \otimes e_{i_2}) = \sum_{j_1, j_2 = 1}^{N} u_{i_1 j_1} u_{i_2 j_2} \otimes T_{p_3} (e_{j_1} \otimes e_{j_2})$$

$$= \sum_{j_1, j_2 = 1}^{N} u_{i_1 j_1} u_{i_2 j_2} \otimes \delta_{j_1 j_2} e_{j_1}$$

$$= \sum_{j=1}^{N} u_{i_1 j} u_{i_2 j} \otimes e_j,$$

---

[5] We did not prove this, but it easily follows from the fact that the corresponding $*$-algebra being commutative, it is of the form $\mathcal{O}(G)$ by Corollary 5.18.

and on the other hand

$$\rho_u \circ T_{p_3}(e_{i_1} \otimes e_{i_2}) = \rho_u(\delta_{i_1 i_2} e_{i_1})$$
$$= \delta_{i_1 i_2} \sum_{j=1}^{N} u_{i_1 j} \otimes e_j.$$

By linear independence, we conclude that for any i_1, i_2, j,

$$u_{i_1 j} u_{i_2 j} = \delta_{i_1 i_2} u_{i_1 j}.$$

Moreover, $T_{p_3}^*$ satisfies the corresponding relations with $u^* = u^t$, yielding Conditions (1) and (3).[6] The converse straightforwardly follows from the same computation.

As for the second point,

$$(\mathrm{id} \otimes T_s) \circ \rho_u(e_i) = \sum_{j=1}^{N} u_{ij} \otimes T_s(e_j)$$
$$= \sum_{j=1}^{N} u_{ij} \otimes 1$$

while $\varepsilon \circ T_s(e_j) = 1 \otimes 1$. Thus,

$$\sum_{j=1}^{N} u_{ij} = 1$$

and using u^* instead, we see that Condition (2) holds. Once again, the converse is straightforward. $\qquad\square$

Remark 3.16 We have included the map T_s in the statement of Exercise 3.5 because it is a nice and important example of a partition map. However, it is redundant because of the orthogonality assumption. Indeed, with the notations of the proof, if $T_\sqcup$ and T_{p_3} are intertwiners, then so is

$$T_\sqcup \circ T_{p_3}^* = T_{\sqcup \circ p_3^*} = T_s.$$

Proposition 3.17 *There are isomorphisms*

$$\mathbb{G}_N(NC) \simeq S_N^+ \text{ and } \mathbb{G}_N(P) \simeq S_N.$$

Proof It follows from Exercise 3.5 that S_N^+ is a partition quantum group and that its category of partitions is generated by p_3 and s. Let us show that this category of partitions is in fact the category NC of all non-crossing partitions. This can be done in two steps.

---

[6] Recall that $u_{ij}^* = u_{ij}$ is part of the axioms of an orthogonal compact matrix quantum group.

1. Let $k \geqslant 1$ and consider the partition

$$p_3^{\otimes k} \circ \left(| \otimes (\sqcap)^{\otimes(k-1)} \otimes | \right). \tag{3.4}$$

This is a partition on $k + 2$ points with only one block. As a consequence, any one-block partition is in the category of partitions $\langle p_3 \rangle$ generated by p_3.

2. Let us prove by induction on n that any non-crossing partition on at most n points is in $\langle p_3 \rangle$. This is clear for $n \leqslant 3$. If it is true for some n, let p be a partition on $n + 1$ points. If p has only one block, then it is in $\langle p_3 \rangle$ by the first point. Otherwise, by Lemma 3.12, there is an interval b in p. Rotating p we can then write it as $b \otimes p'$ for some partition p' on at most n points. The result now follows form the induction hypothesis.

Adding the crossing partition, we see that S_N corresponds to the category of partitions $\mathcal{C}$ generated by p_3 and p_{cross}. One may prove that this is exactly the category P of all partitions either directly or by using p_{cross} to 'uncross' the partitions (this is done in Proposition 6.15). Another solution is to check by hand that for any partition p, f_p is an invariant linear map for some tensor power of the fundamental representation of S_N. Indeed, the fact that f_p is an intertwiner translates into the equation

$$\sum_{j=(j_1,\cdots,j_k)} \delta_p(j) c_{ij} = \sum_{j=(j_1,\cdots,j_k)} \delta_p(i)$$

for all $i = (i_1, \ldots, i_k)$. It is enough to check that it holds when applied to an arbitrary permutation $\sigma \in S_N$. Then, $c_{ij}(\sigma)$ vanishes except with $j_\ell = \sigma(i_\ell)$ for all $1 \leqslant \ell \leqslant k$. Observe now that if $a, b \in \{1, \ldots, k\}$ are two indices such that $a \sim_p b$, then $i_a = i_b$ if and only if $\sigma(i_a) = \sigma(i_b)$. As a consequence, $\delta_p(j) \neq 0$ if and only if $\delta_p(i)$ is, hence the equation holds. We can therefore conclude as in the proof of Proposition 3.15 that $\mathcal{C} = P$. $\qquad\square$

Once again, it follows from this and Theorem 3.9 that the intertwiner spaces of S_N and S_N^+ are given by partition maps, namely

$$\mathrm{Mor}_{S_N^+}\left(P^{\otimes k}, P^{\otimes \ell}\right) = \mathrm{Vect}\left\{T_p \mid p \in NC(k,\ell)\right\},$$

$$\mathrm{Mor}_{S_N}\left(\rho^{\otimes k}, \rho^{\otimes \ell}\right) = \mathrm{Vect}\left\{T_p \mid p \in P(k,\ell)\right\}.$$

In the classical case, this result was proven around the same time by P. Martin in [59] and by V. Jones in [47]. As for the quantum case, it was established by T. Banica in [6].

Remark 3.18 The proof of Proposition 3.17 can be slightly modified to prove as well that NC is a category of partitions. Indeed, we know that

$S_N^+ = \mathbb{G}_N(\langle p_3 \rangle)$ and by the first part of the proof of 3.17, $NC \subset \langle p_3 \rangle$. If now there was a partition $p \in \langle p_3 \rangle$ with a crossing, using the singleton we could remove all the points of p except for the four involved in the crossing. This would entail $p_{\mathrm{cross}} \in \langle p_3 \rangle$, implying that $\mathcal{O}(S_N^+)$ is commutative and therefore contradicting Exercise 1.9. As a consequence, $\langle p_3 \rangle \subset NC$ and we have equality, proving that NC is a category of partitions.

Readers should by now be ready to investigate other examples by themselves. We will therefore conclude this chapter with some exercises introducing additional examples. We start with a variation on O_N^+ and S_N^+.

Exercise 3.6 We denote by $NC_{1,2}$ the set of all non-crossing partitions with blocks of size at most two.

1. Prove that $NC_{1,2}$ is a category of partitions.
2. Prove that $\mathbb{G}_N(NC_{1,2})$ is the quotient of $\mathcal{O}(O_N^+)$ by the relations

$$\sum_{k=1}^{N} U_{ik} = 1 = \sum_{k=1}^{N} U_{kj}$$

   for all $1 \leqslant i, j \leqslant N$.
3. What is the classical group corresponding to the abelianisation of $\mathbb{G}_N(NC_{1,2})$?
4. Prove that $\mathbb{G}_N(NC_{1,2}) \simeq O_{N-1}^+$.

Solution 1. It is clear that $NC_{1,2}$ is stable under horizontal concatenation and reflection. As for vertical concatenation, consider two blocks in a partition p which are merged in its vertical concatenation with another partition q. Then they are merged thanks to a block of the form $\sqcup$. But this means that the size of the new block is at most $2 + 2 - 2 = 2$. In conclusion, $NC_{1,2}$ is stable under vertical concatenation. Eventually, the identity partition is in $NC_{1,2}$ by definition, hence it is a category of partitions.
2. We have seen in Exercise 3.5 that the relations in the statement are exactly those encoded by the singleton partition $s \in NC_{1,2}(1,0)$. What we have to prove is therefore that $NC_{1,2}$ is generated as a category of partitions by NC_2 and s. This can be done by induction with the following hypothesis:

H_n: 'If $p \in NC_{1,2}$ has at most n singletons, then $p \in \langle NC_2, s \rangle$.'

- H_0 holds by definition;

- Assume that H_n holds for some $n \geqslant 0$ and let $p \in NC_{1,2}$ be a partition with $n + 1$ singletons. We can rotate p so that its rightmost point is a singleton, so that it has the form $p' \otimes s$. Since p' has n singletons, it is in $\langle NC_2, s \rangle$ and the proof is therefore complete.

3. Let G be the group corresponding to the abelianisation of $\mathbb{G}_N(NC_{1,2})$. By construction $\mathcal{O}(G)$ is the quotient of $\mathcal{O}(O_N)$ by the relations of Question 2. Therefore, G is the subgroup of O_N consisting in matrices for which the sum of the coefficients on each row and each column is one. Such matrices are called *bistochastic*, hence G is the group of all bistochastic orthogonal matrices.

4. Because T_s is an intertwiner, the vector

$$\xi = \sum_{i=1}^{N} e_i$$

is fixed by the fundamental representation u of $\mathbb{G}_N(NC_{1,2})$ by the same computation as in Exercise 2.5. If we set $e_1 = \xi$ and if $(e_i)_{2 \leqslant i \leqslant N}$ is an orthonormal basis of $\xi^{\perp}$, then the fundamental representation u reads

$$u = \begin{pmatrix} 1 & 0 \\ 0 & v \end{pmatrix}.$$

Because v is an orthogonal $(N-1)$-dimensional representation and its coefficients generate $\mathcal{O}(\mathbb{G}_N(NC_{1,2}))$, there is a surjective $*$-homomorphism $\pi \colon \mathcal{O}(O_{N-1}^+) \to \mathcal{O}(\mathbb{G}_N(NC_{1,2}))$ sending U_{ij} to v_{ij}.

Conversely, consider in the previous basis the matrix

$$U' = \begin{pmatrix} 1 & 0 \\ 0 & U \end{pmatrix} \in M_N(\mathcal{O}(O_{N-1}^+))$$

where U is the fundamental representation of O_{N-1}^+. Then, U' is orthogonal and satisfies, for all $1 \leqslant i, j \leqslant N$,

$$\sum_{k=1}^{N} U'_{ik} = 1 = \sum_{k=1}^{N} U'_{kj}.$$

Thus, there is a surjective $*$-homomorphism $\psi \colon \mathcal{O}(\mathbb{G}_N(NC_{1,2})) \to \mathcal{O}(O_{N-1}^+)$ sending u_{ij} to U'_{ij}. This is an inverse for π, hence the result. $\quad\square$

The next family of quantum groups is of a slightly different flavour and will be studied in detail later on (see Chapter 7).

Exercise 3.7 Let $\mathcal{O}(H_N^+)$ be the quotient of $\mathcal{O}(O_N^+)$ by the relations

$$U_{ij}U_{ik} = 0 = U_{ik}U_{ij} \tag{3.5}$$

for all $1 \leqslant i, j, k \leqslant N$ such that $j \neq k$.

1. Prove that this is the partition quantum group corresponding to the category NC_{even} of all *even* non-crossing partitions, meaning non-crossing partitions such that all blocks have even size.
2. Why is this called the *hyperoctahedral quantum group*?

Solution 1. Let $p_4 = \{\{1, 2, 3, 4\}\} \in NC(2, 2)$ be the partition with only one block. Then,

$$(\mathrm{id} \otimes T_{p_4}) \circ \rho_{U^{\otimes 2}}(e_{i_1} \otimes e_{i_2}) = \sum_{j_1, j_2 = 1}^{N} U_{i_1 j_1} U_{i_2 j_2} \otimes T_{p_4}(e_{j_1} \otimes e_{j_2})$$

$$= \sum_{j_1, j_2 = 1}^{N} U_{i_1 j_1} U_{i_2 j_2} \otimes \sum_{j=1}^{N} \delta_{j_1 j_2} e_j \otimes e_j$$

$$= \sum_{j=1}^{N} U_{i_1 j} U_{i_2 j} \otimes e_j \otimes e_j$$

while

$$\rho_{U^{\otimes 2}} \circ T_{p_4}(e_{i_1} \otimes e_{i_2}) = \delta_{i_1 i_2} \rho_{U^{\otimes 2}}(e_{i_1} \otimes e_{i_1})$$

$$= \delta_{i_1 i_2} \sum_{j_1, j_2 = 1}^{N} U_{i_1 j} U_{i_1 j} \otimes e_{j_1} \otimes e_{j_2}.$$

Identifying the two sums shows on the one hand that $U_{i_1 j} U_{i_2 j} = 0$ if $i_1 \neq i_2$ and on the other hand that $U_{i_1 j_1} U_{i_1 j_2} = 0$ if $j_1 \neq j_2$, which are exactly the relations in the statement. Conversely, if these relations are satisfied then both expressions collapse to

$$\delta_{i_1 i_2} \sum_{j=1}^{N} U_{i_1 j}^2 \otimes e_j \otimes e_j.$$

We have shown that p_4 generates the category of partitions of H_N^+. Note that NC_{even} is indeed a category of partitions, by the same arguments as in Proposition 3.15. Thus, $\langle p_4 \rangle \subset NC_{\text{even}}$ and the converse inclusion follows from the same argument as in Example 3.5: the analogue of Equation (3.4) yields that any one-block partition on an even number of points is in $\langle p_4 \rangle$, and from then on the proof goes by induction.

2. Consider the abelianisation G of H_N^+. Its coefficient functions satisfy Equation (3.5), so that on each row and each column of any element of G, there

is at most one non-zero coefficient. Since G is moreover made of orthogonal matrices, it follows that there is exactly one non-zero coefficient on each row and column, and that this coefficient is -1 or 1. In other words, G is the group of signed permutation matrices, namely permutation matrices where one allows -1 instead of 1 as a coefficient. This group is also known as the *hyperoctahedral group* H_N, hence the name of H_N^+. $\qquad\square$

As before, one can deduce from this a description of the invariants of these compact (quantum) groups. For H_N this recovers a result of K. Tanabe in [66], while for H_N^+ this was first established by T. Banica and R. Vergnioux in [19]: if u (respectively ρ) denotes the fundamental representation of H_N^+ (respectively of H_N), then

$$\mathrm{Mor}_{H_N^+}\left(u^{\otimes k}, u^{\otimes \ell}\right) = \mathrm{Vect}\left\{T_p \mid p \in NC_{\mathrm{even}}(k, \ell)\right\};$$

$$\mathrm{Mor}_{H_N}\left(\rho^{\otimes k}, \rho^{\otimes \ell}\right) = \mathrm{Vect}\left\{T_p \mid p \in P_{\mathrm{even}}(k, \ell)\right\}.$$

The hyperoctahedral groups are important objects because they appear in several contexts. For instance, H_N is the automorphism group of the N-dimensional cube.[7] Moreover, they are basic examples of complex reflection groups, that is to say groups which are generated by reflections in a finite-dimensional vector space. We will encounter these groups and their quantum analogues in Chapter 7.

So far we have only considered categories of partitions which are either non-crossing or symmetric. Even though we will indeed focus on the non-crossing case afterwards, it must be noted that there is a vast world in between these two classes. We will now illustrate this.

Exercise 3.8 Let $\mathcal{A}$ be the quotient of $\mathcal{O}(O_N^+)$ by the relations

$$abc = cba$$

for all $a, b, c \in \{U_{ij} \mid 1 \leqslant i, j \leqslant N\}$.

1. Show that $\mathcal{A}$ has a natural compact quantum group structure, making it a quantum subgroup of O_N^+.
2. Prove that this is the partition quantum group given by the category of partitions generated by the partition (there are three distinct blocks).

---

[7] Let us point out here the surprising fact that H_N^+ is not the quantum automorphism group (see 7.2 for the definition) of the hypercube. The interested reader may refer to [11] for more detail on this phenomenon.

$$p_* = $$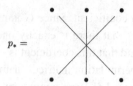

3. What is the classical compact group corresponding to its abelianisation?

Solution 1. By the same argument as in the beginning of the proof of Theorem 3.7, we have to check that the $*$-ideal $\mathcal{I}$ generated by the elements $abc - cba$ for $a, b, c \in \{u_{ij} \mid 1 \leqslant i, j \leqslant N\}$ satisfies

$$\Delta(\mathcal{I}) \subset \mathcal{I} \otimes \mathcal{O}(O_N^+) + \mathcal{O}(O_N^+) \otimes \mathcal{I}.$$

It is, of course, enough to check this on the generators: for $a = U_{i_1 j_1}$, $b = U_{i_2 j_2}$ and $c = U_{i_3 j_3}$:

$$\Delta(abc - cba) = \sum_{k_1,k_2,k_3=1}^{N} U_{i_1 k_1} U_{i_2 k_2} U_{i_3 k_3} \otimes U_{k_1 j_1} U_{k_2 j_2} U_{k_3 j_3}$$

$$- \sum_{k_1,k_2,k_3=1}^{N} U_{i_3 k_1} U_{i_2 k_2} U_{i_1 k_3} \otimes U_{k_1 j_3} U_{k_2 j_2} U_{k_3 j_1}$$

$$= \sum_{k_1,k_2,k_3=1}^{N} U_{i_1 k_1} U_{i_2 k_2} U_{i_3 k_3} \otimes U_{k_1 j_1} U_{k_2 j_2} U_{k_3 j_3}$$

$$- \sum_{k_1,k_2,k_3=1}^{N} U_{i_3 k_3} U_{i_2 k_2} U_{i_1 k_1} \otimes U_{k_1 j_1} U_{k_2 j_2} U_{k_3 j_3}$$

$$+ \sum_{k_1,k_2,k_3=1}^{N} U_{i_3 k_3} U_{i_2 k_2} U_{i_1 k_1} \otimes U_{k_1 j_1} U_{k_2 j_2} U_{k_3 j_3}$$

$$- \sum_{k_1,k_2,k_3=1}^{N} U_{i_3 k_1} U_{i_2 k_2} U_{i_1 k_3} \otimes U_{k_1 j_3} U_{k_2 j_2} U_{k_3 j_1}$$

$$= \sum_{k_1,k_2,k_3=1}^{N} (U_{i_1 k_1} U_{i_2 k_2} U_{i_3 k_3} - U_{i_3 k_3} U_{i_2 k_2} U_{i_1 k_1})$$

$$\otimes U_{k_1 j_1} U_{k_2 j_2} U_{k_3 j_3}$$

$$+ \sum_{k_1,k_2,k_3=1}^{N} U_{i_3 k_3} U_{i_2 k_2} U_{i_1 k_1} \otimes (U_{k_1 j_1} U_{k_2 j_2} U_{k_3 j_3}$$

$$- U_{k_3 j_3} U_{k_2 j_2} U_{k_1 j_1})$$

which lies in $\mathcal{I} \otimes \mathcal{O}(O_N^+) + \mathcal{O}(O_N^+) \otimes \mathcal{I}$.

2. Let us compute the relations corresponding to p_*.

$$\rho_{U^{\otimes 3}} \circ T_{p_*}(e_{i_1} \otimes e_{i_2} \otimes e_{i_3}) = \rho_{U^{\otimes 3}}(e_{i_3} \otimes e_{i_2} \otimes e_{i_1})$$

$$= \sum_{j_1,j_2,j_3=1}^{N} U_{i_3 j_1} U_{i_2 j_2} U_{i_1 j_3} \otimes e_{j_1} \otimes e_{j_2} \otimes e_{j_3}$$

while

$$(\mathrm{id} \otimes T_{p_*}) \circ \rho_{U^{\otimes 3}}(e_{i_1} \otimes e_{i_2} \otimes e_{i_3}) = \sum_{j_1,j_2,j_3=1}^{N} U_{i_1 j_1} U_{i_2 j_2} U_{i_3 j_3} \otimes T_{p_*}$$

$$(e_{j_1} \otimes e_{j_2} \otimes e_{j_3})$$

$$= \sum_{j_1,j_2,j_3=1}^{N} U_{i_1 j_1} U_{i_2 j_2} U_{i_3 j_3} \otimes e_{j_3}$$

$$\otimes e_{j_2} \otimes e_{j_1}).$$

These two quantities are equal if and only if, for all $1 \leqslant i_1, i_2, i_3, j_1,$ $j_2, j_3 \leqslant N$,

$$U_{i_1 j_1} U_{i_2 j_2} U_{i_3 j_3} = U_{i_3 j_3} U_{i_2 j_2} U_{i_1 j_1},$$

hence the result.

3. By definition, $p_* \in P_2$ so that there is an inclusion $\langle p_*, p_{\mathrm{cross}} \rangle \subset P_2$, which translates into the fact that the classical group $\mathbb{G}_N(\langle p_*, p_{\mathrm{cross}} \rangle)$ (corresponding to the abelianisation) contains O_N as a closed subgroup. But conversely, because $\mathcal{O}(\mathbb{G}_N(\langle p_* \rangle))$ is a quotient of $\mathcal{O}(O_N^+)$, its abelianisation is a quotient of $\mathcal{O}(O_N)$, meaning that $\mathbb{G}_N(\langle p_*, p_{\mathrm{cross}} \rangle)$ is a closed subgroup of O_N. Summing up, we have shown that $\mathbb{G}_N(\langle p_*, p_{\mathrm{cross}} \rangle) = O_N$. $\qquad\square$

The previous quantum group is denoted by O_N^* and called the *half-liberated quantum orthogonal group*. It was first introduced in [18] as an example of a genuinely quantum group sitting in between O_N and O_N^+. However, the idea does not work for any non-crossing partition quantum group.

Exercise 3.9 1. Prove that the quotient of $\mathcal{O}(S_N^+)$ by the relations

$$abc = cba$$

for all $a, b, c \in \{p_{ij} \mid 1 \leqslant i, j \leqslant N\}$ is $\mathcal{O}(S_N)$.
2. Give another proof of this result using partitions.
3. What happens for $\mathbb{G}_N(NC_{1,2})$?

Solution 1. For any $1 \leqslant i, j, k, \ell \leqslant N$, we have

$$p_{ij}p_{k\ell} = p_{ij}p_{ij}p_{k\ell}$$
$$= p_{k\ell}p_{ij}p_{ij}$$
$$= p_{k\ell}p_{ij}.$$

In other words, $\mathcal{A}$ is commutative so that there is a surjective $*$-homomorphism $\mathcal{O}(S_N) \to \mathcal{A}$. Because the relations $abc = cba$ hold for any $a, b, c \in \mathcal{O}(S_N)$, the previous map is an isomorphism.

2. As we have seen in Exercise 3.8, $\mathcal{A}$ is the algebra of a partition quantum group with category of partitions generated by that of S_N^+ and p_*. If s denotes the singleton partition, then

$$(| \otimes s \otimes |) \circ p_* \circ (| \otimes s \otimes |)$$

is the crossing partition, so that $\langle NC, p_* \rangle \supset \langle NC, p_{\text{cross}} \rangle = P$ and we recover the previous result.

3. The only thing we used in the previous question was that NC contains the singleton partition. Since $NC_{1,2}$ also contains it, we conclude that $\langle NC_{1,2}, p_* \rangle \supset \langle NC_{1,2}, p_{\text{cross}} \rangle$ so that the quotient is the classical bistochastic group B_N.   □

One may wonder at this point whether the half-liberation only works in the orthogonal case. We will conclude by an exercise showing that in the hyperoctahedral case, there are infinitely many half-liberations.

Exercise 3.10 We denote, for an integer $s \geqslant 3$, by $h_s \in P(0, 2s)$ the partition made of the two blocks $\{1, 3, \cdots, 2s-1\}$ and $\{2, 4, \cdots, 2s\}$. Here is a picture of it:

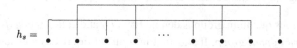

Let $\mathcal{C}_s$ be the category of partitions generated by h_s and p_*, for $s \geqslant 3$.

1. What are h_1 and h_2?
2. Prove that if $\mathcal{C}$ is a category of partitions such that $h_s, h_{s'} \in \mathcal{C}$ for some $1 \leqslant s < s'$, then $h_{s'-s} \in \mathcal{C}$.
3. Show that the partition $p_4 = \{\{1, 2, 3, 4\}\} \in P(4, 0)$ is in $\mathcal{C}_s$ for all $s \geqslant 3$.
4. Prove that for any category of partitions, if $p_4, h_s, h_{s'} \in \mathcal{C}$ for some $s, s' \in \mathbf{N}$, then $h_{s+s'} \in \mathcal{C}$.

5. For a category of partitions $\mathcal{C}$ containing p_4, describe the set $\{s \in \mathbf{N} \mid h_s \in \mathcal{C}\}$.
6. Prove that $\mathcal{O}(\mathbb{G}_N(\mathcal{C}_s))$ is not commutative.
7. Conclude that there exists an infinite set of integers $s \geqslant 3$ such that the categories of partitions $\mathcal{C}_{*,s} = \langle p_*, h_s \rangle$ are pairwise distinct.

Solution 1. By definition, h_1 is the double singleton partition $s \otimes s$, while h_2 is the crossing partition p_{cross}.

2. Simply observe that

$$\left(h_s^* \otimes |^{\otimes(2(s'-s))} \right) \circ h_{s'} = h_{s'-s}.$$

3. This follows from the fact that for $s \geqslant 3$,

$$p_4 = \left(\sqcup^{\otimes(2s-4)} \otimes |^{\otimes 4} \right) \circ h_s.$$

4. Consider first the partition

$$q = \left(|^{\otimes(2s+2)} \otimes \sqcup \otimes \sqcup \otimes |^{\otimes(2s'+2)} \right) \left(|^{\otimes(2s-1)} \otimes p_* \otimes p_* \otimes |^{\otimes(2s'-1)} \right)$$
$$(h_s \otimes p_4 \otimes h_{s'}).$$

This partition consists in two blocks of size $s + s'$. However, it is not $h_{s+s'}$ because the two middle points belong to the same block:

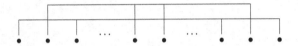

We therefore have to move, say, the left one to the left until the two blocks strictly alternate. To do this, let k_s be the partition obtained by rotating the first s points of h_s (from the left) to the upper row. Then,

$$(k_s \otimes k_s \otimes |^{\otimes 2s'})q = h_{s+s'}.$$

5. Set $s_0 = \min\{s \in \mathbf{N} \mid h_s \in \mathcal{C}\}$. We claim that $h_s \in \mathcal{C}$ if and only if $s \in s_0\mathbf{N}$. Indeed, if $s = as_0 + b$ is the Euclidean division, then by the previous questions $h_{as_0} \in \mathcal{C}$ hence $h_b = h_{s-as_0} \in \mathcal{C}$. Since $b < s_0$ by definition, we must have $b = 0$, concluding the proof.

6. Let r_1 and r_2 be reflections in $\mathbf{C}^2$ with axis forming an angle π/s and consider the diagonal matrix in $M_N(\mathcal{B}(\mathbf{C}^2))$ with diagonal coefficients $r_1, r_2, r_2, \ldots, r_2$. Let us check that these satisfy the defining relations of $\mathcal{O}(\mathbb{G}_N(\mathcal{C}_s))$.

   - The relation corresponding to h_s is equivalent to $(r_i r_j)^s = (r_j r_i)^s$. If $r_i = r_j$ this is clear, and otherwise we have $(r_1 r_2)^s = 1 = (r_2 r_1)^s$;

- The relation corresponding to p_* is $abc = cba$. The only situation in which two generators do not simplify (recall that $r_i^2 = 1$) is when $a = c \neq b$, but in that case the relation is automatically satisfied.

  As a consequence, there exists a surjective $*$-homomorphism

  $$\pi \colon \mathcal{O}(\mathbb{G}_N(\mathcal{C}_s)) \to \mathcal{B}(\mathbf{C}^2)$$

  and since r_1 and r_2 do not commute (because $s \geqslant 3$), $\mathcal{O}(\mathbb{G}_N(\mathcal{C}_s))$ must be non-commutative.

7. Assume that there are only finitely many distinct such categories of partitions, that is to say there exists $s_1, \ldots, s_n \geqslant 3$ such that for all $s \geqslant 3$, there exists $1 \leqslant i \leqslant n$ satisfying $\mathcal{C}_s = \mathcal{C}_{s_i}$. Consider an s which is prime relative to all the integers s_i for $1 \leqslant i \leqslant n$. Then, there exists $1 \leqslant i_0 \leqslant n$ such that $\mathcal{C}_s = \mathcal{C}_{s_{i_0}}$. But then, $h_1 = h_{\gcd(s, s_{i_0})} \in \mathcal{C}_s$. In other words, $\mathcal{C}_s$ contains the double singleton partition $s \otimes s \in P(2, 0)$. But rotating it to $s^* s \in P(1, 1)$ and composing with p_*, we see that $p_{\mathrm{cross}} \in \mathcal{C}_s$. Therefore, $\mathcal{O}(\mathbb{G}_N(\mathcal{C}_s))$ is commutative, a contradiction. $\square$

4

The Representation Theory of Partition Quantum Groups

Our goal in this chapter is to describe the representation theory of the quantum groups O_N^+ and S_N^+. This was first done by T. Banica in [3] and [6] respectively, using Temperley–Lieb algebras and variants thereof. However, the setting of partition quantum groups allows one to take a more general approach. We will therefore give a description of the representation theory of any partition quantum group associated to *non-crossing* partitions and then apply it to our favourite examples. This is the approach developed in [41].

4.1 Projective Partitions

Let us fix once and for all a category of partitions $\mathcal{C}$. Our first task is to find all the irreducible representations of $\mathbb{G}_N(\mathcal{C})$. By Corollary 2.17, we know that it is enough to find irreducible subrepresentations of $u^{\otimes k}$ for all $k \in \mathbf{N}$, which by definition are given by minimal projections in $\mathrm{Mor}_{\mathbb{G}_N(\mathcal{C})}\left(u^{\otimes k}, u^{\otimes k}\right)$.

The good news is that we have a nice generating family of the intertwiner spaces, namely the maps T_p for $p \in \mathcal{C}$. However, this generating family may not be linearly independent, and this is a source of trouble. We will therefore, for the sake of simplicity, rule out the issue in this text thanks to the following result.

Theorem 4.1 *Let N be an integer,*

1. *If $N \geqslant 2$, then the linear maps $(T_p)_{p \in NC_2(k,\ell)}$ are linearly independent for all $k, \ell \in \mathbf{N}$;*

2. *If $N \geqslant 4$ and $\mathcal{C}$ is any category of* non-crossing *partitions, then the linear maps $(T_p)_{p \in \mathcal{C}(k,\ell)}$ are linearly independent for all $k, \ell \in \mathbf{N}$.*

Proof Any proof of this result that we know of relies on non-trivial facts and requires the introduction of several additional combinatorial objects that would

91

take us too far from our subject. We will therefore simply sketch some of the ideas. The strategy is to deduce the second point from the first one.

1. Using the maps $\Phi_{k,\ell}$, it is sufficient to prove that for any $k \in \mathbf{N}$, the vectors $\xi_p = f_p^*$ are linearly independent for all $p \in NC_2(k, 0)$. For this, we can try to show that their Gram determinant is non-zero. Given two partitions p and q,

$$\langle \xi_p, \xi_q \rangle = \sum_{i_1, \cdots, i_k} \sum_{j_1, \cdots, j_k} \langle \delta_p(\mathbf{i}) e_{i_1} \otimes \cdots \otimes e_{i_k}, \delta_q(\mathbf{j}) e_{j_1} \otimes \cdots \otimes e_{j_k} \rangle.$$

$$= \sum_{i_1, \cdots, i_k} \delta_p(\mathbf{i}) \delta_q(\mathbf{i}).$$

The last expression is the number of tuples which are compatible with both p and q. Let us denote by $p \vee q$ the partition obtained by gluing together blocks of p and q having a common point. Then,

$$\delta_p(\mathbf{i}) \delta_q(\mathbf{i}) = \delta_{p \vee q}(\mathbf{i})$$

so that, denoting by $b(\cdot)$ the number of blocks of a partition,

$$\langle \xi_p, \xi_q \rangle = N^{b(p \vee q)}.$$

The determinant of this matrix is known as the *meander determinant* and was computed by P. Di Francesco, O. Golinelli and E. Guitter in [33, section 5.2] (see also [15, theorem 6.1] for another proof), yielding

$$\det \left(N^{b(p \vee q)} \right)_{p,q \in NC_2(k)} = \prod_{i=i}^{k} P_i(N)^{a_{k,i}},$$

where

$$a_{k,i} = \binom{2k}{k-i} - \binom{2k}{k-i-1} + \binom{2k}{k-i-2}$$

and P_i is the ith *dilated Chebyshev polynomial of the second kind*, defined recursively by $P_0(X) = 1$, $P_1(X) = X$ and, for any $i \geqslant 1$,

$$X P_i(X) = P_{i+1}(X) + P_{i-1}(X).$$

One easily checks that the roots of P_i are exactly $\{\cos(2j\pi/i) \mid 0 \leqslant j \leqslant i\} \subset] - 2, 2[$ and the result follows.

2. The idea will be to reduce the problem to the previous case. For that, notice that given a partition $p \in NC_2(2k, 2\ell)$, one can produce another partition $\widehat{p} \in NC(k, \ell)$ by gluing points two by two. Conversely, if $p \in NC(k, \ell)$, then it can be 'doubled' in the following way:

   1. Assume for simplicity that p lies on one line. For each point a of p, draw a point a_ℓ on its left and a_r on its right.

2. Then, connect a_r to b_ℓ if a and b are connected and b is the closest (travelling from left to right in cyclic order) point of the block.
3. Erase the original points and the strings connecting them. Here is an example:

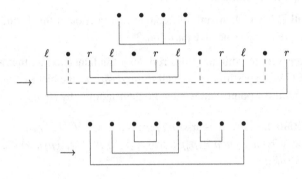

Because of the linear independence proven in the first part, $\xi_p \mapsto \xi_{\widehat{p}}$ yields a well-defined surjective linear map

$$\Phi \colon \text{Vect}\{\xi_p \mid NC_2(2k,0)\} \subset \left(\mathbf{C}^N\right)^{\otimes 2k}$$
$$\to \text{Vect}\{\xi_p \mid p \in NC(k,0)\} \subset \left(\mathbf{C}^{N^2}\right)^{\otimes k}.$$

Now, it follows from [49, proposition 3.1] that, for $p, q \in NC_2(2k)$,

$$\langle \xi_p, \xi_q \rangle = N^k \frac{\langle \xi_{\widehat{p}}, \xi_{\widehat{q}} \rangle}{\|\xi_{\widehat{p}}\| \|\xi_{\widehat{q}}\|}.$$

This means that the Gram matrices of the two families are conjugate by the diagonal matrix with coefficients $N^{k/2} \|\xi_{\widehat{p}}\|$ which is invertible, hence the linear independence for N^2. In particular, for $N = 4$ the maps are linearly independent. If now $N > 4$, restricting the maps to $(\mathbf{C}^4)^{\otimes k}$ (by considering only the first four basis vectors) shows that the family is again linearly independent, hence the result. □

Assuming therefore $\mathcal{C} \subset NC$ and $N \geqslant 4$, how can we build projections in $\text{Mor}_{\mathbb{G}_N(\mathcal{C})}(k, k)$? A natural thing to do is to look first for projections of the form T_p. The partition p should then satisfy $p \circ p = p = p^*$, but one easily sees that this does not yield a projection in general for normalisation reasons, since

$$T_p T_p = N^{\text{rl}(p,p)} T_p.$$

This is, however, easy to fix and leads to the following key definition.

Definition 4.2 A partition p is said to be *projective* if $pp = p = p^*$. Then, there is a multiple S_p of T_p which is an orthogonal projection. The set of projective partitions in $\mathcal{C}(k, k)$ will be denoted by $\mathrm{Proj}_\mathcal{C}(k)$.

Two questions immediately arise:

- Are all T_p's which are proportional to a projection of this form?
- Are there many projective partitions?

The answers to both questions rely on a fundamental fact that we will now explain. Given a partition p, we call *through-blocks* the blocks containing both upper and lower points and we denote their number by $t(p)$.

Proposition 4.3 *Any non-crossing partition $p \in NC(k, \ell)$ can be written in a unique way in the form $p = p_d^* p_u$, where $p_u \in NC(k, t(p))$, $p_d \in NC(\ell, t(p))$ and both satisfy*

1. *All lower points are in different blocks.*
2. *Each lower point is connected to at least one upper point.*
3. *If $i < j$ are lower points and $a(i), a(j)$ are the leftmost upper points connected to i and j respectively, then $a(i) < a(j)$.*

This is called the through-block decomposition *of p.*

Proof This statement is obvious pictorially. For instance,

$$p = $$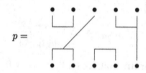

yields

$$p_u = $$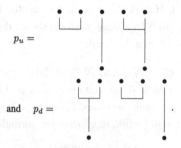

$$\text{and} \quad p_d = $$

This suggests the procedure to build the partitions. Let $b_1, \ldots, b_{t(p)}$ be the through-blocks of p ordered by their leftmost point in the upper row. Then, to build p_u we start with the upper row of p and connect b_i to the ith (starting from the left) lower point. The construction of p_d is similar, using the lower parts of the through-blocks $b'_1, \ldots, b'_n$. The crucial thing is that b_i and b'_i are the two parts of the same through-block. Indeed, if there exist $i < j$ and $k < \ell$ such that b_i is connected to b'_ℓ and b_j to b'_k, then this would produce a crossing. As a consequence, $p_d^* p_u = p$.

To prove uniqueness, simply notice that if $p'^*_d p'_u$ is another through-block decomposition, then the non-through-block parts must coincide, since they are the non-through-blocks of p, and the through-blocks are completely determined by the properties in the statement. $\qquad\square$

Let us show how useful this is by answering the two previous questions in one shot.

Proposition 4.4 *A partition p is projective if and only if there exists a partition r such that $p = r^* r$. As a consequence,*

- T_p *is a multiple of a partial isometry[1] for all p;*
- T_p *is a multiple of a projection if and only if p is projective.*

Proof If p is projective, then $p = p^* p$. Conversely, let r be any partition and let $r = r_d^* r_u$ be its through-block decomposition. The properties of Proposition 4.3 imply that

$$r_d r_d^* = |^{\otimes d} = r_u r_u^*.$$

Thus,

$$r^* r = r_u^* r_d r_d^* r_u = r_u^* r_u$$

and

$$(r_u^* r_u)^* (r_u^* r_u) = r_u^* r_u r_u^* r_u = r_u^* r_u$$

so that $(r^* r)^2 = r^* r$. $\qquad\square$

Even though we will not need it in the sequel, it is possible to work out the precise normalisation of the maps T_p needed to turn them into projections. We will leave the details as an exercise to the reader.

Exercise 4.1 For a partition r, we denote by $\beta(r)$ its number of non-through-blocks and we set

$$S_r = N^{-\beta(r)/2} T_r.$$

---

[1] An operator $V \in \mathcal{B}(H)$ is called a *partial isometry* if $V^* V$ is a projection. It then follows that VV^* is also a projection.

1. Prove that if p is projective, then S_p is a projection.
2. For two partitions p and q, we set

$$\gamma(q,p) = \frac{\beta(p) + \beta(q) - \beta(pq)}{2} - \mathrm{rl}(q,p).$$

(a) Prove that

$$S_q \circ S_p = N^{\gamma(q,p)} S_{q \circ p}$$

(b) Prove that for any partition r, $\beta(r) = \mathrm{rl}(r, r^*) + \mathrm{rl}(r^*, r)$.
(c) Prove that for any partition r, $\gamma(r^*, r) = 0$.
(d) Deduce that S_r is a partial isometry for any partition r.
(e) Conclude that a partition p is projective if and only if S_p is a projection.

Solution   1. We have

$$S_p^2 = N^{-\beta(p)} N^{\mathrm{rl}(p,p)} T_{p \circ p}$$
$$= N^{-\beta(p) + \mathrm{rl}(p,p)} T_p$$
$$= N^{\mathrm{rl}(p,p) - \beta(p)/2} S_p.$$

Consider now a non-through-block in the lower row of p. Because $p^* = p$, there is a corresponding block in the upper row of p and its points exactly match those of p in the concatenation. Therefore, the number of loops equals the number of non-through-blocks in the lower row of p, which by symmetry is $\beta(p)/2$. In other words,

$$\mathrm{rl}(p,p) - \beta(p)/2 = 0,$$

hence S_p is a projection.

2. (a) This follows directly from the definition of S_p and the fact that $T_q \circ T_p = N^{\mathrm{rl}(q,p)} T_{q \circ p}$.

(b) We will use the through-block decomposition of Proposition 4.3, writing $r = r_d^* r_u$. By definition, $\beta(r) = \beta(r_u) + \beta(r_d)$. The following expression of the associativity of composition

$$\left(T_{r_d^*} T_{r_u}\right)\left(T_{r_u^*} T_{r_d}\right) = \left(T_{r_d^*}\left(T_{r_u} T_{r_u^*}\right)\right) T_{r_d}$$

yields

$$\mathrm{rl}(r_u^*, r_d) + \mathrm{rl}(r_d^*, r_u) + \mathrm{rl}(r_d^* r_u, r_u^* r_d) = \mathrm{rl}(r_d^*, r_u r_u^*) + \mathrm{rl}(r_u, r_u^*)$$
$$+ \mathrm{rl}(r_d^* r_u r_u^*, r_d).$$

Now, the properties of the through-block decomposition imply

$$\mathrm{rl}(r_d^*, r_u) = 0 = \mathrm{rl}(r_u^*, r_d),$$

$r_u r_u^* = |^{\otimes t(r)}$ and $r_d^* r_u = r$. Meanwhile the same reasoning as in the first question leads to $\mathrm{rl}(r_u, r_u^*) = \beta(r_u)$. Summing up, we get

$$\mathrm{rl}(r, r^*) = \beta(r_u).$$

Similarly, $\mathrm{rl}(r^*, r) = \beta(r_d)$, hence the result.
(c) Observe that using the previous question we have

$$\beta(r^*, r) = \frac{\beta(r) + \beta(r^*) - \beta(r^*r)}{2} - \mathrm{rl}(r^*, r)$$

$$= \mathrm{rl}(r, r^*) - \frac{\beta(r^*r)}{2}.$$

Letting $r = r_d^* r_u$ be the through-block decomposition of r, we have proven earlier that $\mathrm{rl}(r, r^*) = \beta(r_u)$. We conclude by noticing that $\beta(r^*r) = \beta(r_u^* r_u) = 2\beta(r_u)$.
(d) From what precedes, we have

$$S_r^* S_r = S_{r^*} S_r = N^{\gamma(r^*, r)} S_{r^*r} = S_{r^*r}.$$

Because r^*r is projective, we conclude by the first question that $S_r^* S_r$ is a projection, hence S_r is a partial isometry.
(e) Assume that S_p is a projection. Then, $S_{p^*} = S_p^* = S_p$, and moreover

$$
\begin{aligned}
S_p &= S_p^2 \\
&= S_p^* S_p \\
&= S_{p^* \circ p} \\
&= S_{p \circ p}.
\end{aligned}
$$

Therefore, $p^* = p = p \circ p$ (see Remark 3.2) and p is projective. $\qquad\square$

Using this we can now transfer the usual notions of equivalence and comparison of projections to projective partitions.

Definition 4.5 Let p, q be two projective partitions. Then,

- We say that p is *dominated* by q, and write $p \preceq q$, if $pq = p = qp$;
- We say that p is *equivalent* to q, and write $p \sim q$, if there exists a partition r such that $r^*r = p$ and $rr^* = q$.

All this is nice and encouraging since it shows that partitions encode a structure comparable to that of matrix algebras. However, for a projective partition p, the projection S_p usually fails to be minimal. For instance, $S_{|^{\otimes k}} = \mathrm{Id}_{(\mathbb{C}^N)^{\otimes k}}$. Our next task is therefore to combine projective partitions in order to produce smaller projections. This can be done by exploiting the comparison relation to substract to a given projection S_p all the smaller projections of the form S_q. One may then hope that there is nothing smaller than the resulting projection, namely that it is minimal.

Definition 4.6 Let C be a category of non-crossing partitions and let $p \in C$ be a projective partition. We set

$$R_p = \sup_{q \in C, q \prec p} S_q$$

and

$$P_p = S_p - R_p.$$

Note that it is not even clear that $P_p \neq 0$. In order to prove it, a little linear algebra argument is needed to show that the supremum in the definition is a linear combination of maps T_r with $t(r) < t(p)$.

Proposition 4.7 *The projection R_p is a linear combination of maps T_r with $t(r) < t(p)$. As a consequence, $P_p \neq 0$.*

Proof We first claim that if M is a direct sum of matrix algebras and $(P_i)_{i \in I}$ are orthogonal projections, then $R = \sup_{i \in I} P_i$ is a linear combination of (not necessarily orthogonal) projections $(Q_j)_{j \in J}$ such that for any $j \in J$, there exists $i \in I$ with

$$\mathrm{Im}(Q_j) \subset \mathrm{Im}(P_i).$$

Indeed, there is a basis $(e_\ell)_{1 \leqslant \ell \leqslant s}$ of $\mathrm{Im}(R)$ such that for every $1 \leqslant \ell \leqslant s$, there is an index $i \in I$ such that $e_\ell \in \mathrm{Im}(P_i)$. Complete this basis with an orthonormal basis of the orthogonal complement of $\mathrm{Im}(R)$ and let B be the change-of-basis matrix from this basis to the canonical basis of $\mathbf{C}^n$. This means that

$$R = B^{-1} \left(\sum_{1 \leqslant \ell \leqslant s} E_\ell \right) B,$$

where the (ℓ, ℓ)-th coefficient of E_ℓ is 1 and all the other ones are 0. Setting $Q_\ell = B^{-1} E_\ell B$ for $1 \leqslant \ell \leqslant s$, we get minimal projections summing up to R. Moreover,

$$\mathrm{Im}(Q_\ell) = \mathbf{C} e_\ell \subset \mathrm{Im}(P_i)$$

for some i. Eventually, up to splitting the $P_i's$ into direct sums, we may assume that they belong to one of the blocks of M. Then, $Q_\ell \in M$ for all ℓ and the claim is proven.

Let now $R_p = \sum_\ell Q_\ell$ be the decomposition given by the previous claim applied[2] to the projections $(P_i)_{i \in I} = (S_q)_{q \prec p}$. For each ℓ,

$$Q_\ell = \sum_r \lambda_r S_r,$$

but since its range is contained in the range of some S_q for $q \prec p$, we have

$$Q_\ell = S_q Q_\ell = \sum_r \lambda_r S_{qr}.$$

Because $t(qr) \leqslant t(q) < t(p)$,[3] the proof of the first statement is complete. As for the second one, it follows from the linear independence of Theorem 4.1. □

4.2 From Partitions to Representations

We have just constructed orthogonal projections in the space of self-intertwiners of $u^{\otimes k}$; we can therefore obtain subrepresentations from this in a natural way.

Definition 4.8 For a projective partition $p \in \mathcal{C}(k, k)$, we set

$$u_p = P_p u^{\otimes k} P_p \in \mathcal{L}\left(P_p \left(\mathbf{C}^N\right)^{\otimes k} P_p\right) \otimes \mathcal{O}(\mathbb{G}_N(\mathcal{C}))$$

$$\simeq M_{\mathrm{rk}(P_p)}(\mathbf{C}) \otimes \mathcal{O}(\mathbb{G}_N(\mathcal{C})).$$

The aim of this section is to study these objects in detail. We will use the following simple result several times without further reference.

Lemma 4.9 *The rank of T_p is $N^{t(p)}$. As a consequence,*

- *For any $r_1, r_2 \in P$, $t(r_1 r_2) \leqslant \min(t(r_1), t(r_2))$;*
- *If $q \prec p$ are projective partitions, then $q = p$ if and only if $t(q) = t(p)$.*

Proof Upper non-through-blocks in p yield equations defining the kernel of T_p and consequently have no influence on the rank. If $p \in P(k, \ell)$, the image of T_p is a subspace of $(\mathbf{C}^N)^{\otimes \ell}$. Each lower non-through-block implies that

---

[2] The claim applies here because the span of the partition maps is a subalgebra of $\mathcal{L}((\mathbf{C}^N)^{\otimes k})$ stable under taking adjoints, hence a direct sum of matrix algebras by Corollary A.4.

[3] The inequality follows from the definition but is also a consequence of Lemma 4.9.

some tensor factors reduce to a one-dimensional subspace and each through-block collapses all the tensor factors which are in it to one copy of $\mathbf{C}^N$. Hence, the image is isomorphic to the tensor product of one copy of $\mathbf{C}$ for each lower non-through-block and one copy of $\mathbf{C}^N$ for each through-block, that is, the rank of T_p is $N^{t(p)}$.

The first consequence now follows from the corresponding inequality for the rank of a composition of linear maps, while the second one is clear on the associated projections.   □

4.2.1 Irreducibility

We will now proceed to describe the whole representation theory of $\mathbb{G}_N(\mathcal{C})$ using the representations u_p. The first problem to consider is irreducibility. Before we start, let us make a crucial remark: if p is projective, then $p_d = p_u$ in its through-block decomposition.

Theorem 4.10 *The representation u_p is irreducible for all projective partitions $p \in \mathcal{C}$. Moreover, for any irreducible representation v of $\mathbb{G}_N(\mathcal{C})$, there exists a projective partition $p \in \mathcal{C}$ such that $v \sim u_p$.*

Proof By definition, u_p is irreducible if and only if P_p is minimal, so that we have to prove that

$$P_p \operatorname{Mor}_{\mathbb{G}_N(\mathcal{C})} \left(u^{\otimes k}, u^{\otimes k} \right) P_p = \mathbf{C} P_p$$

and it is, of course, enough to prove that $P_p T_r P_p \in \mathbf{C} P_p$ for all $r \in \mathcal{C}$. Moreover, since $P_p = P_p S_p$, we already know that

$$P_p T_r P_p = P_p S_p T_r S_p P_p \in \mathbf{C} P_p T_{prp} P_p.$$

Note that, by definition, prp is an equivalence between two projective partitions dominated by p. If $t(prp) < t(p)$, then these two partitions are strictly dominated by p by Lemma 4.9, hence T_{prp} is dominated by R_p. It follows that $P_p T_r P_p = 0$. If $t(prp) = t(p)$, consider the partition

$$q = p_u r p_u^*.$$

It has $t(p)$ lower points and $t(p)$ upper points. Moreover,

$$p_u^* q p_u = prp$$

has $t(p)$ through-blocks so that in q, any lower point is connected to exactly one upper point. The only non-crossing partition having this property is $|^{\otimes t(p)}$, hence $q = |^{t(p)}$, implying

$$prp = p_u^* p_u = p.$$

It follows that

$$P_p T_r P_p \in \mathbb{C} P_p T_{prp} P_p = \mathbb{C} P_p T_p P_p = \mathbb{C} P_p.$$

To prove the second part of the statement, it is enough to show that any irreducible subrepresentation of $u^{\otimes k}$ is equivalent to some u_p. By Theorem 2.15, this will follow from the fact that the supremum of the projections P_p is the identity. So, let Q be this supremum and let us prove by induction on the number of through-blocks that it dominates S_p for all projective partitions $p \in \mathcal{C}(k, k)$. If the number of through-blocks of p is minimal, then $P_p = S_p$ is dominated by Q by definition. Assume now that the result holds for all partitions with $t(q) \leqslant n$ and let p have $n + 1$ through-blocks. Then R_p is dominated by Q by the induction hypothesis, as well as P_p by definition. Thus,

$$S_p = P_p + R_p \prec Q,$$

concluding the proof by induction. In particular, it dominates $S_{|\otimes k} = \mathrm{Id}$, hence it is the identity. $\square$

We have now found all the irreducible representations of $\mathbb{G}_N(\mathcal{C})$, but our list is certainly highly redundant. In other words, our second task is to decide whether u_p and u_q are equivalent or not. Once again, this matches perfectly with our previous definitions. Before embarking into the proof, let us clarify a point concerning equivalence.

Lemma 4.11 *Let $\mathcal{C}$ be a category of non-crossing partitions and let $p, q \in \mathrm{Proj}_{\mathcal{C}}(k)$ be projective partitions. If $r \in \mathcal{C}$ is such that $r^* r = p$ and $r r^* = q$, then $r = q_u^* p_u$, where $p = p_u^* p_u$ and $q = q_u^* q_u$ are the through-block decompositions.*

Proof Let $r = r_d^* r_u$ be the through-block decomposition of r. Then, by the properties of Proposition 4.3,

$$p = r^* r = r_u^* r_d r_d^* r_u = r_u^* r_u$$

so that by the uniqueness of the through-block decomposition, $r_u = p_u$. Similarly, $r r^* = q$ implies $r_d = q_d = q_u$, hence the result. $\square$

We can now characterise the equivalence of representations associated to partitions.

Proposition 4.12 *Let p and q be projective partitions in $\mathcal{C}$. Then, $u_p \sim u_q$ if and only if $p \sim q$.*

Proof Assume first that $p \sim q$ and let $r \in \mathcal{C}$ be such that $r^*r = p$ and $rr^* = q$. Let S_r denote a partial isometry proportional to T_r and set

$$V = P_q S_r P_p.$$

By Proposition 4.7, $S_r^* R_q S_r$ is a linear combination of maps $T_{r^*\ell r}$ for $\ell \prec q$. Since

$$t(r^*\ell r) \leqslant t(\ell) < t(q) = t(p),$$

it follows from arguments similar to those of the proof of Theorem 4.10 that $P_p S_r^* R_q S_r P_p = 0$. As a consequence,

$$V^*V = P_p S_r^* S_q S_r P_p \in \mathbf{C} P_p S_{r^*qr} P_p = \mathbf{C} P_p S_p P_p = \mathbf{C} P_p,$$

hence there exists $\lambda \in \mathbf{C}$ such that $V^*V = \lambda P_p$. Because both are positive operators, $\lambda > 0$ and because both have norm one, $\lambda = 1$. A similar argument shows that $VV^* = P_q$, hence V is an equivalence between u_p and u_q.

Conversely, assume that there exists a unitary intertwiner

$$V \in \mathrm{Mor}_{\mathbb{G}_N(\mathcal{C})}\left(u_p, u_q\right) = P_q\, \mathrm{Mor}_{\mathbb{G}_N(\mathcal{C})}\left(u^{\otimes k}, u^{\otimes \ell}\right) P_p$$

and extend it by 0 to a partial isometry $W \in \mathrm{Mor}_{\mathbb{G}_N(\mathcal{C})}\left(u^{\otimes k}, u^{\otimes \ell}\right)$. Then, there exist partitions $r_i \in \mathcal{C}$ such that

$$W = \sum_i \lambda_i T_{r_i}.$$

There is at least one index i such that $P_q T_{r_i} P_p \neq 0$. For such an i, $T_{r_i} P_p \neq 0$, hence $P_p T_{r_i^* r_i} P_p \neq 0$. But as we saw in the proof of Theorem 4.10, this implies that $t(r_i^* r_i) = t(pr_i^* r_i p) = t(p)$, implying that $pr_i^* r_i p = p$. In other words, $r_i^* r_i$ is a projection dominating p with the same rank, hence $r_i^* r_i = p$. Doing the same reasoning for $P_q T_{r_i r_i^*} P_q$ shows that $r_i r_i^* = q$, hence $p \sim q$. $\square$

Using these results, we can refine Theorem 4.10 by giving the exact decomposition of tensor powers of the fundamental representations.

Proposition 4.13 *For any integer k,*

$$u^{\otimes k} \sim \bigoplus_{p \in \mathrm{Proj}_{\mathcal{C}}(k)} u_p.$$

In other words, the multiplicity of an irreducible representation is exactly the cardinality of the equivalence class of the corresponding projective partition.

Proof It follows from the proof of Theorem 4.10 that $u^{\otimes k}$ contains all the representations on the right-hand side and nothing more. However, it is not clear that they are all in direct sum. To prove this, let us first denote by $E_k \subset \mathrm{Proj}_{\mathcal{C}}(k)$ a set of representatives of the equivalence classes of projective partitions and by $n_k(p)$ the cardinality of the equivalence class of p. Consider the map

$$f: \left\{ \begin{array}{ccc} \mathcal{C}(k,k) & \to & \mathrm{Proj}_{\mathcal{C}}(k) \\ r & \mapsto & r^*r \end{array} \right.$$

and note that $f^{-1}(\{p\})$ consists in all the partitions r implementing an equivalence between p and another projective partition q. Thus, $|f^{-1}(\{p\})| = n_k(p)$ and summing up yields the equality

$$|\mathcal{C}(k,k)| = \sum_p n_k(p) = \sum_{p \in E_k} n_k(p)^2.$$

On the other hand, $u^{\otimes k}$ does split into a direct sum of irreducible representations, and each of them is equivalent to u_p for some $p \in E_k$. Thus, there exists integers $\nu_k(p)$, for $p \in E_k$, such that

$$u^{\otimes k} \sim \bigoplus_{p \in E_k} \nu_k(p)u_p$$

so that if we prove that $\nu_k(p) = n_k(p)$ for all p, then we will be done. To see that, let us compute the dimension of the morphism space of both sides of the equality. The left-hand side yields

$$\begin{aligned} \dim \mathrm{Mor}_{\mathbb{G}_N(\mathcal{C})} \left(u^{\otimes k}, u^{\otimes k} \right) &= \dim \left(\mathrm{Vect} \left\{ T_p \mid p \in \mathcal{C}(k,k) \right\} \right) \\ &= |\mathcal{C}(k,k)| \\ &= \sum_{p \in E_k} n_k(p)^2, \end{aligned}$$

because we are considering non-crossing partitions and $N \geqslant 4$, while the right-hand side is

$$\sum_{p \in E_k} \nu_k(p)^2.$$

Since $\nu_k(p) \leqslant n_k(p)$ by definition, we conclude that they are equal. $\qquad \square$

4.2.2 Fusion Rules

The last step to describe the representation theory is to compute the so-called *fusion rules*. This means that for two irreducible representations v and w, we will find all the irreducible subrepresentations of $v \otimes w$. Note that given two projective partitions $p \in \mathcal{C}(k, k)$ and $q \in \mathcal{C}(\ell, \ell)$, the intertwiner space of $u_p \otimes u_q$ is by definition

$$\mathrm{Mor}_{\mathbb{G}_N(\mathcal{C})}(u_p, u_q) = (P_p \otimes P_q) \, \mathrm{Mor}_{\mathbb{G}_N(\mathcal{C})} \left(u^{\otimes (k+\ell)}, u^{\otimes (k+\ell)} \right) (P_p \otimes P_q).$$

A good starting point is therefore to find the projective partitions r such that $(P_p \otimes P_q) \, P_r \neq 0$. For instance, if $r \preceq p' \otimes q$ for some $p' \prec p$, then $(P_p \otimes P_q) \, P_r = 0$. The next proposition states that this is, in fact, enough to characterise the vanishing of this composition.

Proposition 4.14 *Let $X_{\mathcal{C}}(p, q)$ be the set of projective partitions $r \preceq p \otimes q$ such that there is no projective partition $p' \prec p$ or $q' \prec q$ satisfying $r \preceq p' \otimes q$ or $r \preceq p \otimes q'$. Then, there exists a unitary equivalence*

$$u_p \otimes u_q \sim \sum_{r \in X_{\mathcal{C}}(p,q)} u_r.$$

Proof We will proceed in two steps, analysing the operators $(P_p \otimes P_q) \, P_r$.

1. Let us first prove that if $r \notin X_{\mathcal{C}}(p, q)$, then $(P_p \otimes P_q) \, P_r = 0$. Indeed,

$$\begin{aligned} P_p \otimes P_q &= (S_p - R_p) \otimes (S_q - R_q) \\ &= S_p \otimes S_q - (S_p \otimes R_q + R_p \otimes S_q - R_p \otimes R_q) \\ &= S_{p \otimes q} - (A + B - C) \end{aligned}$$

where $A = S_p \otimes R_q$, $B = R_p \otimes S_q$ and $C = R_p \otimes R_q$. Noticing that $AB = BA = C$, we see that

$$R = A + B - C$$

is a projection dominating two commuting projections A and B.[4] In other words, R is the supremum of the projections S_r for all $r \notin X_{\mathcal{C}}(p, q)$. Thus, if $r \notin X_{\mathcal{C}}(p, q)$, then S_r is dominated either by $T_p \otimes R_q$ or by $R_p \otimes T_q$, hence by R, and the same holds for $P_r \prec S_r$. It then follows that $(P_p \otimes P_q) \, P_r = 0$.

2. If now $r \in X_{\mathcal{C}}(p, q)$, P_r is a minimal projection by Theorem 4.10, hence

$$P_r R P_r = \lambda P_r$$

---

[4] Indeed, $R^2 = A^2 + B^2 + C^2 + 2AB - 2AC - 2BC = A + B + C + 2C - 4C = R$ so that R is a projection. Moreover, $RA = A^2 + AB - AC = A + C - C = A$ and similarly for B.

for some $0 \leqslant \lambda \leqslant 1$ (by positivity of the operators involved). We claim that in fact $\lambda < 1$. Indeed, by Proposition 4.7 R is a linear combination of maps S_s where either $(p' \otimes q)s = s$ for some $p' \prec p$ or $(p \otimes q')s = s$ for some $q' \prec q$. As a consequence, $RP_rR = R^*P_rR$ is a linear combination of maps S_s such that $(p' \otimes q')s(p'' \otimes q'') = s$ where at least one of p' or q' is strictly dominated by p (respectively by q) and the same for p'' and q''. If this were equal to $P_r = S_r - R_r$, then S_r would be a linear combination of partition maps satisfying the above property and partition maps appearing in the decomposition of R_r. By Proposition 4.7, none of the partitions involved equals r, hence a contradiction by linear independence. Thus, setting

$$V = (P_p \otimes P_q) \, P_r,$$

we have

$$V^*V = (1 - |\lambda|^2)P_r = \mu^{-2}P_r$$

for some $\mu > 0$. It follows that, setting $W = \mu V$, we get an equivalence between P_r and a subprojection of $P_p \otimes P_q$. Therefore, the range of this subprojection yields a subrepresentation equivalent to u_r.

To conclude, simply notice that, by the two points of the proof,

$$\sup_{r \in X_C(p,q)} (P_p \otimes P_q) \, P_r \, (P_p \otimes P_q) = \sup_{r \preceq p \otimes q} (P_p \otimes P_q) \, P_r \, (P_p \otimes P_q)$$

$$= (P_p \otimes P_q) \, . \qquad \square$$

The previous result means that any irreducible subrepresentation of $u_p \otimes u_q$ is equivalent to u_r for some $r \in X_C(p, q)$, and that any u_r appears at least once up to equivalence. But it does not say that different but equivalent r's need both appear. To see this, we first need a better description of the set $X_C(p, q)$. The intuitive idea is that a partition $r \in X_C(p, q)$ is obtained by 'mixing' some blocks of p with some blocks of q. Concretely, this mixing is done thanks to the following partitions.

Definition 4.15 We denote by $h_\square^k$ the projective partition in $NC(2k, 2k)$ where the ith point in each row is connected to the $(2k - i + 1)$-th point in the same row (i.e. $p_k^*p_k$ with the notations of the proof of Theorem 3.7). If we moreover connect the outer blocks of each row, then we obtain another projective partition in $NC(2k, 2k)$ denoted by $h_\square^k$.

Here is a pictorial representation of these so-called *mixing partitions*:

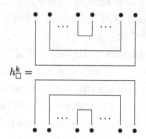

and

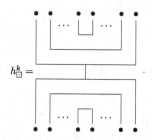

From this, we define binary operations on projective partitions, using $|$ to denote the identity partition:

$$p \,\square^k q = (p_u^* \otimes q_u^*) \left(|^{\otimes t(p)-k} \otimes h_{\square}^k \otimes |^{\otimes t(q)-k} \right) (p_u \otimes q_u),$$

$$p \,\boxdot^k q = (p_u^* \otimes q_u^*) \left(|^{\otimes t(p)-k} \otimes h_{\boxdot}^k \otimes |^{\otimes t(q)-k} \right) (p_u \otimes q_u),$$

where $p = p_u^* p_u$ and $q = q_u^* q_u$ are the through-block decompositions. We are now ready to complete the description of the representation theory of partition quantum groups.

Proposition 4.16 *Let C be a category of non-crossing partitions, let $N \geqslant 4$ be an integer and let $p, q \in C$ be projective partitions. Then, $r \in X_C(p,q) \backslash \{p \otimes q\}$ if and only if there exists $1 \leqslant k \leqslant \min(t(p), t(q))$ such that $r = p \,\square^k q$ or $r = p \,\boxdot^k q$. Moreover,*

$$u_p \otimes u_q = u_{p \otimes q} \oplus \bigoplus_{k=1}^{\min(t(p),t(q))} v_{p \square^k q} \oplus v_{p \boxdot^k q}$$

where $v_r = u_r$ if $r \in C$ and $v_r = 0$ otherwise.

Proof To prove the first assertion, let us consider $r \in X_C(p,q)$ and let us denote by $2a$ and $2b$ the respective number of points of p and q. We start with the partition $p' \in P(a, 0)$ formed by the first a upper points of r. Then, given a block of p', we insert a lower point and connect them if one of the following two conditions is matched:

- The corresponding block in r is a through-block;
- The corresponding block in r is connected with one of the last b upper points.

Starting with the leftmost block and proceeding rightwise, we get a non-crossing partition, $p^\sharp$, satisfying the conditions of Proposition 4.3. The same construction starting with the last b points produces another partition, $q^\sharp$, and we can eventually set

$$h = (p^\sharp \otimes q^\sharp) r (p^{\sharp*} \otimes q^{\sharp*}).$$

Note that, by definition, h is projective. Moreover, there are integers α and β such that $p^{\sharp*} p^\sharp = |^{\otimes \alpha}$ and $q^{\sharp*} q^\sharp = |^{\otimes \beta}$, hence

$$r = (p^{\sharp*} \otimes q^{\sharp*}) h (p^\sharp \otimes q^\sharp).$$

We now have two things to prove. First, that h is of the form of Definition 4.15 and second that $p^\sharp = p_u$ and $q^\sharp = q_u$.

Let us consider two connected upper points of h. They cannot both belong to the first α points, because they are connected to different lower blocks in $p^{\sharp*}$ and hence not connected in r by definition. The same argument works for the last β points. This has two immediate consequences:

- Non-through-blocks in h have size at most two;
- Through-blocks have size two or four.

Moreover, by construction, any upper point of $p^\sharp$ gets connected when composing with r either with a lower point or with another upper point. As a consequence, it cannot yield a singleton in h. The same works for $q^\sharp$, so that we have the following possible blocks:

- A non-through-block of length 2 connecting one of the first a points to one of the last b points;
- An identity partition;
- A through-block with four points connecting one of the first a points to one of the last b points on each row.

Let n be the largest integer such that the point n is not in a through-block. By non-crossingness, no point between $\alpha - n$ and $\beta + n$ can be in a through-block and a straightforward induction on n shows that the restriction of h to $\{\alpha - n, \cdots, \beta + n\}$ is the only possible non-crossing nesting of pairings. Moreover, $\alpha - n - 1$ belongs to a through-block, so that again by non-crossingness, no point $\alpha - n - t$ can be connected to another upper point, forcing them to be connected by an identity partition to the lower row. As a consequence, h is of the desired form.

Observe that, by construction,

$$r \preceq (p^{\sharp*} \otimes q^{\sharp*})(p^{\sharp} \otimes q^{\sharp}) = (p^{\sharp*}p^{\sharp}) \otimes (q^{\sharp*}q^{\sharp}),$$

and consider the composition $(p^{\sharp*}p^{\sharp})p$. Assume that two unconnected points of $p^{\sharp*}p^{\sharp}$ get connected. Then, the same happens in the composition $r(p \otimes q)$ so that the result cannot be r, contradicting the assumption that $r \preceq p \otimes q$. As a consequence,

$$(p^{\sharp*}p^{\sharp})p = (p^{\sharp*}p^{\sharp}),$$

that is, $p^{\sharp*}p^{\sharp} \preceq p$. The same reasoning shows that $q^{\sharp*}q^{\sharp} \preceq q$. If we can prove that both these partitions belong to $\mathcal{C}$, then it will follow by the definition of $X_{\mathcal{C}}(p, q)$ that $p^{\sharp*}p^{\sharp} = p$ and $q^{\sharp*}q^{\sharp} = q$, concluding the proof of the first assertion. Consider a nesting s of b pairings and consider $(|^{\otimes a} \otimes s)(r \otimes |^{\otimes b})(|^{\otimes a} \otimes s)$ as illustrated here:

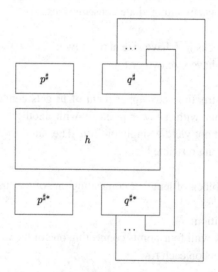

The resulting partition, which belongs to $\mathcal{C}$ by construction, is then precisely $p^{\sharp*}p^{\sharp}$. The same argument works mutatis mutandis for $q^{\sharp*}q^{\sharp}$.

As for the second part of the statement, first notice that as a consequence of the first part, the partitions in $X_C(p, q)$ all have a different number of through-blocks. But it follows from the definition of equivalence that two equivalent projective partitions must have the same number of through-blocks. Thus, the irreducible subrepresentations of $u_p \otimes u_q$ are pairwise non-equivalent, and this implies the direct sum decomposition. $\qquad\square$

Before turning to examples, let us investigate what all this tells us about the simplest representations of a compact matrix quantum groups, namely the ones which are one-dimensional.

Exercise 4.2 Let C be a category of partitions. We introduce the following operations on partitions:

- We denote by $\bar{p}$ the partition obtained by rotating p upside-down,
- We denote by $\tilde{p}$ be the partition obtained by rotating all the lower points of p to the right of its upper points.

Here is an example: for

$$p = \quad$$

we have

$$\bar{p} = \quad$$

and

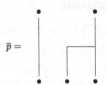

$$\tilde{p} = \qquad :$$

The goal of this exercise is to prove that $\dim(u_p) = 1$ if and only if $t(p) = 0$.

1. Assume that $t(p) = 0$. Compute the ranks of S_p and P_p and conclude.
2. Assume conversely that $\dim(u_p) = 1$ and let us assume by contradiction that $t(p) > 0$.

(a) Prove that

$$p \,\square^{t(p)} \overline{p} = \widetilde{p}^{*} \widetilde{p}$$

and that therefore $u_{p \square^{t(p)} \overline{p}}$ is the trivial representation ε.

(b) Using the formula from Proposition 4.2.9, deduce that

$$p \otimes \overline{p} = \widetilde{p}^{*} \widetilde{p}$$

and conclude.

Solution 1. We first have, by Lemma 4.9,

$$\mathrm{rk}(S_p) = N^{t(p)} = 1.$$

Moreover, since $q \prec p$ implies $t(q) < t(p)$, there is no projective partition strictly dominated by p. As a consequence, $P_p = S_p$ has rank one, hence u_p is one-dimensional.

2. (a) First observe that both partitions have no through-block by definition. Let us number the through-blocks of p from 1 to $t(p)$ (starting from the left). In $p\square^{t(p)}\overline{p}$, we add a string connecting the upper part of the ith through-block to the upper part of the $(t(p) - i)$th through-block of $\overline{p}$. But the latter is nothing but the lower part of the ith through-block of p. Thus, the result is $\widetilde{p}$. The same reasoning works for the lower row, hence the result.

Because $\widetilde{p} \in \mathcal{C}$ since it is a rotation of p, $S_{\widetilde{p}}$ is an equivalence between $u_{\widetilde{p}^{*}\widetilde{p}}$ and the trivial representation.

(b) By the formula for tensor products, $u_p \otimes \overline{u}_p$ contains $u_{p\otimes\overline{p}}$ and $u_{\widetilde{p}^{*}\widetilde{p}}$. Since the three representations involved have dimension one, they must all be equal. As a consequence,

$$t(p) \leqslant t(p \otimes \overline{p}) = t(\widetilde{p}^{*}\widetilde{p}) = 0,$$

a contradiction. $\square$

4.3 Examples

We have now done the hard work. So it is high time to be repaid for our efforts by easily deducing from the previous results the representation theory of our main examples.

4.3.1 Quantum Orthogonal Group

As already mentioned, the representation theory of quantum orthogonal groups was first computed by T. Banica in [3]. The proof relied on the identification of the category $\mathcal{R}(O_N^+)$ of representations of O_N^+ with the *Temperley–Lieb category* $\mathrm{TL}(N)$. With our setting, the result follows from elementary calculations.

Theorem 4.17 (Banica) *For $N \geqslant 2$, the irreducible representations of O_N^+ can be labelled by the non-negative integers in such a way that $u^0 = \varepsilon$, $u^1 = U$ and for any $n \in \mathbf{N}$,*

$$u^1 \otimes u^n = u^{n+1} \oplus u^{n-1}.$$

Proof Recall that if two projective partitions p and q are equivalent, then $t(p) = t(q)$. Assume now that p and q are non-crossing projective pair partitions with $t(p) = t(q)$. Denoting by $p = p_u^* p_u$ and $q = q_u^* q_u$ their through-block decompositions, observe that $p_u, q_u \in NC_2$, so that $r = q_u^* p_u \in NC(2)$. Since $r^* r = p$ and $rr^* = q$, we have proven that the equivalence class of p is given by its number of through-blocks. Setting $u^n = u_{|\otimes n}$ therefore gives all irreducible representations.

Since the empty partition corresponds to the trivial representation, $u^0 = \varepsilon$. Since $NC_2(1,1)$ only contains the identity partition, $P_| = S_| = \mathrm{Id}$ so that $u^1 = U$. Eventually, $t(|\square|^{\otimes n}) = n - 1$ and $|\,\square\,|^{\otimes n} \notin NC_2$, hence the fusion rules. $\qquad\square$

4.3.2 Quantum Permutation Group

The case of quantum permutation groups was studied by T. Banica in [6]. Once again, his strategy relied on a variant of the Temperley–Lieb category. Using non-crossing partition, the proof is almost the same as for O_N^+.

Theorem 4.18 (Banica) *For $N \geqslant 4$, the irreducible representations of S_N^+ can be labelled by the non-negative integers in such a way that $u^0 = \varepsilon$, $P = \varepsilon \oplus u^1$ and for any $n \in \mathbf{N}$,*

$$u^1 \otimes u^n = u^{n+1} \oplus u^n \oplus u^{n-1}.$$

Proof The same argument as for O_N^+ shows that equivalence classes of projective partitions in NC correspond to the number of through-blocks, so that we set $u^n = u_{|\otimes n}$ and $u^0 = \varepsilon$. However, this time $NC(1,1)$ has two elements, namely the identity partition $|$ and the double singleton partition $\{\{1\}, \{2\}\}$. The second one gives a copy of the trivial representation since the singleton partition is an equivalence between it and the empty partition. The announced decomposition of P then follows. Eventually, $t(|\square|^{\otimes n}) = n - 1$ and $t(|\,\square\,|^{\otimes n}) = n$, hence the fusion rules. $\qquad\square$

4.3.3 Quantum Hyperoctahedral Group

The two preceding examples may have given the impression that the number of through-blocks is the only important data of a projective partition. To show that this is not the case, let us consider the quantum hyperoctahedral group H_N^+. Recall that the corresponding category of partitions consists in all even partitions. Let $p_0 \in NC_{\mathrm{even}}(2,2)$ be the partition with only one-block. Then, $t(p_0) = 1$, but

$$r_{p_0}^{|} = (p_0)_u^* \circ | \in NC(1,2)$$

is a block of size three. Thus, u_{p_0} is not equivalent to $u_|$ and it turns out that this parity issue is the only obstruction to equivalence.

Exercise 4.3 Prove that the irreducible representations of H_N^+ can be indexed by words over $\{0,1\}$, with $\varepsilon = u^\emptyset$, $u = u^1$ and $u_{p_0} = u^0$.

Solution Let $p \in NC(k,k)$ and $q \in NC(k',k')$ have only one block. Then, $p_u^* q_u \in NC(k',k)$ is even if and only if k and k' have the same parity. If now $p \in NC_{\mathrm{even}}$, it follows from a straightforward induction that we can write

$$p = p_1 \otimes \cdots \otimes p_{t(p)}$$

where for all $1 \leqslant i \leqslant t(p)$, p_i is a projective partitions with $t(p_i) = 1$. If $\widetilde{p}_i$ is the one-block partition on the same number of points as p_i, then

$$p \sim \widetilde{p}_1 \otimes \cdots \otimes \widetilde{p}_{t(p)}$$

and we conclude that we can label the irreducible representations as in the statement. The last thing to check is that different labels do not yield equivalent partitions. Let $w \neq w'$ be different words and set $p^w = p_{w_1} \otimes \cdots \otimes p_{w_n}$ where $p_1 = |$ and p_0 is as in the statement. Let i be an integer such that $w_i \neq w'_i$. Then, $p_d^{w*} p_u^{w'}$ contains as a subpartition $p_{w_{id}}^* q_{w_{iu}}$ which is a block of size three. Thus, the partitions are not equivalent. $\qquad \square$

We still have to compute the fusion rules and here again a surprise awaits us. Applying Proposition 4.16, we see that

$$u^1 \otimes u^0 = u^{10} \oplus u^1 \oplus \varepsilon,$$
$$u^0 \otimes u^1 = u^{01} \oplus u^1 \oplus \varepsilon.$$

Because $u^{10} \neq u^{01}$, the fusion rules are non-commutative! This is a purely quantum phenomenon. The complete description of the representation theory of H_N^+ was established by T. Banica and R. Vergnioux in [19]. To state it conveniently, we need some notations. Let W be the set of words over $\{0,1\} = \mathbf{Z}_2$ and endow it with the following operations:

- $\overline{w_1 \cdots w_n} = w_n^{-1} \cdots w_1^{-1}$;
- $w_1 \cdots w_n . w_1' \cdots w_m' = w_1 \cdots w_n w_1' \cdots w_m'$;
- $w_1 \cdots w_n * w_1' \cdots w_m' = w_1 \cdots w_{n-1}(w_n + w_1')w_2' \cdots w_m'$.

Exercise 4.4 Prove that the irreducible representations of H_N^+ can be indexed by W in such a way that $\varepsilon = u^\emptyset$, $u = u^1$ and $u_p = u^0$. Moreover, given two words $w, w' \in W$, we have

$$u^w \otimes u^{w'} = \bigoplus_{w=a.z, w'=\overline{z}.b} u^{a.b} \oplus u^{a*b}.$$

Solution The first part of the statement was proven in Exercise 4.3. As for the fusion rules, let us consider two words $w = w_1 \cdots w_n$ and $w' = w_1' \cdots w_m'$ and let $k \leqslant \max(m, n)$. Setting $p_1 = |$ and

$$p_w = p_{w_0} \otimes \cdots \otimes p_{w_n},$$

we see that in $p_w \square^k p_{w'}$, $p_{w_{n-i+1}}$ is glued to $p_{w_i'}$. But if $w_{n-i+1} \neq w_i'$, then this yields a block of size three, which therefore does not belong to NC_{even}. Thus, $p_w \square^k p_{w'} \in NC_{\text{even}}$ if and only if the last k letters of w match the first k letters of w', that is, $w = a.z$ and $w' = \overline{z}.b$ with z of length k. Moreover, the through blocks of $p_w \square p_{w'}$ are just the first $n - i$ through-blocks of p_w followed by the last $m - i$ through-blocks of $p_{w'}$ and this corresponds to $p_{a.b}$. Considering now $p_w \boxdot p_{w'}$, the same decomposition is needed for it to be in NC_{even}, but this time the result contains an extra through-block obtained by gluing $p_{w_{n-k+1}}$ and $p_{w_k'}$. If they have the same parity, then the result has eight points, hence is equivalent to p_0, while it is equivalent to p_1 if they have a different parity. Hence, this is equivalent to p_{a*b}. $\qquad\square$

Knowing the representation theory of a compact quantum group is the first step to study most of its properties. We will illustrate this now with an elementary but important notion named growth.

Exercise 4.5 Let $(\mathbb{G}, u)$ be an orthogonal compact matrix quantum group. For an irreducible representation v, its *length* is the number

$$\ell(v) = \min\{k \mid v \subset u^{\otimes k}\}.$$

Setting

$$S_k = \sum_{v \in \text{Irr}(\mathbb{G}), \ell(v) \leqslant k} \dim(v)^2,$$

$\mathbb{G}$ is said to have *exponential growth with rate r* if there exists $A, A', r > 0$ such that

$$Ar^k \leqslant S_k \leqslant A'r^k$$

for all $k \in \mathbf{N}$. We will consider the case $\mathbb{G} = O_N^+$ for $N > 2$.

1. Let $0 < q < 1$ be such that $q + q^{-1} = N$. Show that for any $n \in \mathbf{N}$,
$$\dim(u^n) = \frac{q^{-n-1} - q^{n+1}}{q^{-1} - q}.$$

2. Prove that for any $n \in \mathbf{N}$,
$$Nq^{-n+1} \leqslant \dim(u^n) \leqslant \frac{q^{-n}}{1 - q^2}.$$

3. Conclude that O_N^+ has exponential growth and compute the rate.

Solution   1. Let us write d_n for $\dim(u^n)$. The fusion rules of Theorem 4.17 yield the recusion formula
$$d_1 d_n = d_{n+1} + d_{n-1}.$$
Because $d_1 = \dim(U) = N$, this is a linear recursion of order 2 with characteristic polynomial $X^2 - NX + 1$. The roots are q and q^{-1}, hence the solutions are of the form
$$Aq^n + Bq^{-n}.$$
Using $d_0 = 1$ and $d_1 = N$, we see that $A + B = 1$ and $qA + q^{-1}B = N$, from which the result follows.

2. Set $a_n = q^n d_n$. Then,
$$\begin{aligned}
\frac{a_{n+1}}{a_n} &= q \frac{d_{n+1}}{d_n} \\
&= \frac{q^{-n} - q^{n+2}}{q^{-n} - q^n} \\
&> 1.
\end{aligned}$$
It follows that the sequence $(a_n)_{n \in \mathbf{N}}$ is increasing, hence each term is larger than the first one and less than the limit.

3. Computing the sums yields
$$(qN)^2 \frac{q^{(-2)n+1}}{1 - q^{-2}} \leqslant S_k \leqslant \frac{1}{(1 - q^2)^2} \frac{q^{(-2)n+1}}{1 - q^{-2}}$$
so that the growth is exponential with rate q^{-2}.   □

Remark 4.19   This notion was first studied in the quantum setting by T. Banica and R. Vergnioux in [20], to which we refer the interested reader for further computations and connections to various problems.

5

Measurable and Topological Aspects

We have taken the bias, throughout the previous chapters, of working exclusively in an algebraic setting. This may seem odd given that the name 'compact quantum groups' suggests that they are *compact*, hence topological objects. Of course, we are focusing on compact matrix groups which can be described algebraically by polynomial equations, but there should still be some kind of topological structure hidden somewhere. Making sense of this topological structure is a non-trivial matter, mainly because it is a *non-commutative* topological structure. More precisely, to any compact matrix quantum group one can associate a non-commutative topological space in the sense of non-commutative geometry, as well as a non-commutative measure space. To explain this, we will therefore need to introduce some elements from the theory of operator algebras and the framework of non-commutative geometry.

5.1 Some Concepts from Non-commutative Geometry

Let us start by thinking about the case of a compact matrix group G. The topology is essentially witnessed by the phenomenon of continuity, hence one may look at the algebra

$$C(G) = C^0(G, \mathbf{C})$$

of continuous complex-valued functions on G. The first remark is that $C(G)$ contains $\mathcal{O}(G)$ as a subalgebra and that the latter is dense with respect to the uniform norm $\| \cdot \|_\infty$ by the Stone–Weierstrass theorem. As a consequence, the normed algebra $(C(G), \| \cdot \|_\infty)$ at least contains all algebraic information about the group G, plus some topological data.

There are two important things to note concerning this normed algebra. First, it is 'nice' in the sense that it is a Banach algebra.[1] Second, the involution

---

[1] A *Banach algebra* is a complete normed algebra A such that $\|xy\| \leqslant \|x\| \|y\|$ for all $x, y \in A$.

115

on $\mathcal{O}(G)$ (given by complex conjugation) extends to $C(G)$, turning it into a *Banach ∗-algebra*.[2] In order to see why this suffices to unravel the topology of G, let us describe an alternate way of recovering the group.

Any point $g \in G$ yields through evaluation a *character*, that is to say a continuous ∗-homomorphism

$$\mathrm{ev}_g \colon C(G) \to \mathbf{C}.$$

Conversely, if η is a character on $C(G)$, it is completely determined by the image of the coefficient functions c_{ij}. Let us denote by g_η the matrix of the images, namely

$$(g_\eta)_{ij} = \eta(c_{ij}).$$

Consider now the ∗-ideal $I \subset \mathbf{C}[X_{ij}, 1 \leqslant i, j \leqslant N]$, such that $\mathcal{O}(G)$ is the quotient of $\mathbf{C}[X_{ij}, 1 \leqslant i, j \leqslant N]$ by I. Because η is a ∗-homomorphism, for any polynomial $P \in I$,

$$P((g_\eta)_{ij}) = P(\eta(c_{ij})) = \eta(P(c_{ij})) = 0,$$

hence g_η is annihilated by I. This means that $g_\eta \in G$. In other words, we have a bijective correspondence between

- the set $\mathrm{Char}(G)$ of characters of $C(G)$,
- the set $\mathrm{Char}_{\mathrm{alg}}(G)$ of characters of $\mathcal{O}(G)$, and
- the elements of G.

The previous construction has a bonus: characters are in particular linear maps of norm at most one, namely points of the unit ball of $C(G)^*$. Clearly, $\mathrm{Char}(G)$ is a closed subset for the weak-∗ topology, hence a compact topological space by the Banach–Alaoglu Theorem (see, for instance, [31, theorem V.3.1]). We are now ready for the *coup de grâce*: the topology induced on $\mathrm{Char}(G)$ by the weak∗-topology coincides, through the set theoretic isomorphism constructed previously, with the original topology on G!

5.1.1 C*-Algebras

Our purpose now is to abstract the crucial features of the Banach ∗-algebra $(C(G), \| \cdot \|_\infty)$ which enable one to recover the topology of G as outlined earlier. It has long been recognised that the correct notion for doing so is that of a C*-algebra, which we will now define.

---

[2] A Banach ∗-algebra is a complex Banach algebra together with an isometric involutive map $x \mapsto x^*$ which is anti-linear and anti-multiplicative.

Definition 5.1 A *C*-algebra* a Banach *-algebra A such that for all $x \in A$,
$$\|x^*x\| = \|x\|^2.$$

Here are the two basic examples of C*-algebras.

Exercise 5.1 Let X be a compact topological space and endow the space $C(X)$ of continuous functions from X to $\mathbf{C}$ with the uniform norm $\| \cdot \|_\infty$. Define furthermore an involution by setting $f^*\colon x \mapsto \overline{f(x)}$. Prove that this yields a C*-algebra.

Solution The fact that the norm is submultiplicative is straightforward and the completeness of $C(X)$ with respect to the uniform norm is a standard result in functional analysis (see, for instance, [31, example III.1.6]). Moreover, the axioms of a Banach *-algebra are readily checked. Eventually, f^*f is the function sending x to $|f(x)|^2$ so that

$$\|f^*f\|_\infty = \sup_{x \in X} |f(x)|^2 = \left(\sup_{x \in X} |f(x)| \right)^2 = \|f\|_\infty^2. \qquad \square$$

Exercise 5.2 Let H be a Hilbert space and let $A \subset \mathcal{B}(H)$ be a closed subalgebra which is stable under taking adjoints. Prove that A is a C*-algebra when endowed with the operator norm and the involution given by adjunction.

Solution Once again, the submultiplicativity of the operator norm and the completeness of $\mathcal{B}(H)$ for it are standard results in functional analysis. Since A is assumed to be closed, it is also complete, hence a Banach algebra. Moreover, checking the axioms of a Banach *-algebra is an easy exercise in Hilbert spaces. As for the last property, recall that for any bounded operator T on a Hilbert space, $\|T^*\| = \|T\|$, so that

$$\|T^*T\| \leqslant \|T^*\|\|T\| = \|T\|^2.$$

To prove the converse equality, simply observe that, for any $x \in H$,

$$\|T(x)\|^2 = \langle T(x), T(x) \rangle = \langle T^*T(x), x \rangle \leqslant \|T^*T\|\|x\|^2. \qquad \square$$

The two fundamental results of the theory are that the general case reduces to the preceding examples. Since this is not our main purpose here, we will not give proofs and refer to the vast literature instead. Of utmost important for us is the commutative case, since it is inspirational for the field of non-commutative geometry (see, for instance, [30] for an overview).

Theorem 5.2 (Gelfand–Naimark) *Let A be a commutative unital C*-algebra. Then, there is a compact topological space X such that $A \simeq C(X)$.*

Proof We refer the reader to [32, theorem 2.1], among other possible sources, for a complete proof. Let us, however, mention that the basic strategy is exactly the one outlined for compact matrix groups G. One considers the weak-$*$ compact set X of characters on A. Evaluation yields a map $A \to C(X)$ and the job is to prove that it is an isomorphism. $\square$

Remark 5.3 The result is even stronger: $X \mapsto C(X)$ is a contravariant equivalence between the category of compact topological spaces with continuous maps and that of commutative unital C*-algebras with $*$-homomorphisms. As a consequence, one can think of general unital C*-algebras as *noncommutative compact spaces*, and this is the intuition behind non-commutative geometry. It is, for instance, possible to adapt, using this idea, tools from algebraic topology, the prominent example being K-theory.

The second main result is that Exercise 5.2 is in fact completely general. We will not use it directly in the sequel, but it has such a strong theoretical interest that we feel compelled to state it.

Theorem 5.4 (Gelfand–Naimark–Segal) *Let A be a C*-algebra. Then, there exists a Hilbert space H such that A embeds into $\mathcal{B}(H)$.*

Proof See [32, theorem 10.7]. The proof relies on the so-called *Gelfand–Naimark–Segal* (GNS in short) construction which will be outlined at the end of Section 5.1.2. $\square$

5.1.2 Von Neumann Algebras

We now have a notion of non-commutative topological space, but what about measure theory? It turns out that the same circle of ideas leads to a satisfying notion of a *non-commutative measure space*. More precisely, consider again the algebra $C(X)$ for a compact space X. If μ is a regular Borel probability measure on X, then we can define a $*$-homomorphism

$$L_\mu \colon C(X) \to \mathcal{B}\left(L^2(X, \mu)\right)$$

given by left multiplication. It is easy to see that this is an embedding if μ has full support. Moreover, it is not very difficult (see Exercise 5.4) to prove that the closure of $L_\mu(C(X))$, with respect to the *strong operator topology*,[3] is

---

[3] This is the topology for which a sequence $(T_n)_{n \in \mathbf{N}}$ of operators converges to T if and only if for any $g \in L^2(X, \mu)$, $\|T_n(g) - T(g)\|$ converges to 0.

exactly the algebra $L^\infty(X, \mu)$ of essentially bounded μ-measurable functions on X.

Now, there are two important remarks to make. First, $L^\infty(X, \mu)$ does not depend on μ (provided it has full support) but only on the corresponding σ-algebra. It therefore reflects the Borel measure space structure of X. Second, $L^\infty(X)$ is a C*-algebra, but it is even better since it is also closed with respect to the strong operator topology, which is strictly weaker in general than the operator norm topology.

Definition 5.5 A *von Neumann algebra* is a C*-algebra A embedded into $\mathcal{B}(H)$ for some Hilbert space H such that it is closed with respect to the strong operator topology.

As for C*-algebras, let us give a commutative example to illustrate the definition.[4]

Exercise 5.3 Let X be a measure space with a finite measure μ.

1. Prove that, for any $f \in L^\infty(X, \mu)$, there is a well-defined bounded linear map $L(f)$ on $L^2(X, \mu)$ such that, for any $g \in L^2(X, \mu)$,

$$L(f)(g) = fg.$$

2. Is L injective?
3. We now want to prove that $A = L(L^\infty(X, \mu))$ is closed in the strong operator topology. Let $(f_n)_{n \in \mathbf{N}}$ be a sequence in $L^\infty(X, \mu)$ such that $(L(f_n))_{n \in \mathbf{N}}$ converges strongly to an operator $T \in \mathcal{B}(L^2(X, \mu))$.
   (a) Prove that there exists a function $g \in L^2(X, \mu)$ such that for all $f \in L^\infty(X, \mu)$, $T(f) = fg$.
   (b) Prove that $g \in L^\infty(X, \mu)$ and conclude.

Solution 1. Simply observe that

$$\|fg\|_2 = \sqrt{\int_X f(x)^2 g(x)^2 \mathrm{d}\mu}$$

$$\leqslant \|f\|_\infty \sqrt{\int_X f(x)^2 g(x)^2 \mathrm{d}\mu}$$

$$= \|f\|_\infty \|g\|_2.$$

---

[4] Note that there is no analogue of Exercise 5.2, since this would just correspond to the definition.

2. We already know from the previous question that $\|L(f)\| \leqslant \|f\|_\infty$. Since, moreover, μ is assumed to be finite, the function $g = 1_{f \geqslant 0} - 1_{f \leqslant 0}$ is square summable. Then,

$$L(f)(g) = \|f\|_\infty = \|f\|_\infty \|g\|_2,$$

hence $\|L(f)\| \geqslant \|f\|_\infty$. In conclusion, L is injective and even isometric.[5]

3. (a) Let $(f_n)_{n \in \mathbf{N}}$ be a sequence in $L^\infty(X, \mu)$ such that $(L(f_n))_{n \in \mathbf{N}}$ converges in the strong operator topology to some $T \in \mathcal{B}(L^2(X, \mu))$. In particular, $(f_n)_{n \in \mathbf{N}}$ converges in $L^2(X, \mu)$ to a function $g = T(1_X)$. Moreover, for any $f \in L^\infty(X, \mu)$,

$$
\begin{aligned}
T(f) &= \lim_{n \to +\infty} L(f_n)f \\
&= \lim_{n \to +\infty} L(f_n f)1_X \\
&= \lim_{n \to +\infty} L(f f_n)1_X \\
&= f \lim_{n \to +\infty} L(f_n)1_X \\
&= f T(1_X) \\
&= fg.
\end{aligned}
$$

(b) To prove that g is essentially bounded, let us denote by C the norm of T and assume by contradiction that $\|g\|_\infty > C$. Then, there exists a measurable subset $Y \subset X$ of strictly positive measure such that $g(x) > C$ for all $x \in Y$. If we now define a function $\tilde{g} \colon X \to \mathbf{C}$ by

$$\tilde{g}(x) = \begin{cases} 1/g(x) & \text{if} \quad x \in Y \\ 0 & \text{if} \quad x \neq Y \end{cases},$$

then $g \in L^\infty(X, \mu)$ so that by the preceding computation,

$$\|T(\tilde{g})\|_2 = \|fg\|_2 = \mu(Y).$$

On the other hand, by definition we have

$$\|T(\tilde{g})\|_2 \leqslant C\|\tilde{g}\|_2 < C\frac{\mu(Y)}{C} = \mu(Y),$$

contradicting the previous inequality.

Summing up, T and $L(g)$ are two bounded operators which coincide on the dense subspace $L^\infty(X, \mu)$ of $L^2(X, \mu)$, hence they are equal. $\square$

---

[5] In fact, any injective *-homomorphism between two C*-algebras is isometric (see, for instance, [32, Corollary 1.8]).

As for C*-algebras, the commutative case is as general as can be (see [32, theorem 14.5] for a proof), provided we add some non-degeneracy assumptions on the way the algebra is embedded into $\mathcal{B}(H)$.

Theorem 5.6 (Gelfand–Naimark) *Let $A \subset \mathcal{B}(H)$ be a commutative von Neumann algebra and assume that there is a vector $\xi \in H$ such that the closure of $A.\xi = \{a(\xi) \mid a \in A\}$ equals H. Then, there exists a compact space X, a regular Borel probability measure μ and an isomorphism*

$$\Phi \colon \mathcal{B}(H) \to \mathcal{B}\left(L^2(X, \mu)\right)$$

such that $\Phi(A) = L^\infty(X, \mu)$.

Remark 5.7 It may help intuition to know that the preceding isomorphism is the conjugation by an isomorphism of the corresponding Hilbert spaces, the latter sending the vector ξ to the indicator function $\mathbf{1}_X$ of X.

Before going further, let us now prove the claims that served as motivations for the introduction of von Neumann algebras in the introduction.

Exercise 5.4 Let X be a compact space and let μ be a regular Borel probability measure on X. We denote by $L_\mu \colon C(X) \to \mathcal{B}(L^2(X, \mu))$ the map sending a function f to the operator of multiplication by f. One can prove, as in the first question of Exercise 5.3, that this is well defined.

1. Prove that $L_\mu(C(X))$ is dense in $L(L^\infty(X, \mu))$ for the strong operator topology.
2. Assume that μ has full support, in the sense that there is no strict closed subset F with $\mu(F) = 1$. Prove that L_μ is injective.

Solution 1. Let $f \in L^\infty(X, \mu)$. In particular, f is in $L^p(X, \mu)$ for all $p \geqslant 1$, hence there exists a sequence $(f_n)_{n \in \mathbf{N}}$ of continuous functions converging to f in $L^\infty(X, \mu)$. Then, for any $g \in C(X)$,

$$
\begin{aligned}
\|(f - f_n)g\|_2^2 &= \int_X (f(x) - f_n(x))^2 g(x)^2 \mathrm{d}\mu(x) \\
&\leqslant \|(f_n - f)\|_4^2 \|g\|_4^2 \\
&\leqslant \|(f_n - f)\|_\infty^2 \|g\|_4^2
\end{aligned}
$$

goes to 0 as n goes to $+\infty$. If now $g \in L^2(X, \mu)$ and $\epsilon > 0$, let $g_\epsilon \in C(X)$ be such that $\|g - g_\epsilon\|_2 < \epsilon$. Then,

$$\|L(f)(g) - L_\mu(f_n)(g)\|_2 = \|fg - f_n g\|_2$$
$$\leqslant \|(f - f_n)g_\epsilon\|_2 + \|(f - f_n)(g - g_\epsilon)\|_2$$
$$\leqslant \|(f - f_n)g_\epsilon\|_2 + \|L_\mu(f) - L_\mu(f_n)\|\|g - g_\epsilon\|_2$$
$$\leqslant \|(f - f_n)g_\epsilon\|_2 + \|f - f_n\|_\infty \|g - g_\epsilon\|_2$$
$$\leqslant \|(f - f_n)g_\epsilon\|_2 + \left(\sup_{n \in \mathbf{N}} \|f - f_n\|_\infty\right)\epsilon.$$

By the first part, there exists N such that the first term is less than ϵ for $n \geqslant N$, hence the result.

2. Let $f \in C(X)$ be such that $L_\mu(f) = 0$ and assume that there exists $x \in X$ such that $f(x) \neq 0$. We may assume without loss of generality that $f(x) > 0$, and by continuity there exists an open subset $U \subset X$ such that $f(x) > 0$ for all $x \in U$. Let us then define a function g in the following way:

$$g(x) = \begin{cases} 1/f(x) & \text{if} \quad x \in U \\ 0 & \text{if} \quad x \notin U \end{cases}.$$

Then, $g \in L^2(X, \mu)$ and

$$0 = L_\mu(f)g = 1_U$$

so that $\mu(U) = 0$. But then, $F = X \setminus U$ is a strict closed subset with measure 1, contradicting the assumption on μ. □

The drawback of von Neumann algebras is, of course, that they require an ambient Hilbert space, which is obtained in the commutative case by choosing a measure. To understand what the analogue of a measure is in the non-commutative world, let us replace μ by the linear form $\varphi_\mu \colon C(X) \to \mathbf{C}$ given by

$$\varphi_\mu(f) = \int_X f \mathrm{d}\mu.$$

Here are three basic observations:

1. The measure μ is positive if and only if $\varphi_\mu(f^*f) = \varphi_\mu(|f|^2) \geqslant 0$ for all $f \in C(X)$;
2. The measure μ has total mass 1 if and only if $\varphi_\mu(1) = 1$;
3. The measure μ has full support if and only if $\varphi_\mu(f^*f) > 0$ for all $f \neq 0$.

Based on these, we can define suitable properties of linear maps to mimic measure theory.

Definition 5.8 Let A be a $*$-algebra. A linear map $\varphi\colon A \to \mathbf{C}$ is said to be

1. *Positive* if $\varphi(x^*x) \geqslant 0$ for all $x \in A$;
2. *Unital* if $\varphi(1) = 1$;
3. *Faithful* if $\varphi(x^*x) > 0$ for all $x \neq 0$.

A positive unital linear map is called a *state*.

Using this, we can build a non-commutative measure space from a $*$-algebra and a state. This is the celebrated *GNS construction*, which we will now explain without proof (see [32, theorem 7.7] for details). Starting with a $*$-algebra A and a state $\varphi\colon A \to \mathbf{C}$, we first define a pre-inner product on A through the formula

$$\langle x, y \rangle = \varphi(xy^*).$$

The problem is that it may be degenerate. To remedy this, consider the *annihilator subspace*

$$\mathcal{N}_\varphi = \{x \in A \mid \varphi(xy^*) = 0, \ \forall \, y \in A\}.$$

Quotienting by this space and completing then yields a bona fide Hilbert space, denoted by $L^2(A, \varphi)$. Moreover, left multiplication yields a $*$-homomorphism

$$L_\varphi\colon A \to \mathcal{B}(L^2(A, \varphi)),$$

and taking the weak operator topology closure of the image yields a von Neumann algebra which deserves to be denoted by $L^\infty(A, \varphi)$. One can furthermore prove that if φ is faithful as a state, then the representation L_φ is faithful.

5.2 The Quantum Haar Measure

Our goal in the present section is to construct a suitable non-commutative topological and measurable structure associated to a compact matrix quantum group $\mathbb{G}$. As outlined in the previous section, the construction of a von Neumann algebra can be done using a state playing the role of a probability measure. Since compact groups have a canonical probability measure, namely the Haar measure, it is natural to wonder whether there is a similar canonical state on a compact matrix quantum group.

5.2.1 A Discrete Detour

A first idea to produce a non-commutative topological structure on a compact matrix quantum group $\mathbb{G}$ could be to consider simply the largest possible C*-algebra containing $\mathcal{O}(\mathbb{G})$ as a dense $*$-subalgebra, which is called the *universal* *enveloping C*-algebra* of $\mathcal{O}(\mathbb{G})$. The existence of such an object is routine to prove (see, for instance, [24, II.8.3]) and the coproduct extends to it by universality. To understand why this is unsatisfying, let us give examples of compact matrix quantum groups of a different type than those encountered up to now.

Let Γ be a finitely generated discrete group,[6] with a generating set $S = \{g_1, \ldots, g_N\}$. Its *group algebra* is the vector space $\mathbf{C}[\Gamma]$ freely spanned by elements a_g for $g \in \Gamma$ with the unique algebra structure such that, for all $g, h \in \Gamma$,

$$a_g a_h = a_{gh}$$

and the unique involution given by

$$a_g^* = a_{g^{-1}}$$

for all $g \in \Gamma$ (note that this is a unital algebra with unit a_e corresponding to the neutral element e of Γ). To see the point of considering that algebra in our context, assume, moreover, that Γ is abelian, and let $\widehat{\Gamma}$ be the set of all homomorphisms from Γ to the circle group

$$\mathbb{T} = \{z \in \mathbf{C} \mid |z| = 1\},$$

called the *Pontryagin dual* of Γ. This is obviously a group, and it is compact when equipped with the topology of pointwise convergence since it then embeds as a closed subspace of $\mathbb{T}^{\Gamma}$. Let us denote it by $\widehat{\Gamma}$. We claim that this is a compact group of matrices. Indeed, consider the group homomorphism

$$\varphi \in \widehat{\Gamma} \mapsto \mathrm{diag}\left(\varphi(a_{g_1}), \ldots, \varphi(a_{g_N})\right) \in \mathrm{GL}_N(\mathbf{C}).$$

It is injective because S is generating, hence an isomorphism onto its image. Moreover, it follows that $a_g \mapsto \varphi(g)$ yields an isomorphism between the group algebra $\mathbf{C}[\Gamma]$ and $\mathcal{O}(\widehat{\Gamma})$.

Based on this, we can define a compact matrix quantum group out of an arbitrary finitely generated discrete group. Let us denote by u_S the matrix

$$u_S = \mathrm{diag}(g_1, \ldots, g_N) \in M_N(\mathbf{C}[\Gamma]).$$

Proposition 5.9 *The pair* $(\mathbf{C}[\Gamma], u_S)$ *forms a compact unitary matrix quantum group in the sense of Definition 1.25.*

---

[6] By this we simply mean a group equipped with the discrete topology and which has a finite subset of generators.

Proof First, it is clear that the coefficients of u_S generate $C[\Gamma]$. Second, we have $u_S^* = u_S^{-1}$ and $\overline{u}_S = u_S^*$ by definition. Eventually, the map

$$\Delta \colon C[\Gamma] \to C[\Gamma] \otimes C[\Gamma]$$

such that

$$\Delta(a_g) = a_g \otimes a_g,$$

which exists by definition of the free vector space $C[\Gamma]$, satisfies the conditions for the coproduct. □

Remark 5.10 Note that such a compact matrix quantum group is orthogonal if and only if the generators can be chosen to have order 2. Thus, we need the theory of unitary compact matrix quantum groups to deal with these objects in general.

By analogy with the abelian case, we will denote this compact matrix quantum group by $\widehat{\Gamma}$ and call it the *dual of* Γ. The universal enveloping C*-algebra of $C[\Gamma]$ is nothing but the universal group C*-algebra $C^*(\Gamma)$. It is well known that this is not the correct operator algebra to look at when investigating geometric or probabilistic properties of Γ (like amenability, property (T) and so on). One should instead consider the reduced C*-algebra $C_r^*(\Gamma)$ and its associated von Neumann algebra $L(\Gamma)$.

To construct these, let us consider the Hilbert space $\ell^2(\Gamma)$ of square summable sequences of complex numbers indexed by the elements of Γ. Denoting by δ_g the sequence which is 1 at g and 0 elsewhere, we can define for any $g \in \Gamma$ a bounded (in fact unitary) operator $\lambda_g \in \mathcal{B}(\ell^2(\Gamma))$ by

$$\lambda_g(\delta_h) = \delta_{gh}.$$

This yields a $*$-homomorphism

$$\lambda \colon C[\Gamma] \to \mathcal{B}(\ell^2(\Gamma))$$

which turns out to be injective. Taking completions of $\lambda(C[\Gamma])$ for the operator norm and the strong operator topology yield respectively $C_r^*(\Gamma)$ and $L(\Gamma)$.

For a general compact matrix quantum group, we do not have a priori an analogue of $\ell^2(\Gamma)$ to build a reduced version. This is where a specific state enters the game by providing an alternate construction. Let us define a state $h \colon C[\Gamma] \to C$ through the formula

$$h(a_g) = \delta_{e,g}.$$

The key fact is that the GNS representation of h exactly yields the previous reduced operator algebras acting on $\ell^2(\Gamma)$.

Exercise 5.5 Prove that the GNS construction applied to h from $\mathbf{C}[\Gamma]$ indeed yields the Hilbert space $\ell^2(\Gamma)$ and the map λ defined earlier.

Solution Let us consider the injective linear map $\Phi\colon \mathbf{C}[\Gamma] \to \ell^2(\Gamma)$ sending g to δ_g. Then,

$$h(gh^*) = h(gh^{-1}) = \langle \delta_g, \delta_h \rangle,$$

so that Φ preserves the (pre-)inner products. We moreover claim that Φ is injective. Indeed, if $x = \sum a_g a \in \mathbf{C}[\Gamma]$ is such that $\Phi(x) = 0$, then because

$$\|\Phi(x)\|^2 = \sum_{g \in \Gamma} |a_g|^2,$$

it follows that all the coefficients vanish, hence $x = 0$. This enables us to identify $(\mathbf{C}[\Gamma], h)$ isometrically with a subspace of $\ell^2(\Gamma)$ (and as a by-product shows that h induces a bona fide inner product). That subspace is obviously dense and $\ell^2(\Gamma)$ is known to be complete, hence it is isomorphic to the Hilbert space obtained from the GNS construction.

If now $g, h \in \Gamma$ and M_g denotes the operator of multiplication by g on $\mathbf{C}[\Gamma]$, we have

$$\Phi \circ M_g \circ \Phi^{-1}(\delta_h) = \Phi\left(M_g(h)\right) = \Phi(gh) = \delta_{gh}$$

so that the identification with $\ell^2(\Gamma)$ identifies the action of the GNS construction with λ. $\qquad\square$

To find an analogue of this for an arbitrary compact matrix quantum group, first note that the irreducible representations of $\widehat{\Gamma}$ are all one-dimensional and given (up to equivalence) by the elements of g. Since h vanishes by definition on all elements $g \neq e$, the following definition is natural.

Definition 5.11 Let $\mathbb{G}$ be a unitary compact matrix quantum group, let $\mathrm{Irr}(\mathbb{G})$ be the set of equivalence classes of irreducible representations of $\mathbb{G}$ and let u^α be a representative of $\alpha \in \mathrm{Irr}(\mathbb{G})$. The *Haar integral* of $\mathbb{G}$ is the linear map $h\colon \mathcal{O}(\mathbb{G}) \to \mathbf{C}$ defined on the basis of coefficients of irreducible representations by

$$h\left(u_{ij}^\alpha\right) = \delta_{\varepsilon,\alpha}.$$

Remark 5.12 The previous definition makes sense because coefficients of irreducible representations span $\mathcal{O}(\mathbb{G})$ and are linearly independent if the representations are not equivalent, so that the span of all coefficients of all non-trivial irreducible representations is a supplement of the line $\mathbf{C}.1$, which is the span of the coefficients of the trivial representation.

The name requires some explanation. To understand where it comes from, first note that for any $\alpha \in \mathrm{Irr}(\mathbb{G})$ and any $1 \leqslant i, j \leqslant \dim(\alpha)$,

$$
\begin{aligned}
(h \otimes \mathrm{id}) \circ \Delta \left(u_{ij}^\alpha\right) &= \sum_{k=1}^{\dim(\alpha)} h(u_{ik}^\alpha) u_{kj}^\alpha \\
&= \delta_{\varepsilon, \alpha}.1 \\
&= h \left(u_{ij}^\alpha\right).1 \\
&= (\mathrm{id} \otimes h) \circ \Delta \left(u_{ij}^\alpha\right).
\end{aligned}
\tag{5.1}
$$

This is an invariance property which is reminiscent of the Haar measure on a classical compact group. In particular, if h' is another linear form on $\mathcal{O}(\mathbb{G})$ satisfying Equation (5.1), then

$$
\begin{aligned}
h(x) &= h(x)h'(1) \\
&= h'\left((h \otimes \mathrm{id}) \circ \Delta(x)\right) \\
&= (h \otimes h') \circ \Delta(x) \\
&= h'(x)h(1) \\
&= h'(x).
\end{aligned}
$$

With this we can give a better justification for the name 'Haar integral':

Exercise 5.6 Prove that if G is a compact matrix group, then the Haar integral on $\mathcal{O}(G)$ coincides with integration with respect to the Haar measure.

Solution Notice first that for any $f \in \mathcal{O}(G)$, $\Delta(f)(x, y) = f(xy)$. If now φ_{Haar} denotes the linear map given by integration with respect to the Haar measure Haar, then

$$
(\varphi_{\mathrm{Haar}} \otimes \mathrm{id}) \circ \Delta(f)(y) = \int_G f(xy) \mathrm{dHaar}(x) = \int_G f(x) \mathrm{dHaar}(x) = \varphi_{\mathrm{Haar}}(f).1
$$

and the same works for $(\mathrm{id} \otimes \varphi_{\mathrm{Haar}}) \circ \Delta$. As explained earlier, this implies that $\varphi_{\mathrm{Haar}} = h$. $\qquad \square$

5.2.2 Positivity of the Haar Integral

The only non-trivial fact about h is that it is positive. This will indeed be the most important result of this chapter. To prove it, let us first give a straightforward, though important, corollary of Theorem 2.15. In the sequel, we denote by B^t the transpose of a matrix B.

Corollary 5.13 *Let* $\mathbb{G}$ *be an orthogonal compact matrix quantum group. For any irreducible unitary representation* v, $\overline{v} = v^{*t}$ *is also an irreducible unitary representation.*

Proof By Corollary 2.17, v is unitarily equivalent to a subrepresentation of $u^{\otimes k}$ for some $k \in \mathbf{N}$. Thus, there exist a unitary matrix B and a unitary representation w satisfying

$$B^* u^{\otimes k} B = \begin{pmatrix} v & 0 \\ 0 & w \end{pmatrix}.$$

As a consequence, v^{*t} is a direct summand of $B^t u^{\otimes k} B^{*t}$, hence is a unitary representation. If V is the subspace of $\left(\mathbf{C}^N\right)^{\otimes k}$ on which v acts, then $\overline{v}$ naturally acts on the conjugate Hilbert space $\overline{V}$. Moreover, T intertwines v with itself if and only if $\overline{T} = T^{*t}$ intertwines $\overline{v}$ with itself. Thus,

$$\dim\left(\mathrm{Mor}_{\mathbb{G}}(\overline{v}, \overline{v})\right) = \dim\left(\mathrm{Mor}_{\mathbb{G}}(v, v)\right) = 1$$

and $\overline{v}$ is irreducible. □

Definition 5.14 Let $\mathbb{G}$ be a compact matrix quantum group and let $v = (v_{ij})_{1 \leqslant i,j \leqslant n}$ be an irreducible representation acting on $V = \mathbf{C}^n$. Then, $\overline{v} = (v_{ij}^*)_{1 \leqslant i,j \leqslant n}$ is called the *conjugate representation* of v. If v is any unitary representation and

$$v = \bigoplus_{i=1}^{n} v^i$$

is a decomposition into irreducible subrepresentations, then its conjugate representation is

$$\overline{v} = \bigoplus_{i=1}^{n} \overline{v}^i.$$

Let us now proceed to prove that the Haar integral is a state. The Haar state for compact matrix quantum groups was first constructed by S. L Woronowicz in [75] in a completely different way: he directly proved the existence of a state satisfying Equation (5.1). Our more algebraic approach rather follows [34].

Theorem 5.15 (Woronowicz) *Let* $(\mathbb{G}, u)$ *be an orthogonal compact matrix quantum group. Then, the Haar integral* h *is a state on* $\mathcal{O}(\mathbb{G})$. *Moreover, it is faithful.*

From now on, h will be called the *Haar state* of $\mathbb{G}$.

Proof It turns out that positivity and faithfulness will be proven in one shot. For the sake of clarity, we proceed in several steps.

1. Let v be any finite-dimensional representation acting on a Hilbert space V. We claim that

$$\widehat{h}(v) = (h(v_{ij}))_{1 \leqslant i,j \leqslant \dim(V)}$$

is the matrix of the orthogonal projection onto the subspace of fixed vectors of ρ_v. Indeed, let W be the subspace of fixed vectors. The proof of the first point in Theorem 2.15 shows that v decomposes along $W \oplus W^{\perp}$ as $v = \varepsilon_W \oplus v_{|W^{\perp}}$, where

$$\varepsilon_W = \mathrm{Id}_W \otimes 1 \in M_{\dim(W)}(\mathbf{C}) \otimes \mathcal{O}(\mathbb{G}) \cong M_{\dim(W)}(\mathcal{O}(\mathbb{G})).$$

Since $v_{W^{\perp}}$ does not contain the trivial representation, $\widehat{h}(v)$ vanishes on $W^{\perp}$, while it acts by the identity on W by definition. Thus, it is the orthogonal projection onto the subspace of fixed vectors.

2. Let us show that for any irreducible representation v acting on V, the vector

$$\xi = \sum_{i=1}^{\dim(V)} e_i \otimes \overline{e}_i \in V \otimes \overline{V}$$

is fixed for $v \otimes \overline{v}$. Indeed,

$$
\begin{aligned}
\rho_{v \otimes \overline{v}}(\xi) &= \sum_{i=1}^{\dim(V)} \sum_{j,k=1}^{\dim(V)} v_{ij} v_{ik}^{*} \otimes e_j \otimes \overline{e}_k \\
&= \sum_{j,k=1}^{\dim(V)} \left(\sum_{i=1}^{\dim(V)} v_{ij} v_{ik}^{*} \right) \otimes e_j \otimes \overline{e}_k \\
&= \sum_{i=1}^{\dim(V)} \sum_{j,k=1}^{\dim(V)} \delta_{j,k} e_j \otimes \overline{e}_k \\
&= 1 \otimes \xi.
\end{aligned}
$$

3. We now claim that for any $\alpha, \beta \in \mathrm{Irr}(\mathbb{G})$,

$$\mathrm{Mor}_{\mathbb{G}} \left(\varepsilon, v^{\alpha} \otimes \overline{v^{\beta}} \right) \simeq \mathrm{Mor}_{\mathbb{G}} \left(v^{\beta}, v^{\alpha} \right).$$

Indeed, if $T \in \mathcal{L} \left(V^{\beta}, V^{\alpha} \right)$, then

$$\xi_T = \sum_{i=1}^{\dim(v^{\beta})} T(e_i) \otimes \overline{e}_i$$

is fixed for $v^\alpha \otimes \overline{v^\beta}$ if and only if T is an intertwiner, where $(e_i)_{1 \leqslant i \leqslant \dim(v^{(\beta)})}$ is an orthonormal basis. To see this, observe that if T is an intertwiner, then

$$
\left(\rho_{v^\alpha \otimes \overline{v^\beta}}\right)(\xi_T) = \sum_{i=1}^{\dim(v^\beta)} \rho_{v^\alpha} \circ T(e_i) \otimes \rho_{\overline{v^\beta}}(\overline{e}_i)
$$

$$
= \sum_{i=1}^{\dim(v^\beta)} (\mathrm{id} \otimes T) \circ \rho_{v^\beta}(e_i) \otimes \rho_{\overline{v^\beta}}(\overline{e}_i)
$$

$$
= (\mathrm{id} \otimes T \otimes \mathrm{id}) \circ \rho_{v^\beta \otimes \overline{v^\beta}}\left(\sum_{i=1}^{\dim(v^\beta)} e_i \otimes \overline{e}_i\right)
$$

$$
= (\mathrm{id} \otimes T \otimes \mathrm{id})\left(\sum_{i=1}^{\dim(v^\beta)} 1 \otimes e_i \otimes \overline{e}_i\right)
$$

$$
= 1 \otimes \xi_T.
$$

As for the converse, it holds by the same computation together with the fact that ρ_{v^β} is injective. In the same way, $\xi \in V^\alpha \otimes \overline{V^\beta}$ is a fixed vector for $v^\alpha \otimes \overline{v^\beta}$, if and only if

$$
T_\xi \colon x \mapsto (\mathrm{id} \otimes \overline{x}^*)(\xi)
$$

is an intertwiner between v^β and v^α, because $\xi_{T_\xi} = \xi$ and $T_{\xi_T} = T$.

4. It follows from the previous points and Schur's Lemma (Lemma 2.14) that

$$
\mathrm{Mor}_G\left(\varepsilon, v^\alpha \otimes \overline{v^\beta}\right) = \mathrm{Mor}_G\left(v^\beta, v^\alpha\right) = \{0\}
$$

if $\alpha \neq \beta$. In particular,

$$
\widehat{h}\left(v^\alpha \otimes \overline{v}^\beta\right) = 0
$$

in that case.

5. For $\alpha = \beta$, the space $\mathrm{Mor}_G(\varepsilon, v^\alpha \otimes \overline{v}^\alpha)$ is one-dimensional (again by Schur's Lemma) and is spanned by

$$
\xi = \sum_{i=1}^{\dim(v^{(i)})} e_i \otimes \overline{e}_i.
$$

Thus, if P_ξ denotes the orthogonal projection onto $\mathbf{C}.\xi$, then

$$
\widehat{h}\left(v_{ij}^\alpha \otimes v_{kl}^{\alpha*}\right) = \langle e_j \otimes \overline{e}_l, P_\xi(e_i \otimes \overline{e}_k)\rangle
$$

$$
= \delta_{ik}\delta_{jl}\frac{1}{\dim\left(v^{(i)}\right)}.
$$

6. We can now conclude. If

$$x = \sum_{\alpha \in \mathrm{Irr}(\mathbb{G})} \sum_{i,j=1}^{\dim(v^\alpha)} \lambda_{i,j}^\alpha v_{ij}^\alpha,$$

then

$$h(xx^*) = \sum_{\alpha \in \mathrm{Irr}(\mathbb{G})} \sum_{i,j=1}^{\dim\left(v^{(\alpha)}\right)} \frac{|\lambda_{i,j}^\alpha|^2}{\dim(v^\alpha)} \geqslant 0$$

and this vanishes if and only if $x = 0$. $\qquad\qquad\square$

Remark 5.16 It follows from the preceding proof that h is *tracial* in the sense that $h(xy) = h(yx)$ for any $x, y \in \mathcal{O}(\mathbb{G})$. See Section C.2 in Appendix C for comments on that property.

Before commenting on the consequences of Theorem 5.15, let us highlight an important fact concerning conjugate representations which was established along the way.

Proposition 5.17 *Let v and w be irreducible representations of an orthogonal compact quantum group $\mathbb{G}$. Then, $w \sim \overline{v}$ if and only if $\mathrm{Mor}_{\mathbb{G}}(\varepsilon, v \otimes w) \neq \{0\}$. Moreover, for any representations w, w_1 and w_2,*

$$\mathrm{Mor}_{\mathbb{G}}(v, w_1 \otimes w_2) \simeq \mathrm{Mor}_{\mathbb{G}}(v \otimes \overline{w}_2, w_1).$$

The latter property is called Frobenius reciprocity.

Proof The first assertion is a restatement of Point (3) in the proof of Theorem 5.15. As for the second one, the proof was done in the particular case where $v = \varepsilon$ in Point (3) in the proof of Theorem 5.15. In the general case, the proof is similar. If V, W_1 and W_2 denote the carrier Hilbert spaces of v, w_1 and w_2 respectively, and if $T: V \to W_1 \otimes W_2$, define $\widetilde{T}: V \otimes \overline{W}_2 \to W_1$ by

$$\left\langle \widetilde{T}(x \otimes \overline{y}_2), y_1 \right\rangle = \langle T(x), y_1 \otimes y_2 \rangle$$

for all $x \in V$, $y_1 \in W_1$ and $y_2 \in W_2$. The result follows from the fact that T is an intertwiner if and only if $\widetilde{T}$ is. This can be seen, for instance, by writing

$$\widetilde{T} = (\mathrm{id}_{W_1} \otimes D_{W_2}) \circ (T \otimes j^{-1}),$$

where D_{W_2} is the duality map and $j: W_2 \to \overline{W}_2$ is the canonical anti-linear isomorphism. $\qquad\square$

Applying the GNS construction to $\mathcal{O}(\mathbb{G})$ and the state h, we get a ∗-homomorphism

$$L_h \colon \mathcal{O}(\mathbb{G}) \to \mathcal{B}(L^2(\mathbb{G}))$$

which is injective because h is faithful. The strong operator topology closure of the range of L_h is then a von Neumann algebra denoted by $L^\infty(\mathbb{G})$, and the norm closure is a C*-algebra denoted by $C_r(\mathbb{G})$. In particular, the ∗-algebra of an orthogonal compact matrix quantum group always embeds into a C*-algebra and this enables one to complete the description in the case where the ∗-algebra is commutative.

Corollary 5.18 *Let $(\mathcal{A}, u)$ be an orthogonal compact matrix quantum group with $\mathcal{A}$ commutative. Then, there exists a closed subgroup G of O_N such that $\mathcal{A} = \mathcal{O}(G)$.*

Proof As explained earlier, $\mathcal{A}$ embeds into its reduced C*-completion A, which is commutative since $\mathcal{A}$ is dense in it. It then follows from Theorem 5.2 that $A = C(X)$ for some compact space X. In particular, if $f \in A$ is non-zero, then there exists $x \in X$ such that $f(x) \neq 0$, so that evaluation at x provides a ∗-homomorphism $A \to \mathbf{C}$ which does not vanish on f. Restricting it to $\mathcal{A}$, we see that the criterion of Question (5) in Exercise 1.10 is satisfied, hence the result. $\square$

Using this, one easily deduces that the abelianisations of $\mathcal{O}(O_N^+)$ and $\mathcal{O}(U_N^+)$ are respectively $\mathcal{O}(O_N)$ and $\mathcal{O}(U_N)$.

Remark 5.19 It is possible to start the theory of compact quantum groups from the non-commutative topological point of view. Starting with a C*-algebra endowed with a coproduct, one then seeks a condition ensuring that this 'compact quantum semigroup' is indeed a group. This was done by S. L Woronowicz in [77]. The reader may also refer to [54] for alternate proofs or to [69, chapters II.4 and II.5] and [60, chapter 1] for detailed accounts. There is also an approach based on von Neumann algebras, which is then a particular case of the definition of a locally compact quantum group; see [51].

5.3 A Glimpse of Non-commutative Probability Theory

5.3.1 Weingarten Calculus

The definition of the Haar state given in the proof of Theorem 5.15 is in a sense explicit since it is given in the basis of coefficients of irreducible representations.

$$h(u_{ij}^{\alpha}) = \frac{\delta_{ij}}{\dim(\alpha)}.$$

However, for the purpose of concrete computations, one sometimes has to compute the Haar state for arbitrary polynomials in the coefficients of the fundamental representation u. There is no practical general formula for this, but if we restrict to partition quantum groups, then it is possible to express the image of these polynomials using the corresponding category of partitions.

When trying to compute $h(u_{i_1 j_1} \cdots u_{i_k j_k})$ for arbitrary indices, the basic idea is to recall that $\widehat{h}(u^{\otimes k})$ is the orthogonal projection onto the subspace of fixed vectors of $u^{\otimes k}$. Therefore, if we have enough data on that subspace, we may be able to derive information on the corresponding orthogonal projection. And it turns out that being a partition quantum group precisely means that we have a combinatorial description of the space $\mathrm{Mor}_{\mathbb{G}}\left(\varepsilon, u^{\otimes k}\right)$, which is the space of fixed vectors of $u^{\otimes k}$.

More precisely, if $\mathcal{C}$ is a category of partitions and $\mathbb{G} = \mathbb{G}_N(\mathcal{C})$, then writing $\xi_p = f_p^*$ we have

$$\mathrm{Mor}_{\mathbb{G}}\left(\varepsilon, u^{\otimes k}\right) = \mathrm{Vect}\left\{\xi_p \mid p \in \mathcal{C}(0,k)\right\}.$$

Now, to describe the orthogonal projection, a convenient tool is the Gram matrix of a basis. This one was already computed in the proof of Theorem 4.1 and we summarise this in a definition. Recall that $p \vee q$ denotes the partition whose blocks are the minimal subsets of $\{1, \ldots, N\}$ which can be written both as unions of blocks of p and as unions of blocks of q.

Definition 5.20 The *Gram matrix* associated to $\mathcal{C}$, k and N is the $|\mathcal{C}(0,k)| \times |\mathcal{C}(0,k)|$-matrix

$$\mathrm{Gr}_N(\mathcal{C},k)$$

with coefficients

$$\mathrm{Gr}(\mathcal{C},k)_{p,q} = N^{b(p \vee q)}$$

for $p, q \in \mathcal{C}(0,k)$. The corresponding *Weingarten matrix* is, if it exists,

$$W_N(\mathcal{C},k) = \mathrm{Gr}_N(\mathcal{C},k)^{-1}.$$

As one may expect from this definition and the title of the section, the final formula for the Haar state will involve W_N rather than Gr_N, so that it will not be completely explicit. We will, however, show in the next section that it is possible to obtain asymptotic estimates on the coefficients of the Weingarten matrix which give (free) probabilistic information on $\mathcal{O}(\mathbb{G})$. The first proof of this result was done by T. Banica and B. Collins for O_N^+ and U_N^+ in [12] and then extended to arbitrary partition quantum groups in [18, theorem 5.4].

Theorem 5.21 (Banica–Collins) *Let $\mathcal{C}$ be a category of partitions and let $N \geqslant$ 4 be an integer. For any $k \in \mathbf{N}$, if $\{\xi_p \mid p \in \mathcal{C}(0, k)\}$ is linearly independent, then*

$$h\left(u_{i_1 j_1} \cdots u_{i_k j_k}\right) = \sum_{p,q \in \mathcal{C}(0,k)} \delta_p(i_1, \cdots, i_k)\delta_q(j_1, \cdots, j_k)W_N(\mathcal{C}, k)_{pq}.$$

(5.2)

Equation (5.2) is called the *Weingarten formula*.

Proof Let us denote by $W \subset \left(\mathbf{C}^N\right)^{\otimes k}$ the subspace of fixed vectors of $u^{\otimes k}$. As already mentioned, the left-hand side of Equation (5.2) is a coefficient of the orthogonal projection P_W onto W. More precisely, using orthogonality of P_W we have

$$
\begin{aligned}
h\left(u_{i_1 j_1} \cdots u_{i_k j_k}\right) &= P_{i_1 \cdots i_k, j_1 \cdots j_k} \\
&= \langle e_{i_1} \otimes \cdots e_{i_k}, P(e_{j_1} \otimes \cdots \otimes e_{j_k})\rangle \\
&= \langle P(e_{i_1} \otimes \cdots e_{i_k}), e_{j_1} \otimes \cdots \otimes e_{j_k}\rangle.
\end{aligned}
$$

Consider now the map $\Phi \colon \left(\mathbf{C}^N\right)^{\otimes k} \to W$ given by

$$\Phi(x) = \sum_{p \in \mathcal{C}(0,k)} \langle \xi_p, x \rangle \xi_p.$$

This is a surjective map but it is not idempotent. Indeed,

$$\Phi(\xi_p) = \sum_{q \in \mathcal{C}(0,k)} \langle \xi_q, \xi_p \rangle \xi_q = \mathrm{Gr}_N(\mathcal{C}, k)\xi_p.$$

In other words, $\Phi = \mathrm{Gr}_N(\mathcal{C}, k) \circ P_W$. Since $N \geqslant 4$, Gr is invertible so that the previous equality readily yields $P_W = W_N(\mathcal{C}, k) \circ \Phi$. The proof now ends with an easy computation,

$$
\begin{aligned}
h\left(u_{i_1 j_1} \cdots u_{i_k j_k}\right) &= \langle W_N(\mathcal{C}, k) \circ \Phi(e_{i_1} \otimes \cdots \otimes e_{i_k}), e_{j_1} \otimes \cdots \otimes e_{j_k}\rangle \\
&= \sum_{p \in \mathcal{C}(0,k)} \langle e_{i_1} \otimes \cdots \otimes e_{i_k}, \xi_p \rangle \langle W_N(\xi_p), e_{j_1} \otimes \cdots \otimes e_{j_k}\rangle \\
&= \sum_{p \in \mathcal{C}(0,k)} \delta_p(i) \langle W_N(\xi_p), e_{j_1} \otimes \cdots \otimes e_{j_k}\rangle \\
&= \sum_{p,q \in \mathcal{C}(0,k)} \delta_p(i)W_N(\mathcal{C}, k)_{pq} \langle \xi_q, e_{j_1} \otimes \cdots \otimes e_{j_k}\rangle \\
&= \sum_{p,q \in \mathcal{C}(0,k)} \delta_p(i)\delta_q(j)W_N(\mathcal{C}, k)_{pq}. \qquad \square
\end{aligned}
$$

We will now give applications of Theorem 5.21 with a probabilistic flavour. This will only be a glimpse of a research direction of its own, which owes much to the combinatorial approach to free probability theory (see, for instance, the book [61]). In particular, the use of *free cumulants* leads to spectacular applications of Weingarten calculus to *non-commutative de Finetti theorems* as in [17] or to asymptotics of random matrices with non-commutative entries *à la Diaconis–Shahshahani*, like in [16].

5.3.2 Spectral Measures

Let us start by making more precise how probability theory enters the stage. Consider an element $x \in L^\infty(\mathbb{G})$ which is self-adjoint. Then, it generates a von Neumann subalgebra $\langle x \rangle \subset L^\infty(\mathbb{G})$ which is commutative, hence isomorphic to $L^\infty(\mathrm{Sp}(x))^7$ by Borel functional calculus (see, for instance, [32, theorem 15.10]). The restriction of h to this algebra is still a state, hence coincides with integration with respect to a Borel probability measure μ_x. We can therefore see x as a random variable and wonder about its *spectral measure* μ_x. We mainly have access to the moments of μ_x, which are given by

$$m_k(\mu_x) = \int_{\mathrm{Sp}(x)} t^k \mathrm{d}\mu_x(t) = h\left(x^k\right).$$

This means that we will compute moments and then try to reconstruct the probability measure. This is not always possible for an arbitrary probability measure, but it will work in our case because all measures are supported on the spectrum of bounded operators, and such spectra are always compact.[8] (See for instance [31, theorem VII.3.6] for a proof.) For later use, let us recall some facts about one of the most important probability distributions in free probability.

Definition 5.22 The *semicircle distribution* (or *Wigner distribution*) μ_{sc} is the probability distribution on $[-2, 2]$ with density

$$\frac{1}{\pi}\sqrt{4 - x^2}$$

with respect to the Lebesgue measure.

---

[7] Here $\mathrm{Sp}(x)$ denotes the spectrum of x, that is to say the set of elements $\lambda \in \mathbb{C}$ such that $x - \lambda.1$ is not invertible.

[8] It follows from the Stone–Weierstrass Theorem that a Borel probability measure on a compact space is determined by its evaluation on polynomials, hence by its moments.

The computation of the moments of μ_{sc} is a standard exercise in undergraduate integration.

Exercise 5.7 Prove that the moments of the semicircle distribution are given by

$$m_{2k+1}(\mu_{\text{sc}}) = 0 \quad \text{and} \quad m_{2k}(\mu_{\text{sc}}) = \frac{1}{k+1}\binom{2k}{k}.$$

Solution Observe that because μ_{sc} has an even density, all its odd moments vanish. We can therefore focus on even moments and compute

$$
\begin{aligned}
m_{2k}(\mu_{\text{sc}}) &= \frac{1}{\pi}\int_{-2}^{2} x^{2k}\sqrt{4-x^2}\,\mathrm{d}x \\
&= \frac{2^{2k+2}}{\pi}\int_{0}^{2}\left(\frac{x}{2}\right)^{2k}\sqrt{1-\left(\frac{x}{2}\right)^2}\,\mathrm{d}x \\
&= \frac{2^{2k+2}}{\pi}\int_{0}^{\pi/2}(\sin(\theta))^{2k}\cos(\theta)^2\mathrm{d}\theta \\
&= \frac{2^{2k+2}}{\pi}\left(\frac{\pi}{2}\frac{(2k)!}{(2^k(k!))^2}-\frac{\pi}{2}\frac{(2k+2)!}{(2^{k+1}((k+1)!))^2}\right) \\
&= \frac{1}{k+1}\binom{2k}{k}.
\end{aligned}
$$

$\square$

5.3.3 Truncated Characters

We will now start by investigating the character of the fundamental representation, that is to say the element

$$\chi = \sum_{i=1}^{N} u_{ii}.$$

Based on our probabilistic intuition, we will use, from now on, the following fancy but suggestive notation for the Haar state: $\int_{\mathbb{G}}$. The moments of χ are easy to compute and do not require Weingarten calculus.

Proposition 5.23 *Let $\mathcal{C}$ be a category of non-crossing partitions and let $N \geqslant 4$. Then,*

$$m_k(\chi) = |\mathcal{C}(k,0)|.$$

Proof Denoting by $\mathrm{Fix}(v)$ the subspace of fixed points of ρ_v, we have by definition

$$m_k(\chi) = \int_{\mathbb{G}_N(\mathcal{C})} \chi^k$$
$$= \mathrm{Tr}\left(\widehat{h}(u^{\otimes k})\right)$$
$$= \dim\left(\mathrm{Fix}\left(u^{\otimes k}\right)\right)$$
$$= \dim\left(\mathrm{Vect}\left\{\xi_p \mid p \in \mathcal{C}(k,0)\right\}\right)$$
$$= |\mathcal{C}(k,0)|,$$

where the last line comes from the linear independence of the partition vectors of Theorem 4.1. $\qquad\square$

Let us illustrate this result in the case of the quantum orthogonal groups.

Exercise 5.8 Let $\mathcal{C} = NC_2$. Prove that the spectral measure of the character is the semicircle distribution.

Solution First, observe that for odd k, $NC_2(k,0) = 0$. We now have to compute the number of non-crossing pair partitions on $2k$ points and this can be done by induction on k. Denoting by $(C_k)_{k\in\mathbb{N}}$ the numbers that we are looking for, we have $C_1 = 1$. Moreover, consider a non-crossing pair partition p on $2k$ points and let ℓ be the point connected to 1. Then, p induces non-crossing pairings on $\{2,\ldots,\ell-1\}$ and $\{\ell+1,\ldots,2k\}$. Conversely, given such pairings one can reconstruct p with the condition that 1 should be connected to ℓ. In other words, we have

$$C_k = \sum_{i=2}^{2k} C_{i-1} C_{2k-i+1}.$$

This uniquely defines the sequence $(C_k)_{k\in\mathbb{N}}$, and it turns out that the moments of the semicircle distribution satisfy this recursion relation. Thus, χ is what is called a *semicircular element*. $\qquad\square$

It is worth comparing this result with the corresponding one for classical orthogonal groups, at least in an asymptotic sense.

Example 5.24 Assume that $\mathcal{C} = P_2$ so that we are considering the classical orthogonal group O_N. We cannot use Proposition 5.23 directly to compute the law of χ for arbitrary N because the linear maps T_p are not linearly independent. Nevertheless, they turn out to be linearly independent for $p \in P_2(k,\ell)$ if $k + \ell \leqslant N$ (see, for instance, the explicit computations in [15]). We can therefore consider the limit of these moments as N goes to infinity, yielding

$$\lim_{N\to+\infty} \int_{O_N} \chi^k d\mu(g) = |P_2(k)| = a_k.$$

Obviously, $a_{2k+1} = 0$ and we have the following recursion formula for a_{2k}: a pair partition on $2k$ points is given by the choice of the point connected to 1 and the choice of a pair partition on $2(k-1)$ points, hence

$$a_{2k} = \sum_{i=2}^{2k} a_{2(k-1)} = (2k-2)a_{2k-2}.$$

This yields $a_{2k} = (2k-2)!!$, where the double factorial symbol means that we are taking the product of all even numbers, which are exactly the moments of the standard Gaussian distribution.

Observe now that $\chi \in \mathcal{O}(O_N)$ is a function on matrices, namely

$$g \mapsto \sum_{i=1}^{N} c_{ii}(g) = \text{Tr}(g).$$

We have therefore shown that the asymptotic law of the trace of a uniformly chosen orthogonal matrix is the standard Gaussian distribution.

Remark 5.25 It is a general fact (which is made formal, for instance, by the Bercovici–Pata bijection [21]) that the semicircle distribution plays in free probability theory a role corresponding to that of the Gaussian distribution in classical probability theory. The previous results therefore support the idea that the quantum orthogonal group can be thought of as a free probabilistic analogue of the classical orthogonal group.

For quantum permutation groups, we can resort to the 'doubling trick' explained in the proof of Theorem 4.1.

Example 5.26 For S_N^+, we have to compute the number of non-crossing partitions on k points. The bijection $p \mapsto \widehat{p}$ used in the proof of Theorem 4.1 shows that

$$|NC(k,0)| = |NC_2(2k,0)|.$$

As a consequence, $\chi_{S_N^+}$ has the same distribution as $\chi_{O_N^+}^2$. This is known as the *free Poisson distribution* (or the *Marschenko–Pastur distribution*) with parameter 1. It is supported on $[0,4]$ and has density

$$\frac{1}{2\pi}\sqrt{\frac{4}{x} - 1}$$

with respect to the Lebesgue measure.

Exercise 5.9 Prove that the Free Poisson distribution indeed has the same moments as the character of the fundamental representation of S_N^+ for $N \geqslant 4$.

Solution It is enough to check that the moment of order k of the free Poisson distribution coincides with the moment of order $2k$ of the semicircle distribution. This follows from a change of variables,

$$
\int_{-2}^{2} x^{2k} \frac{\sqrt{4 - x^2}}{2\pi} \,dx = 2 \int_{0}^{2} x^{2k} \frac{\sqrt{4 - x^2}}{2\pi} \,dx
$$
$$
= 2 \int_{0}^{4} y^k \frac{\sqrt{4 - y}}{2\pi} \frac{dy}{2\sqrt{y}}
$$
$$
= \int_{0}^{4} y^k \frac{\sqrt{4 - y}}{2\pi \sqrt{y}} \,dy
$$
$$
= \int_{0}^{4} y^k \frac{1}{2\pi} \sqrt{\frac{4}{y} - 1} \,dy. \qquad \square
$$

Once again, it is worth comparing this result with its asymptotic counterpart for classical permutations.

Example 5.27 For the classical permutation group S_N, we have

$$
\lim_{N \to +\infty} \int_{S_N} \chi^k \,d\mu(\sigma) = |P(k)| = B_k,
$$

and the numbers B_k are known as the *Bell numbers*. To find a recursion relation, simply observe that a partition on k points is obtained by first choosing $0 \leqslant n \leqslant k-1$ points to connect to the first one, and then choosing any partition of $k - (n + 1)$ points. Thus,

$$
B_k = \sum_{n=0}^{k-1} \binom{k-1}{n} B_{k-1-n}
$$
$$
= \sum_{n=0}^{k-1} \binom{k-1}{k-1-n} B_{k-1-n}
$$
$$
= \sum_{n=0}^{k-1} \binom{k-1}{n} B_n.
$$

It is easy to check that the solution of that recursion is

$$
B_k = \frac{1}{e} \sum_{i=0}^{+\infty} \frac{i^k}{i!},
$$

and these numbers are exactly the moments of the Poisson distribution with parameter 1.

Recall that, as in the case of O_N, χ is the trace function. But the trace of a permutation matrix is the same as its number of fixed points. We have therefore proven that the number of fixed points of a uniformly chosen permutation is asymptotically given by a Poisson distribution with parameter 1.

The quantum hyperoctahedral group is more involved and the distribution of $\chi_{H_N^+}$ is the *free Bessel distribution* introduced in [9]. Instead of proving this, we will, following the work of T. Banica and R. Speicher in [18], try to refine the previous results by considering *truncated characters* in the following sense.

Definition 5.28 Let $\mathcal{C}$ be a category of non-crossing partitions and let $N \geqslant 4$ be an integer. The truncated characters are the elements

$$\chi_t = \sum_{i=1}^{\lfloor tN \rfloor} u_{ii}.$$

for $t \in [0, 1]$.

The previous results can then be improved thanks to the Weingarten formula. This first requires an estimate on the Gram and Weingarten matrices. In the sequel, $O(N^{-1/2})$ means a matrix all of whose coefficients are dominated by $N^{-1/2}$.

Lemma 5.29 *Let $\mathcal{C}$ be a category of partitions, let N, k be integers and let $\Gamma_N(\mathcal{C}, k)$ be the diagonal of $\mathrm{Gr}_N(\mathcal{C}, k)$. Then,*

$$\mathrm{Gr}_N(\mathcal{C}, k) = \Gamma_N(\mathcal{C}, k)^{1/2} \left(\mathrm{Id} + O\left(N^{-1/2} \right) \right) \Gamma_N(\mathcal{C}, k)^{1/2},$$

$$\mathrm{W}_N(\mathcal{C}, k) = \Gamma_N(\mathcal{C}, k)^{-1/2} \left(\mathrm{Id} + O\left(N^{-1/2} \right) \right) \Gamma_N(\mathcal{C}, k)^{-1/2}.$$

Proof To lighten notations we will omit, from now on, $\mathcal{C}$ and k in the computations. The trick is to consider the coefficients of $\Gamma_N^{-1/2} \mathrm{Gr}_N \Gamma_N^{-1/2}$:

$$\left(\Gamma_N^{-1/2} \mathrm{Gr}_N \Gamma_N^{-1/2} \right)_{pq} = \left(\Gamma_N^{-1/2} \right)_{pp} (\mathrm{Gr}_N)_{pq} \left(\Gamma_N^{-1/2} \right)_{qq}$$

$$= N^{b(p \vee q) - \frac{b(p) + b(q)}{2}}.$$

If $p = q$, then the result is 1. Otherwise, there are at least two blocks of p or of q which are merged in $p \vee q$, hence the result is less than $N^{-1/2}$. In other words, the matrix

$$B_N = \Gamma_N^{-1/2} \mathrm{Gr}_N \Gamma_N^{-1/2} - \mathrm{Id}$$

has all its coefficients dominated by $N^{-1/2}$, yielding the first part of the statement.

As for the second part, we have

$$\Gamma_N^{1/2} W_N \Gamma_N^{1/2} = (\mathrm{Id} + B_N)^{-1}$$
$$= \mathrm{Id} + \sum_{n=1}^{+\infty} (-1)^n B_N^n$$
$$= \mathrm{Id} + C_N,$$

and the terms in the sum defining C_N are uniformly dominated by $N^{-1/2}$, yielding the second part. □

We can now give the asymptotics of the moments of truncated characters.

Theorem 5.30 (Banica–Speicher) *Let $N \geqslant 4$, let $\mathcal{C}$ be a category of partitions and let u be the fundamental representation of $\mathbb{G}_N(\mathcal{C})$. Then,*

$$\lim_{N \to +\infty} \int_{\mathbb{G}_N(\mathcal{C})} \chi_t^k = \sum_{p \in \mathcal{C}(k)} t^{b(p)}.$$

Proof For clarity, we will omit k and $\mathcal{C}$ in the notations, since no confusion is possible. Let us first consider the sum of the first ℓ diagonal coefficients. We claim that

$$\int_{\mathbb{G}_N(\mathcal{C})} (u_{11} + \cdots + u_{\ell\ell})^k = \mathrm{Tr}\,(\mathrm{W}_N \mathrm{Gr}_\ell).$$

Indeed, by Theorem 5.21,

$$\int_{\mathbb{G}_N(\mathcal{C})} \left(\sum_{i=1}^{\ell} u_{ii} \right)^k = \sum_{i_1,\cdots,i_k=1}^{\ell} \int_{\mathbb{G}} u_{i_1 i_1} \cdots u_{i_k i_k}$$

$$= \sum_{i_1,\cdots,i_k=1}^{\ell} \sum_{p,q \in \mathcal{C}} \delta_p(i) \delta_q(i) \mathrm{W}_N(p,q)$$

$$= \sum_{p,q \in \mathcal{C}} \left(\sum_{i_1,\cdots,i_k=1}^{\ell} \delta_p(i) \delta_q(i) \right) \mathrm{W}_N(p,q).$$

If both δ_p and δ_q do not vanish on i, then this means that i matches $p \vee q$. Because the indices run over a set of cardinality ℓ, there are therefore $\ell^{b(p \vee q)} = (\mathrm{Gr}_\ell)_{pq}$ tuples yielding a non-zero contribution, hence the result.

Using Lemma 5.29, we can now write (using the fact that $\ell = \lfloor t/N \rfloor$)

$$
\begin{aligned}
\int_{\mathbb{G}_N(\mathcal{C})} \chi_t^k &= \mathrm{Tr}\,(\mathrm{W}_N \mathrm{Gr}_\ell) \\
&= \mathrm{Tr}\left(\Gamma_N^{-1}\Gamma_\ell\right) + \mathrm{Tr}\left(\Gamma_N^{-1/2} C_N \Gamma_N^{-1/2}\Gamma_\ell\right) \\
&\quad + \mathrm{Tr}\left(\Gamma_N^{-1}\Gamma_\ell^{1/2} B_\ell \Gamma_\ell^{1/2}\right) + \mathrm{Tr}(\Gamma_N^{-1/2} C_N \Gamma_N^{-1/2}\Gamma_\ell^{1/2} B_\ell \Gamma_\ell^{1/2}) \\
&= \mathrm{Tr}\left[\left(\Gamma_N^{-1}\Gamma_\ell\right)(\mathrm{Id} + B_\ell + C_N)\right] \\
&\quad + \mathrm{Tr}\left(\Gamma_N^{-1/2} C_N \Gamma_N^{-1/2}\Gamma_\ell^{1/2} B_\ell \Gamma_\ell^{1/2}\right) \\
&= \mathrm{Tr}\left(\Gamma_N^{-1}\Gamma_\ell\right)\left(1 + O(N^{-1/2}) + O(\ell^{-1/2})\right) \\
&\quad + \mathrm{Tr}\left(\Gamma_N^{-1/2} C_N \Gamma_N^{-1/2}\Gamma_\ell^{1/2} B_\ell \Gamma_\ell^{1/2}\right) \\
&= \left(\sum_{p \in \mathcal{C}(k)} \ell^{b(p)} N^{-b(p)}\right)\left(1 + O\left(N^{-1/2}\right)\right) \\
&\quad + \mathrm{Tr}\left(\Gamma_N^{-1/2} C_N \Gamma_N^{-1/2}\Gamma_\ell^{1/2} B_\ell \Gamma_\ell^{1/2}\right) \\
&= \left(\sum_{p \in \mathcal{C}(k)} \left(\frac{\lfloor tN \rfloor}{N}\right)^{b(p)}\right)\left(1 + O\left(N^{-1/2}\right)\right) \\
&\quad + \mathrm{Tr}\left(\Gamma_N^{-1/2} C_N \Gamma_N^{-1/2}\Gamma_\ell^{1/2} B_\ell \Gamma_\ell^{1/2}\right).
\end{aligned}
$$

If we can prove that the second term goes to 0 as N goes to infinity, then the result will be proven. To do this, observe that

$$
\begin{aligned}
\mathrm{Tr}(\Gamma_N^{-1/2} C_N \Gamma_N^{-1/2}\Gamma_\ell^{1/2} B_N \Gamma_\ell^{1/2}) &= \sum_{p,q \in \mathcal{C}(k)} N^{-b(p)/2 - b(q)/2} \\
&\quad (C_N)_{pq}\, \ell^{b(p)/2 + b(q)/2}\, (B_\ell)_{qp} \\
&= \sum_{p,q \in \mathcal{C}(k)} \left(\frac{\ell}{N}\right)^{b(p)/2 + b(q)/2} \\
&\quad (C_N)_{pq}\, (B_\ell)_{qp} \\
&= O\left(\frac{1}{\sqrt{\ell N}}\right) \sum_{p \in \mathcal{C}(k)} \sqrt{\frac{\lfloor tN \rfloor}{N}}^{\,-b(p)/2 + b(q)/2} \\
&\xrightarrow[N \to +\infty]{} 0. \qquad\qquad\qquad\qquad\qquad\square
\end{aligned}
$$

Example 5.31 For $C = NC_2$, the limit of the odd moments vanish, since there is no pair partition on an odd number of points, and we have

$$\lim_{N \to +\infty} \int_{O_N^+} \chi_t^{2k} = \sum_{p \in NC_2(2k)} t^k = t^k \frac{1}{k+1} \binom{2k}{k}.$$

This is the same as the semicircle distribution, except that the radius of the circle has been changed to $2t$ instead of 2. We therefore see that the moments of χ_t smoothly approximate the distribution of χ.

Example 5.32 For $C = NC$, the computation requires the theory of *free cumulants* (see [61]) which we will not introduce here. Let us simply mention that using it, one can prove that χ_t is asymptotically free Poisson with parameter t, that is, has spectral measure

$$(1-t)\delta_0 + \frac{1}{2\pi} \frac{\sqrt{4t - (x-1-t)^2}}{x} \mathrm{d}x.$$

5.3.4 Single Coefficients and Freeness

We will end with another result of the same type, but focusing on the joint behaviour of several elements. More precisely, let us consider all the coefficients u_{ij} of the fundamental representation of O_N^+. Contrary to χ, the spectral measure of u_{ij} is very complicated and depends on N (see [14] for an explicit computation). But the asymptotics can be easily obtained by the Weingarten formula:

Proposition 5.33 *For $N \geqslant 2$, set $x_{ij} = \sqrt{N} U_{ij}$, where U is the fundamental representation of O_N^+. Then,*

$$\lim_{N \to +\infty} \int_{O_N^+} x_{ij}^k = \begin{cases} 0 & \text{if } k \text{ odd} \\ \dfrac{2}{2k+2} \dbinom{k}{k/2} & \text{if } k \text{ even} \end{cases}.$$

In other words, the coefficients are asymptotically semicircular.

Proof Because there is no pair partition on a odd number of points, Theorem 5.21 implies that the odd moments vanish. As for the even ones, using again Theorem 5.21 we have

$$\int_{O_N^+} x_{ij}^{2k} = N^k \sum_{p,q \in NC_2(k)} \delta_p(i, \dots, i)\delta_q(j, \dots, j) (W_N)_{pq}$$

$$= \sum_{p,q \in NC_2(k)} N^k (W_N)_{pq}.$$

By Lemma 5.29,

$$N^k (W_N)_{pq} = N^k \left(\Gamma_N^{-1}\right)_{pq} + \left(\Gamma_N^{-1/2} C_N \Gamma_N^{1/2}\right)_{pq}$$

$$= N^k \left(\Gamma_N^{-1}\right)_{pq} + \left(\Gamma_N^{-1/2}\right)_{pp} \left(\Gamma_N^{-1/2}\right)_{qq} (C_N)_{pq}$$

$$= N^k N^{-(b(p)+b(q))/2} + N^{-(b(p)+b(q))/2} O\left(N^{-1/2}\right).$$

If $p = q$, since we are considering pair partitions on $2k$ points, the number of blocks is k so that the right-hand side equals $1 + O(N^{-1/2})$. If $p \neq q$, the right-hand side equals

$$N^k \left(\Gamma_N\right)_{pq}^{-1} + O(N^{-1/2}) = O(N^{-1/2})$$

which goes to 0 as N goes to infinity. Summing up,

$$\lim_{N \to +\infty} \int_{O_N^+} x_{ij}^k = |NC_2(k, 0)|$$

and the result follows.   □

Remark 5.34 For S_N^+, the computation is trivial. Indeed, since p_{ij} is a projection, $p_{ij}^k = p_{ij}$ for all k so that all moments of order > 0 are equal to N^{-1}.

We can even go further and compute the limit of *mixed moments*, that is to say arbitrary monomials involving the coefficients x_{ij}. In general, in probability theory, it is difficult to compute such mixed moments because of the correlations between the variables. However, the situation is greatly simplified if these correlations vanish, that is, if the variables are independent. In our setting, since the variables do not commute with one another, independence is not the correct notion. Nevertheless, there is a concept of *free independence* which translates the fact that there is 'no correlation'. We will not define this concept here because this would take us too far (we once again highly recommend reading [61]) but simply give a formula for arbitrary joint moments. This requires some terminology.

Definition 5.35 Given a monomial $X = x_{i_1 j_1} \cdots x_{i_k j_k}$ and a partition $p \in \mathcal{P}(k)$, we will say that p *matches* X if, for any points ℓ_1 and ℓ_2 connected by p, we have $i_{\ell_1} = i_{\ell_2}$ and $j_{\ell_1} = j_{\ell_2}$. The set of partitions matching X will be denoted by $\mathcal{P}(X)$.

Proposition 5.36 *We have, for any monomial* $X = x_{i_1 j_1} \cdots x_{i_k j_k}$,

$$\lim_{N \to +\infty} \int_{O_N^+} X = \left\{ \begin{array}{ll} 0 & if \quad k \ odd \\ |\mathcal{P}(X) \cap NC_2(k)| & if \quad k \ even \end{array} \right. .$$

Proof The proof starts, as for Proposition 5.33, using Theorem 5.21. The odd moments vanish, so let us consider a moment of length $2k$:

$$\int_{O_N^+} X = \sum_{p,q \in NC(2k)} \delta_p(i) \delta_q(j) (W_N)_{pq}.$$

The same computation as in the proof of Proposition 5.33 shows that as N goes to infinity, all the terms in the right-hand side vanish except for those with $p = q$. We therefore have

$$\lim_{N \to +\infty} \int_{O_N^+} X = \sum_{p \in NC_2(k)} \delta_p(i) \delta_p(j).$$

To conclude, simply notice that both δ-functions do not vanish, if and only if p matches X. $\qquad\square$

Part III

FURTHER EXAMPLES AND APPLICATIONS

6

A Unitary Excursion

6.1 The Classification of Categories of Non-crossing Partitions

Now that we have a firm grasp on the theory of orthogonal compact matrix quantum groups, we may start to feel frustrated by the restrictions that the orthogonality assumption entails. Indeed, there are not so many categories of partitions: combining the results of [18] and [73] yields

Theorem 6.1 (Banica–Speicher, Weber) *There are exactly seven categories of non-crossing partitions.*

This leaves few examples to study. The obvious way to go further is, of course, to consider unitary quantum groups, which were briefly alluded to in Definition 1.25. It turns out that the results of the previous chapters carry on to this more general setting with little modifications, including the combinatorial aspects. But before turning to this, we will give a proof of Theorem 6.1. This will illustrate the basic techniques in the classification program for partition quantum groups. Because this is a long proof involving various arguments, we will split it into several subsections.

6.1.1 The Proof

Our strategy to prove the classification theorem will be to introduce invariants for categories of partitions and refine them until we can show that they exhaust all possibilities.

6.1.1.1 The Block Size

We start with a very natural notion which is fundamental in all works on classifications of categories of partitions.

Definition 6.2 The *block size* of a category of partitions $\mathcal{C}$ is the set $\mathrm{BS}(\mathcal{C})$ of all sizes of blocks of partitions in $\mathcal{C}$.

Thanks to that tool, we can already split the classification into four separate cases. All along the proofs of this section, we will use an elementary but fundamental technique which we now explain. Let $\mathcal{C}$ be a category of partitions and let $p \in \mathcal{C}$. Consider two neighbouring points of p and remove them, merging the blocks to which they belong. Then, the resulting partition is again in $\mathcal{C}$. Indeed, if the points where the ith and $(i + 1)$-th and p is assumed to lie on the upper row, then the operation simply amounts to taking

$$p \circ \left(|^{\otimes i-1} \otimes \sqcup \otimes |^{\otimes k-i-1} \right),$$

where k is the number of points of p. For short, we will say that the new partition has been obtained by *capping* and use this without further reference hereafter.

Let us now prove a lemma which will be useful in understanding the invariant BS.

Lemma 6.3 *Let $\mathcal{C}$ be a category of partitions and assume that it contains a partition with a block of size at least three. Then, the partition*

$$p_4 = \{\{1, 2, 3, 4\}\} \in NC(4, 0)$$

is in $\mathcal{C}$.

Proof Let p be a partition with a block b of size at least three and let p' be a rotation of p such that the endpoints of the upper row are two consecutive points of b. (By this we mean that between these two points there is no point of b.) Then, the two endpoints of the upper row are connected to at least one point in the lower row since b has at least three points, hence the partition $q = p'^* \circ p'$ has the form where b' is some partition lying on one line. To conclude, simply observe that the following is a rotation of q:

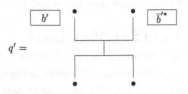

and that $q' \circ q'^* \in \mathcal{C}$ is a rotation of p_4. □

Proposition 6.4 *There are four possible values for* $\mathrm{BS}(\mathcal{C})$, *namely* $\{2\}$, $\{1,2\}$, $2\mathbf{N}$ *and* $\mathbf{N}$. *Moreover,*

1. *If* $\mathcal{C}$ *is non-crossing and* $\mathrm{BS}(\mathcal{C}) = \{2\}$, *then* $\mathcal{C} = NC_2$;
2. *If* $\mathcal{C}$ *is non-crossing and* $\mathrm{BS}(\mathcal{C}) = 2\mathbf{N}$, *then* $\mathcal{C} = NC_{even}$.

Proof Let $p \in \mathcal{C}$, let b be a block of p and let b' be a block of p'. Let q be a rotation of p such that all the points of b are in the upper row and the extreme points of the upper row are in b and let q' be a similar rotation of p'. If k is the number of upper points of q and k' is the number of upper points of q', then

$$(q \otimes q') \circ (|^{\otimes k-1} \otimes \sqcap \otimes |^{k'-1})$$

has a block whose size is the sum of the sizes of b and b' minus 2 since it is obtained by merging one point of b with one point of b'. As a consequence, if $x, y \in \mathrm{BS}(\mathcal{C})$, then $x + y - 2 \in \mathrm{BS}(\mathcal{C})$. In particular, if $\mathcal{C}$ contains a partition with a block of size at least three, then it contains a partition with a block of size four by Lemma 6.3 and it then follows that for any $x \in \mathrm{BS}(\mathcal{C})$,

$$x + 4 - 2 = x + 2 \in \mathrm{BS}(\mathcal{C}),$$

that is, $2\mathbf{N} \subset \mathrm{BS}(\mathcal{C})$. If all the partitions in $\mathcal{C}$ have even size, we have equality. Otherwise, let p be a partition with a block of odd size. If that block is a singleton, then using it and p_4 we can produce a partition with a block of size three in $\mathcal{C}$. As a consequence, $x + 3 - 2 = x + 1 \in \mathrm{BS}(\mathcal{C})$ as soon as $x \in \mathrm{BS}(\mathcal{C})$, so that $\mathrm{BS}(\mathcal{C}) = \mathbf{N}$. Otherwise, consider an odd block such that all partitions between its points have an even number of blocks (such a block exists by finiteness of the number of blocks). Capping all the intermediate points, we end up with a block of even size at least three, which can in turn be reduced by capping to a block of size three and we once again conclude that $\mathrm{BS}(\mathcal{C}) = \mathbf{N}$.

As for the second part, if $\mathrm{BS}(\mathcal{C}) = \{2\}$ and $\mathcal{C}$ is non-crossing then $\mathcal{C} \subset NC_2$ and we already proved in 3.15 that this implies $\mathcal{C} = NC_2$. If now $\mathrm{BS}(\mathcal{C}) = 2\mathbf{N}$ and $\mathcal{C}$ is non-crossing, then $\mathcal{C} \subset NC_{even}$. Let, moreover, p be a partition with a block of size four. Because all the other blocks have even size, we can remove them by capping and conclude that $\mathcal{C}$ contains p_4. As we have seen in Exercise 3.7, this implies $\mathcal{C} = NC_{\text{even}}$, concluding the proof. $\square$

6.1.1.2 The Block Number

To go further in the classification, it is crucial to observe that the remaining cases are exactly the category of partitions whose partitions contain singletons. To distinguish between them, let us introduce one example which we have not seen before.

Definition 6.5 We denote by NC' the set of all non-crossing partitions with an even number of odd blocks.

This is a category of partitions[1] which is distinct from NC because it does not contain the singleton partition $s = \{\{1\}\}$. The distinction between the two can be captured by the following invariant.

Definition 6.6 The *block number* of a category of partitions $\mathcal{C}$ is the set $\text{BN}(\mathcal{C})$ of numbers of odd blocks in the partitions of $\mathcal{C}$.

Once again, there are few possible values for this invariant. Before proving this and extending the classification, let us isolate for convenience a result concerning yet another category of non-crossing partitions.

Lemma 6.7 *Let $NC_{1,2}$ be the set of all partitions with blocks of size at most two. Then, this is the category of partitions generated by the singleton s.*

Proof It is clear that the defining property of $NC_{1,2}$ is preserved under reflection and horizontal concatenation. The stability under vertical concatenation can be proven by induction exactly as in the proof of Proposition 3.15. Let us now prove by induction that if $\mathcal{C}$ is the category of partitions generated by the singleton, then $\mathcal{C} = NC_{1,2}$.

- If $p \in NC_{1,2}$ has one point, then it is in $\mathcal{C}$ by definition.
- Assuming now that all partitions of size at most k in $NC_{1,2}$ are in $\mathcal{C}$, let us consider a partition of size $k + 1$. If it only has even blocks, then it is in $NC_2 \subset \mathcal{C}$. Otherwise, we can rotate it so that it becomes $s \otimes q$, where s denotes the singleton partition. Applying the induction hypothesis to q then concludes the proof. $\square$

We are now ready to investigate the properties of $\text{BN}(\mathcal{C})$.

---

[1] This can be proven in the same way as Lemma 3.14, observing that merging two blocks does not change the parity of the number of odd blocks.

Proposition 6.8 *There are three possible values for* $\mathrm{BN}(\mathcal{C})$, *namely* $\{0\}$, $2\mathbf{N}$ *and* $\mathbf{N}$. *Moreover,*

1. *If* $\mathcal{C}$ *is non-crossing and* $\mathrm{BN}(\mathcal{C}) = 0$, *then* $\mathcal{C} = NC_2$ *or* $\mathcal{C} = NC_{even}$;
2. *If* $\mathcal{C}$ *is non-crossing and* $\mathrm{BN}(\mathcal{C}) = \mathbf{N}$, *then* $\mathcal{C} = NC$ *or* $\mathcal{C} = NC_{1,2}$.

Proof Taking horizontal concatenations shows that $\mathrm{BN}(\mathcal{C})$ is a sub-semigroup of $\mathbf{N}$. Moreover, when capping two points from different blocks, then either the number of odd blocks stays the same (if we merge an odd block with an even one or two even ones together) or the number of odd blocks is reduced by two (if we merge two odd blocks). This means that if $x \in \mathrm{BN}(\mathcal{C})$ is at least two, then $x - 2 \in \mathrm{BN}(\mathcal{C})$ and the first assertion follows.

By definition, $\mathrm{BN}(\mathcal{C}) = \{0\}$ if and only if $\mathcal{C} \subset NC_{even}$, hence the first point by Proposition 6.4. As for the second one, let $p \in \mathcal{C}$ be a partition with an odd number of odd blocks. By capping, we can remove all even blocks and reduce all odd blocks to singletons. We can, moreover, pair the singletons until only one remains. Thus, $\mathcal{C}$ contains the singleton partition and we conclude by Lemma 6.7 that $NC_{1,2} \subset \mathcal{C}$. If $\mathrm{BS}(\mathcal{C}) = \{1, 2\}$, we must have equality. If instead $\mathrm{BS}(\mathcal{C}) = \mathbf{N}$, let $q \in \mathcal{C}$ contain a block of size three. Using the singleton, we can remove all the points of q except for that block and we conclude by Exercise 3.5 that $\mathcal{C} = NC$. $\qquad\square$

6.1.1.3 The Interval Number

We are now entering the subtle part of the classification, that is, the case $\mathrm{BN}(\mathcal{C}) = 2\mathbf{N}$. For the moment, we have at least two categories of partitions which are not characterised by the values of BS and BN, namely NC' and $NC'_{1,2} = NC_{1,2} \cap NC'$. If we tried to prove that these are the only remaining possibilities, we would be led to consider the category of partitions generated by the double singleton partition $s \otimes s$. This is contained in $NC'_{1,2}$ but it turns out that it is not sufficient to generate it. An easy way of seeing this is through a third invariant.

Definition 6.9 The *interval number* of a category of partitions $\mathcal{C}$ is the set $\mathrm{IN}(\mathcal{C})$ of the number of points between two consecutive points of the same block in a partition of $\mathcal{C}$.

Consider now the so-called *positioner partition*:

$$p_\bullet = \quad \underset{\bullet\ \ \bullet\ \ \bullet\ \ \bullet}{\lceil\quad\ \rceil} \quad .$$

It is in $NC'_{1,2}$, but not in $NC^\sharp_{1,2} = \langle s \otimes s \rangle$ (we will prove this in a minute), so that these two categories of partitions are different. Moreover, it translates into an interesting 'commutation property' for singletons.

Lemma 6.10 *Let p be a partition on k points. Then, any partition obtained from p by moving singletons is in $\langle p, p_\bullet \rangle$.*

Proof Assume that p lies on one line and consider the following rotation of $p_\bullet$:

$$q_\bullet = \quad \begin{matrix} \bullet & \bullet \\ | & \\ \bullet & \bullet \end{matrix} \quad .$$

If now the ith point of p is a singleton that we want to move, then

$$\left(|^{\otimes i-1} \otimes q_\bullet \otimes |^{\otimes k-i-1} \right) \circ p$$

is the same partition as p except that the singleton has been moved to the $(i-1)$-th position. Iterating this one can move the singleton anywhere, hence the result. $\square$

Using this, we can describe the properties of $\mathrm{IN}(\mathcal{C})$.

Proposition 6.11 *There are two possible values for $\mathrm{IN}(\mathcal{C})$, namely $2\mathbf{N}$ and $\mathbf{N}$. Moreover,*

1. *If $\mathrm{IN}(\mathcal{C}) = 2\mathbf{N}$ and $\mathrm{BS}(\mathcal{C}) = \{1,2\}$, then $\mathcal{C} = NC^\sharp_{1,2}$;*
2. *If $\mathrm{IN}(\mathcal{C}) = \mathbf{N}$, $\mathrm{BN}(\mathcal{C}) = 2\mathbf{N}$ and $\mathrm{BS}(\mathcal{C}) = \{1,2\}$, then $\mathcal{C} = NC'_{1,2}$.*

Proof By capping we see that if $x \in \mathrm{IN}(\mathcal{C})$ is at least three, then $x - 2 \in \mathrm{IN}(\mathcal{C})$. Moreover, we claim that $\mathrm{IN}(\mathcal{C})$ is a semigroup. To prove this, let us consider $p, q \in \mathcal{C}$ and two points in each which are consecutive points of a block and which are separated by respectively x and y points. By rotating, we may assume that these two points are the first and last points of the upper rows. But then, considering $p \otimes q$ and capping the middle points, the first and last points of the upper row are now connected and there are between them $x + y$ points which are not connected to the extreme ones, hence the claim.

If $\mathrm{BS}(\mathcal{C}) = \{1,2\}$, then consider a partition on one line containing a singleton. If it is not s, then by capping points which are not singletons and then singletons together we end up with s or $s \otimes s$ and if the original partition was s, the same conclusion holds. As a consequence, $NC^\sharp_{1,2} \subset \mathcal{C}$. Let us prove by induction on the number of points that if $\mathrm{IN}(\mathcal{C}) = 2\mathbf{N}$, then the inclusion is an equality.

- If $p \in \mathcal{C}$ has two points, then either $p = \sqcup$ or $p = s \otimes s$, hence the result.
- Assume now that all partitions in $\mathcal{C}$ on less than $2k$ points are in $NC_{1,2}^{\sharp}$. If p has $2(k+1)$ points (note that $\mathrm{IN}(\mathcal{C}) = 2\mathbf{N}$ forces $\mathrm{BN}(\mathcal{C}) = 2\mathbf{N}$), there are two possibilities:
  - If there is no singleton in p, then $p \in NC_2 \subset NC_{1,2}^{\sharp}$;
  - Otherwise, there exist up to rotation two connected points with only singletons between them. There is an even number of such singletons, hence up to a rotation we have $p = s \otimes s \otimes q$ with $q \in \mathcal{C}$ having $2k$ points, and we conclude by induction.

Eventually, if $\mathrm{BN}(\mathcal{C}) = 2\mathbf{N}$ and $\mathrm{BS}(\mathcal{C}) = \{1,2\}$, then $NC_{1,2}^{\sharp} \subset \mathcal{C} \subset NC_{1,2}'$ by definition. We will therefore simply prove that there is no category of partitions strictly in between the two extreme ones. To do this, let p be a partition in $NC_{1,2}' \setminus NC_{1,2}^{\sharp}$. Then, there are two connected points in p with an odd number of singletons between them. By capping, we can reduce the number of points of p to four while still having two of them connected with an odd number of points between them which do not belong to the same block, and the only possibility is then the positioner partition $p_\bullet$. Now if $p \in NC_{1,2}'$, using Lemma 6.10 we can move all its singletons to the left, producing a partition of the form $s^{2k} \otimes q \in NC_{1,2}^{\sharp}$, with $q \in NC_2$. This means that $\langle NC_{1,2}^{\sharp}, p_\bullet \rangle = NC_{1,2}$, hence the result. $\qquad\square$

6.1.1.4 Conclusion of the Proof

There is one case left, namely when $\mathrm{BS}(\mathcal{C}) = \mathbf{N}$ but $\mathrm{BN}(\mathcal{C}) = 2\mathbf{N}$. Surprisingly, in a sense, $\mathrm{IN}(\mathcal{C})$ makes no difference in that case.

Proposition 6.12 *Let $\mathcal{C}$ be a category of non-crossing partitions such that* $\mathrm{BS}(\mathcal{C}) = \mathbf{N}$ *and* $\mathrm{BN}(\mathcal{C}) = 2\mathbf{N}$. *Then,* $\mathcal{C} = NC'$.

Proof It is clear that $NC_{1,2}^{\sharp} \subset \mathcal{C} \subset NC'$. Let p be a partition in $\mathcal{C}$ with a block of odd size. We can cap points which do not belong to that block until only singletons remain. Then, we can cap again until the block has three points. There are then one or three singletons. In the first case, we get $s \otimes b_3$, where b_3 is the one-block partition with three points. In the second case, we have

$$q = \quad \text{ } \quad \text{ }$$

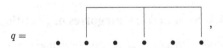

and capping the last two points of q yields the positioner partition $p_\bullet$. Using Lemma 6.10, we therefore deduce that $s^{\otimes 3} \otimes b_3 \in \mathcal{C}$ which by capping in turn yields $s \otimes b_3 \in \mathcal{C}$. Note that composing $s \otimes b_3$ with $s \otimes s$ yields $p_\bullet$, so that both partitions are in $\mathcal{C}$. We are now ready to prove that $\mathcal{C} = NC'$ by induction on the number of points.

- If $p \in NC'$ has two points, then it is obviously in $\mathcal{C}$.
- Assume now that all partitions in NC' on at most $2k$ points are in $\mathcal{C}$. If p has $2(k + 1)$ points, we will split into three cases:
  - If p contains at least two singletons, then they can be moved by Lemma 6.10 so that p becomes of the form $s \otimes s \otimes q$ with $q \in NC'$ and we conclude by induction that $p \in \mathcal{C}$.
  - If there is one singleton in p, then there exist two neighbouring points which are connected (up to rotation). Using the following partition,

    which is a rotation of $s \otimes b_3$, we can collapse these two neighbouring points at the price of adding a singleton. We now have two singletons and can apply the first step.
  - If there are no singletons, we can apply the previous step twice.   □

The combination of Propositions 6.4, 6.8, 6.11 and 6.12 now proves Theorem 6.1, and the list of all categories of non-crossing partitions is
$$NC_2, NC_{1,2}, NC'_{1,2}, NC^\sharp_{1,2}, NC_{\text{even}}, NC' \text{ and } NC.$$
We even get a little more since we have an explicit invariant.

Corollary 6.13 *The triple* $(\mathrm{BS}(\mathcal{C}), \mathrm{BN}(\mathcal{C}), \mathrm{IN}(\mathcal{C}))$ *is a complete invariant of categories of non-crossing partitions.*

Remark 6.14 The quantum groups $\mathbb{G}_N(NC'_{1,2})$, $\mathbb{G}_N(NC^\sharp_{1,2})$ and $\mathbb{G}_N(NC')$ can be described in terms of the other ones using direct products and free products; see Exercise 6.5.

6.1.2 Symmetric Categories of Partitions

Before turning to other matters, let us briefly explain how one can use the classification of categories of non-crossing partitions to obtain a classification of symmetric categories of partitions. Remember that a category of partitions is symmetric if it contains the crossing partition p_{cross}, and that this is equivalent to the corresponding partition quantum group being classical. We will rely on the following result.

Proposition 6.15 *Let C be a symmetric category of partitions. Then,*

$$C = \langle C \cap NC, p_{\text{cross}} \rangle.$$

Proof Let us start with some simple observations. To any permutation $\sigma \in S_k$ there corresponds a partition $p_\sigma \in P_2(k, k)$ obtained by connecting the ith upper point with the $\sigma(i)$-th lower point. Moreover, $p_\sigma \circ p_\tau = p_{\sigma\tau}$ so that this yields a group isomorphic to S_k. Now,

$$|^{\otimes i-1} \otimes p_{\text{cross}} \otimes |^{\otimes k-i-1} = p_{(i,i+1)}$$

and these transpositions generate S_k, hence any symmetric category of partitions contains p_σ for all $\sigma \in S_k$ and all $k \in \mathbf{N}$.

Let now $p \in C$. We will prove by induction on the number of crossings that $p \in \langle C \cap NC, p_{\text{cross}} \rangle$.

- If p is non-crossing, the result is clear.
- Assume that the result holds for all partitions with at most c crossings and consider a partition p with $c + 1$ crossings. Recall that these are quadruples $k_1 < k_2 < k_3 < k_4$ such that k_1 is connected to k_3 but not to k_2, and k_2 is connected to k_4. Let us focus on one of these crossings and let σ be the transposition exchanging k_2 and k_3. Then, $p_\sigma \circ p$ has c crossings, hence belongs to $\langle C \cap NC, p_{\text{cross}} \rangle$ by the induction hypothesis. Composing with $p_{\sigma^{-1}} \in \langle p_{\text{cross}} \rangle$ concludes the proof. $\square$

We are now ready for the classification in the symmetric case.

Corollary 6.16 *There are exactly six symmetric categories of partitions, namely*

$$P_2, P_{1,2}, P'_{1,2}, P_{\text{even}}, P' \text{ and } P.$$

Proof One simply computes $\langle C, p_{\text{cross}} \rangle$ for C ranging over the seven categories of non-crossing partitions. It is clear that this gives the following correspondence:

$$
\begin{aligned}
NC_2 &\longrightarrow P_2 \\
NC_{1,2} &\longrightarrow P_{1,2} \\
NC'_{1,2} &\longrightarrow P'_{1,2} \\
NC_{\text{even}} &\longrightarrow P_{\text{even}} \\
NC' &\longrightarrow P' \\
NC &\longrightarrow P
\end{aligned}
$$

The only case which is not obvious is $NC_{1,2}^\sharp$. Observe that using p_{cross}, we can move one singleton in $s \otimes s \otimes \sqcap$ so that it becomes $p_\bullet$. Thus,

$$\langle NC_{1,2}^\sharp, p_{\text{cross}} \rangle = \langle NC_{1,2}', p_{\text{cross}} \rangle = P_{1,2}. \qquad \square$$

6.2 Coloured Partitions

In this section we will give the unitary generalisations of all the main results of the past chapters. Because the proofs are very similar, we will only indicate the important differences with the orthogonal case when needed. Let us first note that Theorem 2.15 was already stated for unitary compact matrix quantum groups because the proof indeed works in that setting.

Now, in order to understand how the partition description should be modified to cover unitary quantum groups, let us go back to the duality map. Recall that for a Hilbert space V, we defined it as the map $D_V \colon V \otimes V \to \mathbf{C}$ given by

$$D_V(x \otimes y) = \langle x, \overline{y} \rangle.$$

If $\mathbb{G}$ is an *orthogonal* compact matrix quantum group with fundamental representation $u \in \mathcal{O}(\mathbb{G}) \otimes \mathcal{L}(V)$, we have seen that D_V is an intertwiner between $u \otimes u$ and the trivial representation ε. However, the proof crucially relies on the fact that $u_{ij}^* = u_{ij}$. If this is not the case, then the result fails.

To see what should be the correct statement in general, let us write the duality operator differently as $D_V \colon V \otimes \overline{V} \to \mathbf{C}$:

$$D_V(x \otimes \overline{y}) = \langle x, y \rangle.$$

In this way, the question is, given a representation of $\mathbb{G}$ on V, whether there is a natural representation acting on $\overline{V}$. This was answered in Definition 5.14: one should consider the conjugate representation $\overline{u}$ whose coefficients are u_{ij}^*. By the axioms of a unitary compact matrix quantum group, this is again a unitary representation and

$$D_V \in \mathrm{Mor}_{\mathbb{G}}(u \otimes \overline{u}, \varepsilon).$$

Remark 6.17 Another way to understand this is that in the orthogonal case, we implicitly used the fact that the anti-linear isomorphism $j \colon V \to \overline{V}$ sending x to $\overline{x}$ intertwines u with itself, so that we can equivariantly identify V with its conjugate space. This is not true any more in general.

The main point of the preceding discussion is to stress that since we no longer assume $u = \overline{u}$, we should not focus tensor products of u alone, but on tensor products of u and $\bar{u}$. In order to make this precise, we need some shorthand notations. Let $n \in \mathbf{N}$ be an integer and let $w = w_1 \cdots w_n$ be a word over $\{\circ, \bullet\}$. Setting $u^\circ = u$ and $u^\bullet = \bar{u}$, we write

$$u^w = u^{w_1} \otimes \cdots \otimes u^{w_n},$$

acting on $V^w = V^{w_1} \otimes \cdots \otimes V^{w_n}$. We can now state the corollary of Theorem 2.15 which is crucial for the partition approach.

Corollary 6.18 *Let* $\mathbb{G} = (\mathcal{O}(\mathbb{G}), u)$ *be a unitary compact matrix quantum group. Then, any irreducible representation is equivalent to a subrepresentation of* u^w *for some word* w *on* $\{\circ, \bullet\}$.

Based on this, it is natural to try to extend the partition picture by distinguishing points corresponding to u and $\bar{u}$. This is easily done as follows.

Definition 6.19 A *coloured partition* is a partition with the extra data of a colour on each point. The set of coloured partitions is denoted by $P^{\circ \bullet}$. A coloured partition is said to be non-crossing if the underlying uncoloured partition is non-crossing. The set of non-crossing coloured partitions is denoted by $NC^{\circ \bullet}$.

Here are two examples, the first one being crossing while the second one is not.

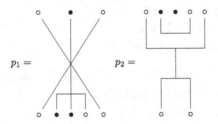

To define the linear map corresponding to a coloured partition, let us call the word formed by the colours of the upper points of p (read from left to right) its *upper colouring*, while the colours of the lower points (again read from left to right) are its *lower colouring*. Then, for a partition p with upper colouring w and lower colouring w', we define a linear map

$$T_p \colon V^w \to V^{w'}$$

by Equation (3.1) as in the uncoloured case.

Based on this, we can clarify the notion of a category of coloured partitions. There are two differences with the uncoloured case. The first one is that it is only possible to compose q with p (in that order) if the upper colouring of q matches the lower colouring of p. The second one is that the partition corresponding to the duality map for $\mathbf{C}^N$ is, in this setting, the partition

$$D_{\circ\bullet} = \begin{array}{cc} \circ & \bullet \\ \lfloor\underline{}\rfloor \end{array}$$

while the duality map $D_{\bullet\circ}$ for $\overline{\mathbf{C}}^N$ is the same one with the colours exchanged. Moreover, the identity map of $\mathbf{C}^N$ is given by the partition

$$\mathrm{id}_\circ = \begin{array}{c} \circ \\ | \\ \circ \end{array}$$

and the identity $\mathrm{id}_\bullet$ of $\overline{\mathbf{C}}^N$ is given by the same partition but coloured in black. For convenience, given a set $\mathcal{C}$ of partitions, we will denote by $\mathcal{C}(w, w')$ the subset of all partitions in $\mathcal{C}$ with upper colouring w and lower colouring w'.

Definition 6.20 A *category of coloured partitions* $\mathcal{C}$ is a collection of sets of partitions $\mathcal{C}(w, w')$ for all words w and w' on $\{\circ, \bullet\}$ such that

1. If $p \in \mathcal{C}(w, w')$ and $q \in \mathcal{C}(w'', w''')$, then $p \otimes q \in \mathcal{C}(w.w'', w'.w''')$ (where $\cdot$ denotes the concatenation of words);
2. If $p \in \mathcal{C}(w, w')$ and $q \in \mathcal{C}(w', w'')$, then $q \circ p \in \mathcal{C}(w, w'')$;
3. If $p \in \mathcal{C}(w, w')$, then $p^* \in \mathcal{C}(w', w)$;
4. $\mathrm{id}_\circ \in \mathcal{C}(\circ, \circ)$ and $\mathrm{id}_\bullet \in \mathcal{C}(\bullet, \bullet)$;
5. $D_{\circ\bullet} \in \mathcal{C}(\circ\bullet, \emptyset)$ and $D_{\bullet\circ} \in \mathcal{C}(\bullet\circ, \emptyset)$.

Remark 6.21 It is not difficult to prove (see the following rotation operation) that assuming $\mathrm{id}_\circ \in \mathcal{C}$, we automatically have $\mathrm{id}_\bullet \in \mathcal{C}$. We have nevertheless included both for symmetry.

We have not mentioned crossing partitions so far. One has to be a little careful here, since there are several ways of colouring the crossing partition $\{\{1, 3\}, \{2, 4\}\}$. Considering the corresponding relations, we see that if we colour, for instance, all the points in white, then this yields

$$u_{ij} u_{k\ell} = u_{k\ell} u_{ij}$$

for all $1 \leqslant i, j, k, \ell \leqslant N$. This, of course, implies the corresponding relation for the adjoints, but does not tell us anything about commutation between u_{ij} and u_{ij}^*. In order to get everything, the simplest way is to include all possible colourings.

Definition 6.22 A category of partitions $\mathcal{C}$ is said to be *symmetric* if it contains all coloured versions of the crossing partition p_{cross}.

We can now give the coloured analogue of Theorem 3.9.

Theorem 6.23 *Let N be an integer and let $\mathcal{C}$ be a category of coloured partitions. Then, there exists a unique unitary compact matrix quantum group $\mathbb{G} = (\mathcal{O}(\mathbb{G}), u)$, where u has dimension N, such that for any words w, w' on $\{\circ, \bullet\}$,*

$$\text{Mor}_{\mathbb{G}}(u^w, u^{w'}) = \text{Vect}\,\{T_p \mid p \in \mathcal{C}(w, w')\}\,.$$

Moreover, $\mathbb{G}$ is a classical group if and only if $\mathcal{C}$ is symmetric. The compact quantum group $\mathbb{G}$ will be denoted by $\mathbb{G}_N(\mathcal{C})$ and called the partition quantum group *associated to N and $\mathcal{C}$.*

To complete the picture, let us clarify the rotation operation in the coloured setting. Assume that we want to rotate a white point in the upper row to the lower row. We can certainly use the partition $D_{\bullet\circ}$ to do this in the following way:

$$p \mapsto (\text{id}_{\bullet} \otimes p) \circ (D_{\bullet\circ}^* \otimes \text{id}_*^{\otimes k}),$$

where the $*$ index is either $\circ$ or $\bullet$. However, the point has been changed to a black one in the process. As a consequence, rotations are allowed in categories of coloured partitions on condition that the colour of the rotated point is changed. The following is a pictorial description of this operation.

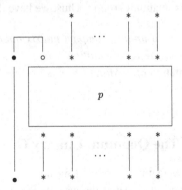

As one may expect, this setting contains the one without colours that we have been working in up to now. Here is a formal statement of that fact.

Exercise 6.1 Let $\mathcal{C}$ be a category of coloured partitions and assume that the partition $\text{id}_{\circ\bullet}$, which is the coloured version of $|$ in $P(\circ, \bullet)$, is in $\mathcal{C}$. Prove that the set C' of uncoloured partitions obtained by removing the colours from C is a category of partitions, and that

$$\mathbb{G}_N(C) = \mathbb{G}_N(C').$$

Solution First note that by applying $\mathrm{id}_{\circ\bullet}$ and its rotated version $\mathrm{id}_{\bullet\circ}$, we can turn black points of any partition in C into white points. Thus, if $p \in C'$ is a partition such that one of its coloured versions is in C, then p with all points coloured in white is also in C. As a consequence, given $p, q \in C'$, there are coloured versions in C which can be composed, hence $q \circ p \in C'$. The other axioms being trivially satisfied, C' is a category of partitions.

Applying the definition, we see that $T_{\mathrm{id}_{\circ\bullet}}$ intertwines u with $\overline{u}$ if and only if $u_{ij} = u_{ij}^*$ for all i, j, hence $\mathbb{G}_N(C)$ is an orthogonal compact matrix quantum groups. In particular, it is completely determined by the sets

$$\mathrm{Mor}_{\mathbb{G}_N(C)}(u^{\otimes k}, u^{\otimes \ell}) = \mathrm{Vect}\left\{T_p \mid p \in C(\circ^k, \circ^\ell)\right\}$$
$$= \mathrm{Vect}\left\{T_p \mid p \in C'(k, \ell)\right\}$$

and the result follows from the uniqueness in Theorem 3.7. $\square$

Note that conversely, if C' is a category of uncoloured partition, then the category of coloured partitions obtained by taking all the possible colourings of elements of C' satisfies $\mathbb{G}_N(C) = \mathbb{G}_N(C')$.

Let us end with a comment concerning conjugate representations. The proof of Corollary 5.13 used the orthogonality of the fundamental representation u, but all that is really needed is the fact that $\overline{u}$ is unitary, which holds for any unitary compact matrix quantum group.[2] Thus, we have the following:

Corollary 6.24 *Let $\mathbb{G}$ be a unitary compact matrix quantum group. For any unitary representation v, the representation $\overline{v}$ is again unitary and is called the* conjugate representation *of v. Moreover, v is irreducible if and only if $\overline{v}$ is irreducible.*

6.3 The Quantum Unitary Group

Let us explore the new features exhibited by unitary compact matrix quantum groups by working out in detail the case of the quantum unitary group U_N^+ from Definition 1.27. This first requires identifying its category of partitions. Because U_N^+ is the largest unitary compact matrix quantum group, in

---

[2] This is precisely the reason behind the strange extra assumption in the definition.

the sense that all the other ones correspond to quotients of $\mathcal{O}(U_N^+)$ for some N, its corresponding category of partitions must be the smallest possible one. Proving this is a good exercise for practicing with coloured partitions.

Exercise 6.2 Let $\mathcal{U}$ be the set of all non-crossing coloured partitions in pairs such that:

- If the two points of a block are in the same row, then they have different colours;
- If the two points of a block are in different rows, then they have the same colour.

Prove that this is a category of partitions and that it is generated by $D_{\circ\bullet}$.

Solution The fact that $\mathcal{U}$ is stable under reflection and horizontal concatenation is straightforward. Observe also that it is by definition stable under rotations. As for vertical concatenation, it is enough to prove it for $q \circ p$ with p lying on the lower row, and the proof then goes by induction on the length of p.

- If p has length two, then the result is clear;
- Otherwise, consider an interval in p, which is of the form $D_{\circ\bullet}$ or $D_{\bullet\circ}$ by definition of $\mathcal{U}$. If the corresponding points of q are also connected, we can remove all of them without changing the result of the composition, hence $q \circ p \in \mathcal{U}$ by the induction hypothesis. Otherwise, we can remove the points and furthermore connect the two points of q connected to these two points. Because the two original points had different colours, the two new connected points have the same colour if they lie on different rows, while they will have different colours if they lie on the same row. In both cases, we get a new partition in $\mathcal{U}$ and the composition equals $q \circ p$, hence we can conclude once again by the induction hypothesis.

Thus, $\mathcal{U}$ is a category of partitions.

Let us now prove that any partition in $\mathcal{U}$ is in the category of partitions $\mathcal{U}'$ generated by $D_{\circ\bullet}$. For this we will first prove by induction on the number of points that for any partition in $NC_2^{\circ\bullet}$ lying on one line, there are two consecutive points which are connected. This is clear for a partition on two points. Let us assume that it is also true for all partitions on at most $2n$ points for some $n \in \mathbf{N}$, and let p be a partition on $2(n+1)$ points. Numbering the points from 1 to $2n + 2$ from left to right, point 1 must be connected to another one, say i.

- If $i = 2$, then we are done;
- Otherwise, the points $2, \cdots, i-1$ can only be connected to another point in the same interval by non-crossingness. Thus, the restriction of p to these points is a subpartition which is in $NC_2^{\circ\bullet}$. By induction, there are two consecutive points connected there, concluding the proof.

Back now to the original problem, let $p \in \mathcal{U}$ and assume, up to rotation, that the first two points are connected. Then, either $p = D_{\circ\bullet} \otimes q$ or $p = D_{\bullet\circ} \otimes q$. In the first case, we have

$$q = (D_{\circ\bullet}^* \otimes \mathrm{id}_* \otimes \cdots \otimes \mathrm{id}_*) \circ p \in \mathcal{U}$$

with id_* being suitably coloured to match the colours of p, and similarly in the second case. A straightforward induction based on this observation then yields the result.   □

The second thing we need is the generalisation of the results of Chapter 4. The notion of projective partition still makes sense, and the only thing which must be adapted is the through-block decomposition of Proposition 4.3. The question here simply is, what should be the colour of the lower points of p_d and p_u? An examination of the proofs shows that this does not matter as long as they are all the same. Indeed, the point is only to be able to compose p_d^* with p_u, or to compose these with mixing partitions. As a consequence, we can simply decide once and for all to colour all those points in white, as well as those of the mixing partitions. Then, the proofs carry on verbatim to yield

Theorem 6.25 *Let $\mathcal{C}$ be a category of coloured partitions and let $N \geqslant 4$ be an integer. Then, the irreducible representations of $\mathbb{G}_N(\mathcal{C})$ can be indexed by the projective partitions of $\mathcal{C}$ in such a way that*

- $u_p \sim u_q$ *if and only if* $p \sim q$;
- $u^w = \displaystyle\bigoplus_{p \in \mathcal{C}(w,w)} u_p.$

Moreover,

$$u_p \otimes u_q = u_{p \otimes q} \oplus \bigoplus_{k=1}^{\min(t(p),t(q))} v_{p \,\square^k\, q} \oplus v_{p \,\boxdot^k\, q}$$

where $v_r = u_r$ if $r \in \mathcal{C}$ and $v_r = 0$ otherwise.

We can now describe the representation theory of U_N^+. For convenience, we will first define the abstract object indexing the equivalence classes of irreducible representations. Let M be the set of all words over $\{\circ, \bullet\}$. It has a natural monoid structure given by the concatenation of words, but in order to deal with conjugate representations, we also need a conjugation operation $w \mapsto \overline{w}$ on M. Setting $\overline{\circ} = \bullet$ and $\overline{\bullet} = \circ$, we define

$$\overline{w_1 \cdots w_k} = \overline{w}_k \cdots \overline{w}_1.$$

Theorem 6.26 (Banica) *For $N \geqslant 2$, the irreducible representations of U_N^+ can be labelled by the elements of M in such a way that $u^\emptyset = \varepsilon$, $u^\circ = V$ and for any $w \in M$,*

$$\overline{u^w} = u^{\overline{w}}.$$

Moreover, for any $w, w' \in M$,

$$u^w \otimes u^{w'} = \sum_{w=az, w'=\overline{z}b} u^{ab}.$$

Proof We define a map $\Phi \colon M \to \mathrm{Irr}(U_N^+)$ sending $w = w_1 \cdots w_k$ to the equivalence class of

$$p_{\mathrm{id}_{w_1} \otimes \cdots \otimes \mathrm{id}_{w_k}}.$$

By construction, $\Phi(\circ) = u$ and $\Phi(\bullet) = \overline{u}$ and we will prove that Φ is a bijection.

1. We first prove that Φ is onto. Let $p \in \mathcal{U}$ be a projective partition. A straightforward induction shows that it can be written as

$$p = (b_0^* b_0) \otimes p_1 \otimes (b_1^* b_1) \otimes p_2 \otimes \cdots \otimes p_k \otimes (b_k^* b_k),$$

where $p_i \in \{\mathrm{id}_\circ, \mathrm{id}_\bullet\}$ and b_i lies on one line. Now, any two connected points in b_i have different colours because $p \in \mathcal{U}$, hence $b_i \in \mathcal{U}$. Composing with

$$r = b_0^* \otimes | \otimes b_1^* \otimes | \otimes \cdots \otimes | \otimes b_k^*$$

therefore yields an equivalence in $\mathcal{U}$ between p and

$$q = p_1 \otimes p_2 \otimes \cdots \otimes p_k.$$

Setting w_i to be the colour of p_i, u_p is then in the equivalence class of $\Phi(w_1 \cdots w_k)$, proving surjectivity.

2. Consider now $w, w' \in M$ such that $\Phi(w) \sim \Phi(w')$ and notice first that they must have the same number of through-blocks, hence w and w' must have the same number of letters. Let r_i be the identity partition with the colour w_i on the upper row and w_i' on the lower row. Then,

$$r = r_1 \otimes \cdots \otimes r_k$$

implements the equivalence between $\Phi(w)$ and $\Phi(w')$ so that $r \in \mathcal{U}$. It then follows from the definition of $\mathcal{U}$ that $w_i = w_i'$ for all i, hence $w = w'$. In other words, Φ is injective.

3. As for the fusion rules, first note that $\Phi(w) \,\boxdot\, \Phi(w')$ is never in $\mathcal{U}$ since it has a block of size four. Moreover, in $\Phi(w)\square^n\Phi(w')$, the last point in the upper row of $\Phi(w)$ is connected to the first one in the upper row of $\Phi(w')$, hence they must have different colours. Similarly, the point next to the last one in the upper row of $\Phi(w)$ gets connected to the point next to the last one in the upper row of $\Phi(w')$, hence they must have different colours, and so on. Therefore, for this partition to be in $\mathcal{U}$, the last n letters of w must coincide (in reverse order) with the opposite of the first n ones of w'. In other words, we must have $w = az$ and $w' = \overline{z}b$ with z of length n. In that case, all the non-through-blocks can be removed up to equivalence, yielding a representation equivalent to $\Phi(ab)$, as claimed.

4. Eventually, for $u^w \otimes u^{w'}$ to contain the trivial representation $u^\emptyset$, one must have $w' = \overline{w}$. Hence, by Proposition 5.17, $\overline{u}^w = u^{\overline{w}}$ and the proof is complete.

$\square$

Further examples of unitary compact quantum groups will be studied in detail in Chapter 7, but we would like to give a slightly different example right now. As we will see later on (Exercise 6.7), this is connected to the hyperoctahedral quantum group H_N^+.

Exercise 6.3 Let $\mathcal{C}_{\mathrm{alt}}$ be the set of all non-crossing coloured even partitions such that, once rotated on one line, the colours alternate. We admit that this is a category of partitions (see Exercise 7.2 for a proof).

1. Prove that $\mathcal{C}_{\mathrm{alt}}$ is generated by the following partition:

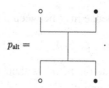

2. Deduce from this a description of $\mathcal{O}(\mathbb{G}_N(\mathcal{C}_{\mathrm{alt}}))$ as a quotient of $\mathcal{O}(U_N^+)$.

Solution 1. Let us first prove that all alternating one-block partitions are in $\langle p_{\mathrm{alt}}\rangle$ by induction on the number of points. An alternating partition on two points is a rotation of $D_{\circ\bullet}$ and an alternating partition on four points is a rotation of p_{alt}, hence they are all in $\langle p_{\text{alt}} \rangle$. Let us therefore assume that all one-block partitions of $\mathcal{C}_{\text{alt}}$ on at most $2n$ points are in $\langle p_{\text{alt}} \rangle$ for some n and let p_n be the alternating one-block partition in $NC^{\circ\bullet}((\circ\bullet)^n, (\circ\bullet)^n)$. Then,

$$p_{n+1} = (p_n \otimes p_{\text{alt}})(\text{id}_*^{\otimes(n-1)} \otimes p_{\text{alt}} \otimes \text{id}_\circ) \in \langle p_{\text{alt}} \rangle,$$

proving the claim (the other one-block partitions can be obtained from p_{n+1} by rotations). The result now follows by induction following the standard strategy of isolating an interval.

2. It follows from the preceding question that $\mathcal{O}(\mathbb{G}_N(\mathcal{C}_{\text{alt}}))$ is the quotient of $\mathcal{O}(U_N^+)$ by the relations coming from the fact that $T_{p_{\text{alt}}}$ is an intertwiner. Let us compute these explicitly:

$$(\text{id} \otimes T_{p_{\text{alt}}}) \circ \rho_{u \otimes \bar{u}}(e_{i_1} \otimes \bar{e}_{i_2}) = \sum_{j_1, j_2 = 1}^{N} u_{i_1 j_1} u_{i_2 j_2}^* \otimes T_{p_{\text{alt}}}(e_{j_1} \otimes \bar{e}_{j_2})$$

$$= \sum_{j=1}^{N} u_{i_1 j} u_{i_2 j}^* \otimes e_j \otimes \bar{e}_j$$

while

$$\rho_{u \otimes \bar{u}} \circ T_{p_{\text{alt}}}(e_{i_1} \otimes \bar{e}_{i_2}) = \delta_{i_1 i_2} \rho_{u \otimes \bar{u}}(e_{i_1} \otimes \bar{e}_{i_1})$$

$$= \delta_{i_1 i_2} \sum_{j_1, j_2 = 1}^{N} u_{i_1 j_1} u_{i_2 j_2}^* e_{j_1} \otimes e_{j_2}.$$

Therefore, we have that $u_{i_1 j_1} u_{i_1 j_2}^* = 0$ for $j_1 \neq j_2$ and that $u_{i_1 j} u_{i_2 j}^* = 0$ for $i_1 \neq i_2$. Conversely, these relations imply that $T_{p_{\text{alt}}}$ intertwines $u \otimes \bar{u}$ with itself, hence $\mathcal{O}(\mathbb{G}_N(\mathcal{C}_{\text{alt}}))$ is the quotient of $\mathcal{O}(U_N^+)$ by the relations $V_{ij} V_{i'j}^* = 0$, $V_{ij} V_{ij'}^* = 0$ for all $1 \leqslant i, i', j, j' \leqslant N$ such that $i \neq i'$ and $j \neq j'$. $\qquad \square$

6.4 Making Things Complex

Classically, one may think of U_N as a complex version of O_N, even though this is not the complexification in the sense of Lie or algebraic groups.[3] At least, they play the same role of largest possible compact matrix group respectively

---

[3] The complexification of O_N in that sense is the complex orthogonal group $O(N, \mathbf{C})$, that is, the group of all complex matrices with inverse equal to their transpose. This is an interesting locally compact matrix group, but it is not compact, hence not suited for us.

in the unitary and orthogonal setting. In the quantum case, there is a precise way of describing U_N^+ as a complexification of O_N^+. This relies on an important procedure called the *unital free product* construction. Since it is interesting in its own right, we will devote a bit of time to introducing it in detail.

6.4.1 The Unital Free Product Construction

Usually, a free product construction takes a family of objects and builds the 'largest possible one' generated by the data. In particular, no relation between the original objects should hold in the free product except for those required by the structure of the objects themselves.

6.4.1.1 The Algebra Case

Let us focus for the moment on algebras. There is a standard way to build a free product of algebras using tensor algebras: this is simply the quotient of the tensor algebra[4] of the direct sum $A_1 \oplus A_2$ by the ideal generated by the elements $a \otimes a' - aa'$ and $b \otimes b' - bb'$ for all $a, a' \in A_1$ and $b, b' \in A_2$. Note, however, that if A_1 and A_2 are unital, then the inclusions of the original algebras into the free product will not be unital algebra homomorphisms. This is, of course, easy to remedy by further quotienting by the ideal generated by $1 - 1_A$ and $1 - 1_B$. It turns out, however, that this description of the unital free product of unital algebras is not practical when it turns to studying compact quantum groups. We will therefore use another construction which, though equivalent, is a bit more involved.

We start with two unital algebras, A_1 and A_2, and fix a direct sum decomposition of vector spaces:

$$A_1 = \mathbf{C}.1_{A_1} \oplus A_1' \text{ and } A_2 = \mathbf{C}.1_{A_2} \oplus A_2'.$$

Associated to this decomposition are the projections p_i on $\mathbf{C}.1_{A_i}$ parallel to A_i' for $i = 1, 2$. The idea is that we have taken out the units to be able to easily identify them, so that we can focus on the rest. Intuitively, an element of the free product should be a sum of a multiple of the unit and a polynomial in the elements of A_1' and A_2'. Here is how to make this precise: we set

$$A = \mathbf{C}.1_A \oplus \bigoplus_{k=1}^{+\infty} \bigoplus_{i_1 \neq \cdots \neq i_k \in \{1,2\}} A_{i_1}' \otimes A_{i_2}' \otimes \cdots \otimes A_{i_k}'.$$

---

[4] Do not worry if you do not know what a tensor algebra is; we will not need it in the sequel.

This makes sense as a vector space, and we claim that it can be endowed with a unital algebra structure in a natural way. For the unitality, we simply declare that the generator of the first summand is a multiplicative unit. The rest of the definition of the product is subtler. For convenience, let us set, for $1 \leqslant k < +\infty$,

$$A^{(k)} = \bigoplus_{i_1 \neq \cdots \neq i_k \in \{1,2\}} A'_{i_1} \otimes A'_{i_2} \otimes \cdots \otimes A'_{i_k}.$$

Definition 6.27 Given $a = a_1 \otimes \cdots \otimes a_k \in A^{(k)}$ and $b = b_1 \otimes \cdots \otimes b_\ell \in A^{(\ell)}$, we define their product inductively as follows:

- If $i_k \neq i_\ell$ (i.e. if a_k and b_1 do not belong both to A_1 or both to A_2), then

$$ab = a_1 \otimes \cdots \otimes a_k \otimes b_1 \otimes \cdots \otimes b_\ell \in A^{(k+\ell)};$$

- Otherwise, if $i \in \{1, 2\}$ is such that $a_k, b_1 \in A_i$, we set

$$ab = a_1 \otimes \cdots \otimes a_{k-1} \otimes (a_k b_1 - p_i(a_k b_1)) \otimes b_2 \otimes \cdots \otimes b_\ell$$
$$+ p_i(a_k b_1)(a_1 \otimes \cdots \otimes a_{k-1})(b_2 \otimes \cdots \otimes b_\ell),$$

where the second term is already defined by induction.

It is straightforward to check that this bilinear operation endows A with the structure of a unital algebra and that the inclusions $A_1 \hookrightarrow A$ and $A_2 \hookrightarrow A$ are unital algebra homomorphisms. But to convince the reader that the definition is sound, the best way is to establish the unavoidable universal property that it should satisfy.

Exercise 6.4 Let B be a unital algebra and let $\pi_1 : A_1 \to B$ and $\pi_2 : A_2 \to B$ be unital algebra homomorphisms. Prove that there exists a unique unital algebra homomorphism $\pi : A \to B$ such that $\pi_{|A_1} = \pi_1$ and $\pi_{|A_2} = \pi_2$.

Solution Because π_1 and π_2 are in particular linear maps, we can extend them to $A^{(k)}$ for each $k \geqslant 1$ by using $\pi_{i_1} \otimes \cdots \otimes \pi_{i_k}$ on each summand. As for the unit, since π_1 and π_2 are both unital, it is coherent to set $\pi(1_A) = 1_B$. The map π defined in that way certainly restricts correctly to the summands and we only have to check that it is an algebra homomorphism. This directly follows by induction on the total length of the two elements we multiply, using the definitions and the fact that π_1 and π_2 are algebra homomorphisms. $\square$

Remark 6.28 The preceding universal property implies that A does not depend, up to unital isomorphism, on the decompositions of A_1 and A_2

which were fixed in the beginning. In fact, it shows that it coincides with the construction using tensor algebras given at the beginning of this section.

We will denote the algebra A by $A_1 * A_2$ and call it the *unital free product* of A_1 and A_2. Note that if A_1 and A_2 are $*$-algebras, then there is a unique associated $*$-algebra structure on $A_1 * A_2$ which turns the inclusions into $*$-homomorphisms, namely $1^* = 1$ and

$$(a_1 \otimes \cdots \otimes a_k)^* = a_k^* \otimes \cdots \otimes a_1^*.$$

We can now use this definition to extend the notion to the setting of quantum groups.

6.4.1.2 The Compact Matrix Quantum Group Case

Let $(\mathbb{G}, u)$ and $(\mathbb{H}, v)$ be two unitary compact matrix quantum groups. Form the unital free product $\mathcal{A} = \mathcal{O}(\mathbb{G}) * \mathcal{O}(\mathbb{H})$ using the decompositions

$$\mathcal{O}(\mathbb{G}) = \mathbf{C}.1_{\mathcal{O}(\mathbb{G})} \oplus \ker(h_\mathbb{G}) \text{ and } \mathcal{O}(\mathbb{H}) = \mathbf{C}.1_{\mathcal{O}(\mathbb{H})} \oplus \ker(h_\mathbb{H}),$$

where $h_\mathbb{G}$ and $h_\mathbb{H}$ denote the Haar states of $\mathbb{G}$ and $\mathbb{H}$ respectively, and consider the matrix

$$w = \begin{pmatrix} u & 0 \\ 0 & v \end{pmatrix} \in M_{\dim(u)+\dim(v)}(\mathcal{A}).$$

Applying Exercise 6.4 to the coproducts

$$\Delta_\mathbb{G} \colon \mathcal{O}(\mathbb{G}) \to \mathcal{O}(\mathbb{G}) \otimes \mathcal{O}(\mathbb{G}) \subset \mathcal{A} \otimes \mathcal{A},$$

$$\Delta_\mathbb{H} \colon \mathcal{O}(\mathbb{H}) \to \mathcal{O}(\mathbb{H}) \otimes \mathcal{O}(\mathbb{H}) \subset \mathcal{A} \otimes \mathcal{A},$$

yields a $*$-homomorphism $\Delta \colon \mathcal{A} \to \mathcal{A} \otimes \mathcal{A}$ such that

$$\Delta(w_{ij}) = \sum_{k=1}^{\dim(u)+\dim(v)} w_{ik} \otimes w_{kj}.$$

Since, moreover, both w and $\overline{w}$ are unitary by construction, and their coefficients generate $\mathcal{A}$, we have built a new unitary compact matrix quantum group.

Definition 6.29 The unitary compact matrix quantum group $(\mathcal{O}(\mathbb{G}) * \mathcal{O}(\mathbb{H}), w)$ is called the *free product* of $(\mathbb{G}, u)$ and $(\mathbb{H}, v)$ and denoted by $\mathbb{G} * \mathbb{H}$.

Remark 6.30 It is an easy exercise to check that if Γ_1, Γ_2 are discrete groups, then

$$\widehat{\Gamma}_1 * \widehat{\Gamma}_2 \simeq \widehat{\Gamma_1 * \Gamma_2},$$

where $\Gamma_1 * \Gamma_2$ denotes the free product of groups (see, for instance, [27, definition E.8] for the definition). For that reason, many authors write $\mathbb{G}\hat{*}\mathbb{H}$ for the free product of two compact quantum groups. We have, however, chosen to drop the hat to lighten notations, since we will not consider the discrete setting hereafter.

In the remainder of this section, we want to prove that U_N^+ is a complexification of O_N^+ in a free product sense. The proof of this requires knowledge of the representation theory of free products in terms of the representation theory of the factors. This was first described by S. Wang in [70]. For practical reasons, we will start by defining a complete family of representatives of the irreducible representations.

Definition 6.31 Let $(\mathbb{G}, u)$ and $(\mathbb{H}, v)$ be unitary compact matrix quantum groups and let $\varepsilon_\mathbb{G}$ and $\varepsilon_\mathbb{H}$ denote their respective trivial representations. A word $w = \alpha_1\alpha_2\cdots\alpha_k$ on

$$(\mathrm{Irr}(\mathbb{G}) \setminus \{\varepsilon_\mathbb{G}\}) \amalg (\mathrm{Irr}(\mathbb{H}) \setminus \{\varepsilon_\mathbb{H}\}),$$

where $\amalg$ denotes the disjoint union of sets, is said to be *reduced* if

- $\alpha_{i+1} \in \mathrm{Irr}(\mathbb{G})$ if $\alpha_i \in \mathrm{Irr}(\mathbb{H})$;
- $\alpha_{i+1} \in \mathrm{Irr}(\mathbb{H})$ if $\alpha_i \in \mathrm{Irr}(\mathbb{G})$.

If w is a reduced word, then

$$u^w = u^{\alpha_1} \otimes \cdots \otimes u^{\alpha_k},$$

is said to be a *reduced representation*.

Theorem 6.32 (Wang) *Let $(\mathbb{G}, u)$ and $(\mathbb{H}, v)$ by unitary compact matrix quantum groups. Then, reduced representations are irreducible and pairwise inequivalent. Moreover, any non-trivial irreducible representations of $\mathbb{G} * \mathbb{H}$ is equivalent to a reduced representation.*

Proof We will proceed stepwise to prove by induction on the length of w that

H_k: 'All reduced representations given by reduced words of length less than or equal to k define pairwise inequivalent irreducible representations.'

1. Words of length one are just irreducible representations of $\mathbb{G}$ and $\mathbb{H}$ so that words of length one yield irreducible representations. Moreover, a representation of $\mathbb{G}$ cannot be equivalent to a representation of $\mathbb{H}$. Indeed, if this was the case, then the coefficients would have the same linear span by point

(2) of Theorem 2.15. But the decomposition we took to define the unital free product implies that these coefficients lie respectively in $\ker(h_{\mathbb{G}})$ and $\ker(h_{\mathbb{H}})$, hence are linearly independent. Thus, H_1 holds.

2. Assume now that H_k holds and consider a reduced word $w = \alpha_1 \cdots \alpha_{k+1}$. Then, by Frobenious reciprocity (see Proposition 5.17),

$$\mathrm{Mor}_{\mathbb{G}*\mathbb{H}}\left(u^w, u^w\right) \simeq \mathrm{Mor}_{\mathbb{G}*\mathbb{H}}\left(u^w \otimes \overline{u}^{\alpha_{k+1}}, u^{\alpha_1} \otimes u^{\alpha_2} \otimes \cdots \otimes u^{\alpha_k}\right).$$

We know again by Proposition 5.17 that there is a decomposition

$$u^{\alpha_{k+1}} \otimes \overline{u}^{\alpha_{k+1}} = \varepsilon \oplus u^{\beta_1} \oplus \cdots \oplus u^{\beta_n}$$

into irreducible subrepresentations with $u^{\beta_i} \neq \varepsilon$ for all $1 \leqslant i \leqslant n$. Now, the coefficients of $u^{\alpha_1} \otimes u^{\alpha_2} \otimes \cdots \otimes u^{\beta_i}$ belong to $\mathcal{A}^{(k+1)}$, hence are linearly independent from those of $u^{\alpha_1} \otimes u^{\alpha_2} \otimes \cdots \otimes u^{\alpha_k}$ that lie in $\mathcal{A}^{(k)}$. As a consequence,

$$\mathrm{Mor}_{\mathbb{G}*\mathbb{H}}\left(u^{\alpha_1} \otimes u^{\alpha_2} \otimes \cdots \otimes u^{\beta_i}, u^{\alpha_1} \otimes u^{\alpha_2} \otimes \cdots \otimes u^{\alpha_k}\right) = \{0\}.$$

Thus,

$$\mathrm{Mor}_{\mathbb{G}*\mathbb{H}}\left(u^w, u^w\right) \simeq \mathrm{Mor}_{\mathbb{G}*\mathbb{H}}\left(u^{\alpha_1} \otimes u^{\alpha_2} \otimes \cdots \otimes u^{\alpha_k},\right.$$
$$\left. u^{\alpha_1} \otimes u^{\alpha_2} \otimes \cdots \otimes u^{\alpha_k}\right)$$

which is one-dimensional by H_k, hence u^w is irreducible.

3. Consider now two words w and w' of length $k + 1$. There are two cases.

- If α_i and α_i' are not representations of the same quantum group for some $1 \leqslant i \leqslant k + 1$, then the coefficients of u^w and $u^{w'}$ belong to different summands of $\mathcal{A}^{(k+1)}$, hence the representations are inequivalent by point (2) of Theorem 2.15.

- Otherwise, we have

$$\mathrm{Mor}_{\mathbb{G}*\mathbb{H}}\left(u^w, u^{w'}\right) \simeq \mathrm{Mor}_{\mathbb{G}*\mathbb{H}}\left(u^w \otimes \overline{u}^{\alpha_{k+1}'}, u^{\alpha_1'} \otimes \cdots \otimes u^{\alpha_k'}\right).$$

Proceeding as for irreducibility, we split $u^{w_{k+1}} \otimes \overline{u}^{\alpha_{k+1}}$ into a sum of irreducibles. All the terms yield null morphism spaces except perhaps the trivial representation, which appears in the decomposition if and only if $w_{k+1} = \alpha_{k+1}$. This then reduces the problem to two words of length k and we conclude by H_k.

To conclude the proof of the statement, simply observe that the coefficients of reduced representations, together with 1 which is the coefficient of the trivial representation, span $\mathcal{O}(\mathbb{G} * \mathbb{H})$. Thus, by Theorem 2.15, any irreducible representation is equivalent to a reduced one. $\qquad\square$

Let us illustrate the free product construction by an example coming from the classification of partition quantum groups done in Section 6.1.

Exercise 6.5 For $N \geqslant 2$, we consider the compact quantum group $\mathbb{G}_N(NC_{1,2}^\sharp)$ together with its fundamental representation u and set, for $1 \leqslant i \leqslant N$,

$$\gamma_i = \sum_{j=1}^N u_{ij},$$

$$\lambda_j = \sum_{i=1}^N u_{ij}.$$

1. Prove that $\gamma_i = \lambda_j$ for all $1 \leqslant i, j \leqslant N$. We will denote by γ that element.
2. Prove that $\gamma^2 = 1$.
3. Deduce from this that there is an orthonormal basis B of $\mathbf{C}^N$ and, setting $v = B^{-1}uB$, a surjective $*$-homomorphism

$$\pi \colon \mathcal{O}(O_{N-1}^+) * \mathcal{O}(\mathbf{Z}_2) \to \mathcal{O}(\mathbb{G}_N(NC_{1,2}^\sharp))$$

such that $\pi(U_{ij}) = v_{ij}$ for all $1 \leqslant i, j \leqslant N$ and $\pi(g) = \gamma$, where g is the generator of $\mathcal{O}(\mathbf{Z}_2)$.
4. Prove that the preceding $*$-homomorphism is an isomorphism.

Solution 1. Let $s \in P(1, 0)$ be the singleton partition. We will translate the fact that T_{s^*s} interwtines u with itself. On the one hand,

$$(\mathrm{id} \otimes T_{s^*s}) \circ \rho_u(e_i) = \sum_{j=1}^N u_{ij} \otimes T_{s^*s}(e_j)$$

$$= \sum_{j=1}^N u_{ij} \otimes \left(\sum_{k=1}^N e_k\right)$$

$$= \sum_{k=1}^N \gamma_i \otimes e_k$$

and on the other hand

$$\rho_u \circ T_{s^*s}(e_i) = \sum_{j=1}^N \rho_u(e_j)$$

$$= \sum_{j=1}^N \sum_{k=1}^N u_{jk} \otimes e_k$$

$$= \sum_{k=1}^N \lambda_k \otimes e_k.$$

Combining both yields, for all $1 \leqslant k \leqslant N$, $\gamma_i = \lambda_k$.

2. We compute, recalling that u is orthogonal,

$$\gamma^2 = \frac{1}{N} \sum_{i=1}^{N} \gamma_i^2$$

$$= \sum_{i=1}^{N} \sum_{j,k=1}^{N} u_{ij} u_{ik}$$

$$= \frac{1}{N} \sum_{j,k=1}^{N} \sum_{i=1}^{N} u_{ij} u_{ik}$$

$$= \frac{1}{N} \sum_{j,k=1}^{N} \delta_{jk}$$

$$= 1.$$

3. This follows from the universal property of free products. Indeed, set

$$\xi = \sum_{i=1}^{n} e_i$$

and complete ξ to an orthonormal basis B. By definition, we have

$$\rho_u(\xi) = \gamma \otimes \xi$$

so that

$$B^{-1} u B = \begin{pmatrix} w & 0 \\ 0 & \gamma \end{pmatrix}.$$

Moreover, the restriction of u to $\xi^{\perp}$, w, is again orthogonal, so that $\mathcal{O}(\mathbb{G}_N(NC_{1,2}^{\sharp}))$ is generated by the coefficients of an $(N-1) \times (N-1)$ orthogonal matrix and a generator of $\mathcal{O}(\mathbf{Z}_2)$, whence the result.

4. Consider the fundamental representation V of $O_{N-1}^+ * \mathbf{Z}_2$ with coefficients $V_{ij} = U_{ij}$ for $1 \leqslant i, j \leqslant N$ and $V_{iN} = \delta_{iN} \gamma$ and let us set $W = BVB^{-1}$. This is an orthogonal matrix, hence there is a surjective $*$-homomorphism

$$\psi \colon \mathcal{O}(O_N^+) \to \mathcal{O}(O_{N-1}^+) * \mathcal{O}(\mathbf{Z}_2)$$

sending U_{ij} to W_{ij} for all $1 \leqslant i, j \leqslant N$. Moreover, we have for all $1 \leqslant i, j \leqslant N$ that

$$\sum_{k=1}^{N} W_{ik} = \sum_{k=1}^{N} W_{kj}.$$

These relations precisely mean that $T_{s^* s}$ intertwines W with itself, so that ψ factors through a $*$-homomorphism

$$\widetilde{\psi}\colon \mathcal{O}(\mathbb{G}_N(NC_{1,2}^\sharp)) \to \mathcal{O}(O_{N-1}^+) * \mathcal{O}(\mathbf{Z}_2).$$

It is clear on the generators that $\widetilde{\psi}$ is inverse to π, hence the result. $\qquad\square$

6.4.2 Free Complexification

We can now define a general notion of complexification of an orthogonal compact matrix quantum group, based on the unital free product construction. Let us denote from now on by $\mathbb{T}$ the circle group, namely the group of complex numbers of modulus one. If z denotes the inclusion $\mathbb{T} \hookrightarrow \mathbf{C}$, then z is a fundamental (one-dimensional) representation of $\mathbb{T}$, since it generates the $*$-algebra $\mathcal{O}(\mathbb{T})$ of trigonometric polynomials. We will always use this description of $\mathbb{T}$ as a unitary compact matrix quantum group.

Proposition 6.33 *Let $(\mathbb{G}, u)$ be an orthogonal compact matrix quantum group and let $(\mathcal{O}(\mathbb{T}), z)$ be the circle group. Then, the subalgebra $\mathcal{A}$ of $\mathcal{O}(\mathbb{G} * \mathbb{T})$ generated by $u_{ij}z$ for all $1 \leqslant i, j \leqslant N$ yields a unitary compact matrix quantum group when endowed with the fundamental representation $uz = (u_{ij}z)_{1 \leqslant i,j \leqslant N}$.*

Proof First observe that $(uz)^* uz = u^t u = \mathrm{Id} = uz(uz^*)$ and that the same holds for $\overline{uz} = u^t z$. Moreover, by definition of the coproduct on the free product,

$$\Delta(u_{ij}z) = \Delta(u_{ij})\Delta(z)$$

$$= \left(\sum_{k=1}^N u_{ik} \otimes u_{kj} \right)(z \otimes z)$$

$$= \sum_{k=1}^N u_{ik}z \otimes u_{kj}z.$$

Since by construction the coefficients of uz generate $\mathcal{A}$ as an algebra, the proof is complete. $\qquad\square$

We will set $\mathcal{A} = \mathcal{O}(\widetilde{\mathbb{G}})$, write $\widetilde{u} = uz$ and call $(\widetilde{\mathbb{G}}, \widetilde{u})$ the *free complexification* of $\mathbb{G}$. We now have everything in hand to make sense of U_N^+ as a complexification of O_N^+. This result is originally due to T. Banica in [4].

Theorem 6.34 (Banica) *There is an isomorphism*

$$\widetilde{O}_N^+ \simeq U_N^+.$$

Proof Because $\widetilde{O}_N^+$ is a unitary compact matrix quantum group generated by a fundamental representation $W = Uz$ of dimension N, there exists by the universal property of $\mathcal{O}(U_N^+)$ a surjective $*$-homomorphism

$$\pi \colon \mathcal{O}\left(U_N^+\right) \to \mathcal{O}\left(\widetilde{O}_N^+\right)$$

sending V_{ij} to W_{ij} for all $1 \leqslant i, j \leqslant N$. Our strategy will be to prove that π induces a bijection between equivalence classes of irreducible representations of both compact matrix quantum groups, and then use this to conclude that π itself must be an isomorphism. We will proceed stepwise.

1. We will use the notations of Theorem 4.17 and write $u_n = u_{|\otimes n}$ for a representative of the nth equivalence class of irreducible representations of O_N^+. Recall in particular that $U = u_1$. Let us denote by $\mathcal{S} \subset \mathrm{Irr}(O_N^+ * \mathbb{T})$ the set of all irreducible representations of the form

   $$z^{[\epsilon_0]-} u_{n_1} z^{\epsilon_1} \cdots z^{\epsilon_{p-1}} u_{n_p} z^{[\epsilon_p]+}$$

   with $[-1]_- = -1, [-1]_+ = 0, [1]_- = 0, [1]_+ = 1$ and $\epsilon_{i+1} = (-1)^{n_{i+1}+1} \epsilon_i$. We claim that $\mathcal{S}$ is stable under taking tensor products with $u_1 z$ and $\overline{u_1 z} = z^{-1} u_1$. Indeed,

   - $(u_1 z) \otimes (u_{2n'_1} \cdots) = u_1 z u_{2n'_1} \cdots$,
   - $(u_1 z) \otimes (z^{-1} u_{2n'_1} z \cdots) = (u_{2n'_1-1} z \cdots) \oplus (u_{2n'_1+1} z \cdots)$,
   - $(u_1 z) \otimes (u_{2n'_1+1} z \cdots) = u_1 z u_{2n'_1+1} z \cdots$,
   - $(u_1 z) \otimes (z^{-1} u_{2n'_1+1} \cdots) = (u_{2n'_1} \cdots) \oplus (u_{2n'_1+2} \cdots)$

   and similarly for $z^{-1} u_1$. Since $\mathcal{S}$ moreover contains $u_1 z$ and its conjugate, it contains representatives of all the equivalence classes of irreducible representations of $\widetilde{O}_N^+$.

2. We will now use the indexing of irreducible representations of U_N^+ from Theorem 6.26, which is given by the set M of all words over $\{\circ, \bullet\}$. We define a map $\Phi \colon M \to \mathcal{S}$ inductively on the length of the words in the following way:

   - $\Phi(\circ) = u_1 z$, $\Phi(\bullet) = z^{-1} u_1$;
   - If $w = (\circ\bullet)^\ell \bullet w'$, then $\Phi(w) = u_{2\ell} \Phi(\bullet w')$;
   - If $w = (\bullet\circ)^\ell \circ w'$, then $\Phi(w) = z^{-1} u_{2\ell} z \Phi(\circ w')$;
   - If $w = (\circ\bullet)^\ell \circ w'$ with w' not starting with $\bullet$, then $\Phi(w) = u_{2\ell+1} z \Phi(\circ w')$;
   - If $w = (\bullet\circ)^\ell \bullet w'$ with w' not starting with $\circ$, then $\Phi(w) = z^{-1} u_{2\ell+1} \Phi(\bullet w')$.

This is indeed a bijection, with inverse inductively defined by

- $\Phi^{-1}(u_1 z) = \circ$, $\Phi^{-1}(z^{-1} u_1) = \bullet$;
- If $w = u_{2\ell} w'$, then $\Phi^{-1}(w) = (\circ\bullet)^\ell \Phi^{-1}(w')$;
- If $w = z^{-1} u_{2\ell} z w'$, then $\Phi^{-1}(w) = (\bullet\circ)^\ell \Phi^{-1}(w')$;
- If $w = u_{2\ell+1} z w'$, then $\Phi^{-1}(w) = (\bullet\circ)^\ell \circ \Phi^{-1}(w')$;
- If $w = z^{-1} u_{2\ell+1} w'$, then $\Phi^{-1}(w) = (\circ\bullet)^\ell \bullet \Phi^{-1}(w')$.

3. One easily checks by induction that the following properties hold:

- $\Phi(\overline{w}) = \overline{\Phi(w)}$;
- $\Phi(\circ \otimes w) = \Phi(\circ) \otimes \Phi(w)$;
- $\Phi(\bullet \otimes w) = \Phi(\bullet) \otimes \Phi(w)$.

As a consequence, any element of $\mathcal{S}$ can be obtained by taking tensor products of $u_1 z$ and $z^{-1} u_1$. In other words, $\mathcal{S} = \mathrm{Irr}(\widetilde{O}_N^+)$.

4. Note that because π sends V to W coefficientwise, we have

$$\pi \circ \Delta_{U_N^+} = \left(\Delta_{\widetilde{O}_N^+} \otimes \Delta_{\widetilde{O}_N^+} \right) \circ \pi.$$

As a consequence, if $u = (u_{ij})_{1 \leqslant i,j \leqslant \dim(u)}$ is a representation of U_N^+, then

$$\widehat{\pi}(u) = (\pi(u_{ij}))_{1 \leqslant i,j \leqslant \dim(u)}$$

is a representation of $\widetilde{O}_N^+$. Moreover, we have the following straightforward properties:

- $\widehat{\pi}(\overline{u}) = \overline{\widehat{\pi}(u)}$;
- $\widehat{\pi}(u \otimes v) = \widehat{\pi}(u) \otimes \widehat{\pi}(v)$;
- $\widehat{\pi}(u^\circ) = u_1 z$ and $\widehat{\pi}(u^\bullet) = z^{-1} u_1$.

Because the map Φ defined earlier satisfies the same properties when applied to the representations u^w for $w \in M$, we have that, for any $w \in M$,

$$[\widehat{\pi}(u^w)] = \Phi(w)$$

so that, in particular, $\widehat{\pi}(u^w)$ is non-trivial if w is not the empty word.

5. Let us notice that by the previous point, if u_{ij}^w is a coefficient of the non-trivial irreducible representation u^w, then

$$\pi(u_{ij}^w) = \widehat{\pi}(u^w)_{ij}$$

is a coefficient of the non-trivial irreducible representation $\widehat{\pi}(u^w)$. Thus, composing with the Haar state of $\widetilde{O}_N^+$ yields

$$h_{\widetilde{O}_N^+} \circ \pi(u_{ij}^w) = 0 = h_{U_N^+}(u_{ij}^w)$$

and because the coefficients of irreducible representations form a linear basis of $\mathcal{O}(U_N^+)$, this implies that $h_{U_N^+} = h_{\widetilde{O}_N^+} \circ \pi$. We can now conclude: if $x \in \ker(\pi)$, then

$$h_{U_N^+}(x^*x) = h_{\tilde{O}_N^+} \circ \pi(x^*x)$$
$$= h_{\tilde{O}_N^+}\left(\pi(x)^*\pi(x)\right)$$
$$= 0$$

and by faithfulness of the Haar state $h_{U_N^+}$, $x = 0$. Thus, π is injective, hence an isomorphism. □

A consequence of Theorem 6.34 is that the fundamental character $\chi^{U_N^+}$ of U_N^+ has the same moments with respect to $h_{U_N^+}$ as the product $\chi^{O_N^+}z$ with respect to $h_{O_N^+ * \mathbb{T}}$. Note, however, that these are not moments of a classical random variable, because $\chi^{U_N^+}$ is not a *normal element* in the sense that it does not commute with its adjoint $\left(\chi^{U_N^+ *}\right)$. One sometimes speaks, of a *non-commutative random variable* in that case and the term moments is often specified as *-moments* since one needs to consider all monomials in the variable and its adjoint.

It follows from Exercise 6.2 that if we set $\chi^\circ = \chi^{U_N^+}$ and $\chi^\bullet = \chi^{U_N^+ *}$, then, for any word $w = w_1 \cdots w_k$ on $\{\circ, \bullet\}$,

$$\int_{U_N^+} \chi^{w_1} \cdots \chi^{w_k} = |\mathcal{U}(\emptyset, w)|.$$

A non-commutative random variable with these *-moments is called a *circular element*.

It is natural to wonder at this point whether we can produce more examples of unitary compact matrix quantum groups through the free complexification operations. Here follow a few exercises in that direction. We start with the quantum permutation group.

Exercise 6.6 Prove that $\tilde{S}_N^+ = S_N^+ * \mathbb{T}$.

Solution Recall that the fundamental representation P of S_N^+ splits as $P = u^1 \oplus \varepsilon$. This means that the matrix Pz is block diagonal with one block being just z. But then, the corresponding coefficient of Pz is z, so that $\mathcal{O}(\tilde{S}_N^+)$ contains z. Thus, it also contains $\bar{z}$ and $P = Pz\bar{z}$. In other words, it contains both $\mathcal{O}(S_N^+)$ and $\mathcal{O}(\mathbb{T})$ so that it is the whole free product. □

Next, we consider the quantum hyperocathedral group, and it turns out that the free complexification operation recovers a construction introduced earlier in Exercise 6.3.

Exercise 6.7 Prove that $\mathbb{G}_N(C_{\text{alt}})$ is the free complexification of H_N^+.

Solution By Exercise 6.3, $\mathcal{O}(\mathbb{G}_N(\mathcal{C}_{\text{alt}}))$ is the quotient of $\mathcal{O}(U_N^+)$ by the ideal I generated by the elements $V_{ij}V_{i'j}^*$ and $V_{ij}V_{ij'}^*$, for all $1 \leqslant i, i', j, j' \leqslant N$ with $i \neq i'$ and $j \neq j'$. Moreover, by the universal property of free products applied to the canonical surjection $\mathcal{O}(O_N^+) \to \mathcal{O}(H_N^+)$ and to id$: \mathcal{O}(\mathbb{T}) \to \mathcal{O}(\mathbb{T})$, there is a surjective $*$-homomorphism

$$\phi: \mathcal{O}(O_N^+) * \mathcal{O}(\mathbb{T}) \to \mathcal{O}(H_N^+) * \mathcal{O}(\mathbb{T})$$

whose kernel is generated by the elements $U_{ij}U_{i'j}$ and $U_{ij}U_{ij'}$ for all $1 \leqslant i, i', j, j' \leqslant N$ with $i \neq i'$ and $j \neq j'$ (these are the defining relations of $\mathcal{O}(H_N^+)$). Because it sends the fundamental representation to the fundamental representation, ϕ restricts to a surjective $*$-homomorphism $\mathcal{O}(\widetilde{O}_N^+) \to \mathcal{O}(\widetilde{H}_N^+)$ still denoted by ϕ.

We now claim that if $\pi: \mathcal{O}(U_N^+) \to \mathcal{O}(\widetilde{O}_N^+)$ is the $*$-isomorphism of Theorem 6.34, then $\pi(I) = \ker(\phi)$. This follows from the fact that

$$(Uz)_{ij}(Uz)_{k\ell}^* = U_{ij}z\bar{z}U_{k\ell} = U_{ij}U_{k\ell}.$$

As a consequence, $\phi \circ \pi$ induces an isomorphism between $\mathcal{O}(U_N^+)/I = \mathcal{O}(\mathbb{G}_N(\mathcal{C}_{\text{alt}}))$ and $\mathcal{O}(\widetilde{H}_N^+)$ and the proof is complete. $\qquad\square$

One may use this result to compute the representation theory of $\mathbb{G}_N(\mathcal{C}_{\text{alt}})$, or use the theory of Chapter 4. In any case, the result is intricate to write down and we will therefore not do it. The reader may refer to [63] where this was first done, or to [38] for an alternate approach closer to our setting.

We conclude with another natural question: what happens if we complexify a compact matrix quantum group which is already a complexification? We will work out the most natural case, which is that of U_N^+.

Exercise 6.8 Prove that $\widetilde{U}_N^+ = U_N^+$.

Solution Let us write $W = Vz$ for the fundamental representation of $\widetilde{U}_N^+$. First, by the universal property there is a surjective $*$-homomorphism

$$\pi: \mathcal{O}(U_N^+) \to \mathcal{O}(\widetilde{U}_N^+)$$

sending V to W. Second, we can build an inverse to π by exploiting the universal property of the free product. Indeed, let $\phi_1: \mathcal{O}(U_N^+) \to \mathcal{O}(U_N^+)$ be the identity map and let $\phi_2: \mathcal{O}(\mathbb{T}) \to \mathcal{O}(U_N^+)$ be the unique $*$-homomorphism sending z to 1. Then, there exists a unique $*$-homomorphism

$$\phi: \mathcal{O}(U_N^+) * \mathcal{O}(\mathbb{T}) \to \mathcal{O}(U_N^+)$$

which coincides with ϕ_1 and ϕ_2 on the factors. Because ϕ sends by definition W to V, it is inverse to π and the proof is complete. $\qquad\square$

6.4.3 Tensor Complexification

If one applies the free complexification operation starting with a classical group, the result will be a non-commutative $*$-algebra, hence not classical any more. There is, however, a natural way to modify the construction so that commutativity of the $*$-algebra is preserved, and that is to replace the free product with the tensor product. Even though this will not be used in the sequel, we will now for the sake of completeness explain that so-called *tensor complexification*.

We start by describing the tensor product operation at the level of quantum groups. Let $(\mathbb{G}, u)$ and $(\mathbb{H}, v)$ be unitary compact matrix quantum groups and form the tensor product $\mathcal{A} = \mathcal{O}(\mathbb{G}) \otimes \mathcal{O}(\mathbb{H})$. We can define the tensor product of the coproducts

$$\Delta_{\mathbb{G}} \otimes \Delta_{\mathbb{H}} \colon \mathcal{A} \to \mathcal{O}(\mathbb{G}) \otimes \mathcal{O}(\mathbb{G}) \otimes \mathcal{O}(\mathbb{H}) \otimes \mathcal{O}(\mathbb{H})$$

but it does not have the correct target algebra. The problem can easily be remedied by setting

$$\Delta = (\mathrm{id} \otimes \Sigma \otimes \mathrm{id}) \circ \Delta_{\mathbb{G}} \otimes \Delta_{\mathbb{H}} \colon \mathcal{A} \to \mathcal{A} \otimes \mathcal{A},$$

where $\Sigma \colon \mathcal{O}(\mathbb{G}) \otimes \mathcal{O}(\mathbb{H}) \to \mathcal{O}(\mathbb{H}) \otimes \mathcal{O}(\mathbb{G})$ is the *flip map* sending $x \otimes y$ to $y \otimes x$. One can then check directly that this coproduct is compatible with the representation

$$w = \begin{pmatrix} u \otimes 1 & 0 \\ 0 & 1 \otimes v \end{pmatrix}$$

so that $(\mathcal{A}, w)$ is a unitary compact matrix quantum group.

Definition 6.35 The unitary compact matrix quantum group $(\mathcal{O}(\mathbb{G}) \otimes \mathcal{O}(\mathbb{H}), w)$ is called the *direct product* of $\mathbb{G}$ and $\mathbb{H}$ and denoted by $\mathbb{G} \times \mathbb{H}$.

Remark 6.36 As in Remark 6.30, it is straightforward to prove that given two finitely generated discrete groups Γ_1 and Γ_2, there is an isomorphism

$$\widehat{\Gamma}_1 \times \widehat{\Gamma}_2 \cong \widehat{\Gamma_1 \times \Gamma_2},$$

so that $\widehat{\times}$ could also be a reasonable notation.

Even though this is not needed in the sequel, let us work out the representation theory of direct products as computed by S. Wang in [71].

Theorem 6.37 (Wang) *Let $(\mathbb{G}, u)$ and $(\mathbb{H}, v)$ be unitary compact matrix quantum groups. Then, the representations*

$$u^\alpha \otimes u^\beta,$$

for (α, β) *in* $\mathrm{Irr}(\mathbb{G}) \times \mathrm{Irr}(\mathbb{H})$ *not both trivial, form a complete set of equivalence classes of irreducible and pairwise inequivalent non-trivial representations of* $\mathbb{G} \times \mathbb{H}$.

Proof If $\beta \neq \beta'$, then by Frobenius reciprocity (Proposition 5.17)

$$\mathrm{Mor}_{\mathbb{G} \times \mathbb{H}} \left(u^\alpha \otimes u^\beta, u^{\alpha'} \otimes u^{\beta'} \right) \simeq \mathrm{Mor}_{\mathbb{G} \times \mathbb{H}} \left(u^\alpha \otimes u^\beta \otimes \overline{u}^{\beta'}, u^{\alpha'} \right)$$
$$\simeq \bigoplus_{\gamma \subset \beta \otimes \overline{\beta}'} \mathrm{Mor}_{\mathbb{G} \times \mathbb{H}} \left(u^\alpha \otimes u^\gamma, u^{\alpha'} \right).$$

If there is a non-zero element in the latter space then, by irreducibility, there exists $\gamma \subset \beta \otimes \overline{\beta}'$ such that $u^{\alpha'}$ is a subrepresentation of $u^\alpha \otimes u^\gamma$. Thus, the coefficients $u_{ij}^{\alpha'} \otimes 1$ are linear combinations of coefficients $u_{ab}^\alpha \otimes u_{cd}^\gamma$. But since $\gamma \neq \varepsilon$, 1 is not in the linear span of the coefficients u_{cd}^γ, hence a contradiction.

The same argument applies to the case $\alpha \neq \alpha'$, so that the representations in the statement are pairwise inequivalent. Moreover, by the same argument,

$$\mathrm{Mor}_{\mathbb{G} \times \mathbb{H}} \left(u^\alpha \otimes u^\beta, u^\alpha \otimes u^\beta \right) \simeq \mathrm{Mor}_{\mathbb{G} \times \mathbb{H}} \left(u^\alpha \otimes u^\beta \otimes \overline{u}^\beta, u^\alpha \right)$$
$$\simeq \mathrm{Mor}_{\mathbb{G} \times \mathbb{H}} \left(u^\alpha, u^\alpha \right) \oplus \bigoplus_{\gamma \subset \beta \otimes \overline{\beta}', \gamma \neq \varepsilon}$$
$$\mathrm{Mor}_{\mathbb{G} \times \mathbb{H}} \left(u^\alpha \otimes u^\gamma, u^\alpha \right)$$
$$= \mathrm{Mor}_{\mathbb{G} \times \mathbb{H}} \left(u^\alpha, u^\alpha \right)$$

and the latter space is one-dimensional, proving the irreducibility of $u^\alpha \otimes u^\beta$.

Eventually, because the coefficients of all these representations generate $\mathcal{O}(\mathbb{G} \otimes \mathbb{H})$, we have found all irreducible representations. $\square$

Noticing that $\mathcal{O}(\mathbb{G}) \otimes \mathcal{O}(\mathbb{H})$ is commutative as soon as both $\mathcal{O}(\mathbb{G})$ and $\mathcal{O}(\mathbb{H})$ are, the recipe to complexify a classical group is now clear. For convenience, let us simply write ab for $a \otimes b$ when the context is unambiguous.

Exercise 6.9 Let $(\mathbb{G}, u)$ be an orthogonal compact matrix quantum group. Prove that the subalgebra $\mathcal{A}$ of $\mathcal{O}(\mathbb{G} \times \mathbb{T})$ generated by $u_{ij}z$ yields a unitary compact matrix quantum group when endowed with the fundamental representation $uz = (u_{ij} \otimes z)_{1 \leqslant i, j \leqslant N}$.

Solution First, observe that $(uz)^* uz = u^t u = \mathrm{Id} = uz(uz^*)$ and that the same holds for $\overline{uz} = u^t z$. Moreover, by definition of the coproduct on the tensor product,

$$\Delta(u_{ij}z) = \Delta(u_{ij})\Delta(z) = \left(\sum_{k=1}^N u_{ik} \otimes u_{kj} \right) (z \otimes z) = \sum_{k=1}^N u_{ik}z \otimes u_{kj}z.$$

Since by construction the coefficients of uz generate $\mathcal{A}$, the proof is complete. $\qquad\square$

We will write[5] $\mathcal{A} = \mathcal{O}(\overline{\mathbb{G}})$ and call this unitary matrix quantum group the *tensor complexification* of $\mathbb{G}$. Tensor complexification naturally appears in the classification of all unitary non-crossing partition quantum groups (see [68]). Unfortunately, despite that usefulness in the genuinely quantum setting, the construction turns out to be of little interest in the classical case.

Exercise 6.10 Let $G \subset U_N$ be a classical compact group of matrices. We identify the group $\mathbb{T}$ of complex numbers with $\{z.\,\mathrm{Id} \mid z \in \mathbb{T} \subset U_N\}$ and we denote by H the image of the group homomorphism

$$\pi \colon G \times \mathbb{T} \to U_N$$

sending (g, z) to gz.

1. We denote as usual by $(c_{ij})_{1 \leqslant i,j \leqslant N}$ the matrix coefficient functions on $M_N(\mathbf{C})$. Prove that the $*$-subalgebra of $\mathcal{O}(G \times \mathbb{T})$ generated by the functions $c'_{ij} = c_{ij} \circ \pi$ is $\mathcal{O}(\overline{G})$.
2. Conclude that $\overline{G} = H$.

Proof 1. Simply observe that

$$\begin{aligned}
c'_{ij}(g, z) &= c_{ij} \circ \pi(g, z) \\
&= c_{ij}(gz) \\
&= c_{ij}(g)z,
\end{aligned}$$

so that $c'_{ij} = c_{ij}z$.

2. Let us denote by Φ the map induced by π at the level of $*$-algebras, that is to say $\Phi \colon f \mapsto f \circ \pi$. Since we have just shown that the range of Φ is $\mathcal{O}(\overline{G})$, it is enough to prove that it is injective. But this is clear: if $f \in \ker(\Phi)$, then $f(gz) = 0$ for any $g \in G$ and $z \in \mathbb{T}$, hence $f = 0$. $\qquad\square$

Applying this to the orthogonal group O_N, we see that $\overline{O}_N$ is the group of matrices of the form zg with g orthogonal and $z \in \mathbb{T}$. Observing that $(-1).(-\mathrm{id}) = 1.\mathrm{id}$, we see that the map π is not surjective in that case. In fact, we have just found its kernel: if $zg = \mathrm{id}$, then z must be real, hence $z \in \{\pm 1\}$ which only leaves the possibility $g \in \{\pm \mathrm{id}\}$. As a conclusion, $\overline{O}_N$ is isomorphic to the quotient of $O_N \times \mathbb{T}$ by the subgroup $\{(\mathrm{id}, 1), (-\mathrm{id}, -1)\}$. It may not be clear that this cannot be isomorphic to U_N, but this will follow indirectly from another argument; see Remark 8.37.

---

[5] This notation is not standard in any respect, but we simply needed something different from the tilda used for the free complexification.

7

Further Examples

Now that we have adapted our combinatorial tools to the unitary setting, we can use them to study an interesting and important family of examples, called *quantum reflection groups*. To understand the origin of the definition, let us first discuss classical reflection groups. We have already encountered two examples of classical finite groups generated by elements of order two, which can be thought of as 'reflections' in the general sense: the permutation group S_N and the hyperoctahedral group H_N. Recall that the second one was made of permutation matrices, but with coefficients allowed to be ± 1. This can be generalised under the name of *reflection groups*.

7.1 Quantum Reflection Groups

7.1.1 Liberation of Reflection Groups

Let us start with the formal definition of reflection groups. A complex-valued matrix will be called *monomial* if it has exactly one non-zero coefficient on each row and column.

Definition 7.1 Let N and s be integers. The reflection group H_N^s is the group of all monomial matrices in $M_N(\mathbb{C})$ with coefficients being sth roots of unity.

Our purpose in this section is to find a definition of a quantum version of reflection groups. The problem of finding the correct defining relations of the algebra $\mathcal{O}(H_N^s)$ which would also make sense in a non-commutative context is not an easy one. This kind of problem was in fact one of the main motivations for the introduction of partition quantum groups in [18]. We will follow the main ideas of this work by using the following strategy: first describe the intertwiners of reflection groups with coloured partitions, then restrict to non-crossing partitions to produce a quantum object.

It turns out that for H_N^s, a partition description of the intertwiner spaces was given by K. Tanabe in [66], though in a different language. We will restate this in terms of coloured partitions using the following two ones. First, we define $q_\circ \in NC^{\circ\bullet}(2,2)$ to be the one-block partition with all points coloured in white:

$$q_\circ = \qquad \qquad .$$

Second, we define $p_s \in NC^{\circ\bullet}(s,0)$ to be the one-block partition with all blocks coloured in white:

$$p_s = \qquad \qquad .$$

Let us prove that this is sufficient to recover the reflection groups H_N^s as partition quantum groups.

Theorem 7.2 (Tanabe) *Let* $s, N \in \mathbf{N}$ *and let* $\mathcal{C}_{s,class}$ *be the* symmetric category *of coloured partitions generated by* $q_\circ$, *and* p_s. *Then, there is an isomorphism*

$$\mathbb{G}_N(\mathcal{C}_{s,class}) \simeq H_N^s$$

Proof By definition, $\mathbb{G}_N(\mathcal{C}_{s,class})$ is a group of unitary matrices in $M_N(\mathbf{C})$ whose coefficients satisfy some extra relations given by the partitions $q_\circ$ and p_s. Let us start with $q_\circ$ and compute

$$(\mathrm{id} \otimes T_{q_\circ}) \circ \rho_{u \otimes u}(e_{i_1} \otimes e_{i_2}) = \sum_{j_1, j_2 = 1}^{N} u_{i_1 j_1} u_{i_2 j_2} \otimes T_{q_\circ}(e_{j_1} \otimes e_{j_2})$$

$$= \sum_{j=1}^{N} u_{i_1 j} u_{i_2 j} \otimes e_j \otimes e_j$$

while

$$\rho_{u \otimes u} \circ T_{q_\circ}(e_{i_1} \otimes e_{i_2}) = \delta_{i_1 i_2} \sum_{k=1}^{N} \rho_{u \otimes u}(e_{i_1} \otimes e_{i_1})$$

$$= \delta_{i_1 i_2} \sum_{j_1, j_2 = 1}^{N} u_{i_1 j_1} u_{i_1 j_2} \otimes e_{j_1} \otimes e_{j_2}.$$

These two expressions are equal if and only if

$$u_{ij}u_{ij'} = 0 = u_{ij}u_{i'j}$$

for any $1 \leqslant i, i', j, j' \leqslant N$ such that $i \neq i'$ and $j \neq j'$. This precisely means that the matrices in $\mathbb{G}_N(\mathcal{C}_{s,\text{class}})$ have only one non-zero coefficient on each row. Because the columns of a unitary matrix form an orthonormal basis, there is also only one non-zero coefficient on each column, and these coefficients have modulus one.

We now turn to p_s and compute

$$
\begin{aligned}
(\text{id} \otimes T_{p_s}) \circ \rho_{u^{\otimes s}}(e_{i_1} \otimes \cdots \otimes e_{i_s}) &= \sum_{j_1,\cdots,j_s=1}^{N} u_{i_1 j_1} \cdots u_{i_s j_s} \\
&\qquad \otimes T_{p_s}(e_{j_1} \otimes \cdots \otimes e_{j_s}) \\
&= \sum_{j=1}^{s} u_{i_1 j} \cdots u_{i_s j} \otimes e_j \otimes \cdots \otimes e_j
\end{aligned}
$$

while

$$\rho_\varepsilon \circ T_{p_s}(e_{i_1} \otimes \cdots \otimes e_{i_s}) = \delta_{i_1 \cdots i_s}.$$

These two quantities are equal if and only if

$$\sum_{j=1}^{N} u_{ij}^s = 1.$$

Given that only one coefficient in the sum can be non-zero, this is equivalent to the fact that for each matrix in $\mathbb{G}_N(\mathcal{C}_{s,\text{class}})$, the non-zero coefficient of each row is an sth root of unity, hence the result. $\qquad\square$

Remark 7.3 In his work [66], K. Tanabe deals with the more general family of complex reflection groups $G(m, p, n)$, of which H_N^s is the particular case of $G(s, s, N)$. Defining compact matrix quantum group analogues of the full family $G(m, p, n)$ is, to this date, still an open problem (see [42] for an example in the special case of $G(2, 1, 4)$). However, the complete classification of non-crossing unitary partition quantum groups (see [68]) shows that it cannot be done in this setting.

We can now illustrate one of the interesting features of the partition approach to groups and quantum groups. Indeed, as a direct by-product of Theorem 7.2, we obtain a definition of a quantum version of the reflection groups.

Definition 7.4 Let s, N be integers and set $C_s = \langle q_\circ, p_s \rangle$. The partition quantum group $\mathbb{G}_N(C_s)$ is called a *quantum reflection group* and denoted by H_N^{s+}.

By definition, the abelianisation map is a surjection from $\mathcal{O}(H_N^{s+})$ to $\mathcal{O}(H_N^s)$ mapping the fundamental representation to the fundamental representation, hence H_N^s is indeed the classical version of H_N^{s+}. As a first attempt to get some grasp on these objects, let us work out some very basic examples.

Exercise 7.1 Prove that H_N^{1+} and H_N^{2+} are isomorphic to partition quantum groups which were already studied earlier.

Solution Let $s = 1$, so that p_s is the white singleton partition $s_\circ$. Then,

$$D_{\circ\circ} = (p_1 \otimes p_1)q_\circ \in C_1$$

and rotating this yields the identity partition with different colours on the rows. Thus, by Exercise 6.1 we can forget the colours in C_1. Since it contains a four-block and a singleton, it contains a three block. Thus, $C_1 = NC^{\circ\bullet}$ by Proposition 3.17 and it follows that

$$H_N^{1+} = S_N^+.$$

Let now $s = 2$, so that $p_2 = D_{\circ\circ}$ with both points coloured in white. Rotating it yields the identity partition with different colours, hence by Exercise 6.1 we can forget the colours. Since C_2 contains the four-block and only partitions with even blocks, $C_2 = NC_{\text{even}}^{\circ\bullet}$ by Exercise 3.7, so that

$$H_N^{2+} = H_N^+. \qquad \square$$

7.1.2 Representation Theory

Our goal is now to compute the representation theory of quantum reflection groups. However, our definition of the categories of partitions C_s is inconvenient for that purpose, because it only gives us generators. It would be better to have an intrinsic characterisation of the partitions in C_s. We will present such a characterisation now, which requires some vocabulary.

Definition 7.5 For a coloured partition $p \in P^{\circ\bullet}(k, 0)$, let $c_u(p)$ (respectively $c_\ell(p)$) be the difference between the number of white and black points in its upper row (respectively its lower row). The *colour sum* of p is then defined as

$$c(p) = c_u(p) - c_\ell(p).$$

We will characterise the partitions in C_s by the colour sum of their blocks, but for practical reasons it is simpler to define another set of partitions, and then to prove equality.

Definition 7.6 We define $\mathcal{D}_s$ to be the set of all non-crossing partitions $p \in NC^{\circ\bullet}$ such that for each block $b \subset p$,

$$c(b) = 0 \mod s.$$

We only defined $\mathcal{D}_s$ as a set, but we will need to make sure that it is in fact a category of partitions.

Lemma 7.7 *The set $\mathcal{D}_s$ is a category of partitions.*

Proof The defining property of $\mathcal{D}_s$ is preserved by reflection and horizontal concatenation and is satisfied by $D_{\circ\bullet}$, $D_{\bullet\circ}$, $\mathrm{id}_\circ$ and $\mathrm{id}_\bullet$, so that all we have to prove is stability under vertical concatenation. To do this, we will first prove the result for the set P_s consisting in all partitions (possibly crossing) such that for each block $b \subset p$, $c(b) = 0 \mod s$. Consider two partitions $p, q \in P_s$ such that the composition $q \circ p$ makes sense. We will proceed by induction on the total number of blocks $b(p) + b(q)$.

- If $b(p) + b(q) = 2$, then $b(p) = 1 = b(q)$ and the lower colouring of p equals the upper colouring of q. In particular, we have the following equalities modulo s,
$$c_u(p) = c_\ell(p) = c_u(q) = c_\ell(q)$$
so that $q \circ p \in P_s$.
- If $b(p) + b(q) > 2$, assume first that there are two blocks of p (respectively two blocks of q) which get merged in the composition. Then, $q \circ p = q \circ \widetilde{p}$ (respectively $\widetilde{q} \circ p$)) where $\widetilde{p}$ (respectively $\widetilde{q}$) is obtained by merging the two given blocks. Because $\widetilde{p}$ (respectively $\widetilde{q}$) is in P_s by definition and the total number of blocks has been decreased by one, we can conclude by the induction hypothesis.
- If $b(p) + b(q) > 2$ and no blocks get merged in the composition, then $b(p) = b(q)$ and each block of p is combined with exactly one block of q. The argument of the first step of the present proof then shows that the composition of these two blocks is in P_s, hence $q \circ p \in P_s$.

Now that we have proven that P_s is a category of partitions, it suffices to note that

$$\mathcal{D}_s = P_s \cap NC^{\circ,\bullet}. \qquad \square$$

Before going further, let us mention a similar result concerning the category of partitions C_{alt} introduced at the end of Section 6.3.

Exercise 7.2 Prove that C_{alt} is a category of partitions.

Solution Note that we cannot use an argument involving crossings because using them we can change alternating blocks into non-alternating ones. We will therefore rather use the same strategy as in Lemma 3.14. Only the stability under vertical concatenation is not obviously satisfied, and up to rotations it is enough to prove it for $p \in C_{\text{alt}}(0, \ell)$ and $q \in C_{\text{alt}}(\ell, m)$. We can then work by induction on ℓ.

If $\ell = 2$, then $p \in \{D_{\circ\bullet}, D_{\bullet\circ}\}$ and we are therefore removing two points of q while merging the corresponding blocks. If the two points were in the same block, the resulting one is still alternating. If not, let us first merge the two blocks without removing points. Since the two points which now become neighbours in the block have different colours, the merged block is alternating. Removing two consecutive points therefore again yields an alternating block.

If the result holds up to some ℓ, assume that p has $\ell + 2$ points and that $q \circ p$ has a crossing. Then, p has an interval and removing two neighbouring points in that interval preserves the alternating structure. Doing the corresponding operation on q (removing the points and merging the corresponding blocks) also does as explained in the previous point, and the composition remains unchanged. We then conclude by induction. $\square$

Because $c(q_\circ) = 0$ and $c(p_s) = s$, we have $C_s \subset D_s$ by definition. We claim that this inclusion is an equality. To keep things clear, we will first work with $q_\circ$ alone.

Lemma 7.8 *Let D_∞ be the set of all non-crossing partitions $p \in NC^{\circ\bullet}$ such that for each block $b \subset p$, $c(b) = 0$. Then,*

$$\langle q_\circ \rangle = D_\infty$$

Proof Let us first observe that D_∞ is a category of partitions by the same argument as in the proof of Lemma 7.7. As a consequence, because $q_\circ \subset D_\infty$, we have an inclusion of the left-hand side into the right-hand side.

As for the converse, let us first show that it holds for one-block partitions by induction on the number of points.

- For partitions with two points the result is clear.
- Let us now assume that the result holds for one-block partitions on at most $2n$ points (note that for the colour sum to vanish, there must be an even

number of points) and consider a one-block partition p on $2(n + 1)$ points. Up to rotation, we may assume that p lies on one line and that its first two points (from the left) are coloured by $\circ\bullet$. Let p' be the partition formed by the last $2n$ points of p. By induction, since

$$c(p') = c(p) = 0,$$

$p' \in \mathcal{D}_\infty$, hence $p' \in \langle q_\circ \rangle$ by the induction hypothesis. But then,

$$p = (q \otimes p')(\mathrm{id}_\circ \otimes q' \otimes \mathrm{id}_*^{\otimes n}),$$

where id_* denotes suitably coloured versions of the identity partition and q' is one of the following partitions:

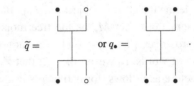

If we prove that these two partitions are in $\langle q_\circ \rangle$, then we will be done. The second one is just the upside-down rotation of $q_\circ$. As for the first one, we have

$$\widetilde{q} = (\mathrm{id}_\circ \otimes D_{\circ\bullet} \otimes \mathrm{id}_\bullet)\,(q_\circ \otimes q_\bullet)\,(\mathrm{id}_\circ \otimes D_{\circ\bullet}^* \otimes \mathrm{id}_\bullet) \in \langle q_\circ \rangle,$$

hence the result.

We can now prove the equality by induction on the number of points. For partitions on at most four points, the result is clear. If it is true for partitions on at most $2n$ points, let $p \in \mathcal{D}_\infty$ be a partition on $2(n + 1)$ points. If it has only one block, then it is in $\langle q_\circ \rangle$ by the first part of the proof. Otherwise, there is an interval q in p and rotating yields a partition of the form $q \otimes p'$. It follows from the definition of $\mathcal{D}_\infty$ that both q and p' are again in $\mathcal{D}_\infty$, hence the same holds for p, concluding the proof. $\square$

The proof of the equality $\mathcal{C}_s = \mathcal{D}_s$ now follows easily by arguments similar to those of Lemma 7.8.

Proposition 7.9 *For any $1 \leqslant s < +\infty$, $\mathcal{C}_s = \mathcal{D}_s$.*

Proof Let us first prove that any one-block partition with only white points of $\mathcal{D}_s$ is in $\mathcal{C}_s$. Such a partition must be of the form p_{ns} for some integer n. But

$$p_{ns} = \left(p_s^{\otimes n}\right)\left(\mathrm{id}_\circ^{\otimes s-1} \otimes q_\circ \otimes \mathrm{id}_\circ^{\otimes s-2} \otimes q_\circ \otimes \cdots \otimes q_\circ \otimes \mathrm{id}_\circ^{\otimes s-1}\right) \in \mathcal{C}_s.$$

If now p is a one-block partition which is not of the previous form up to rotation, we can rotate it on one line so that the colours of the first two points are $\circ\bullet$. If p' is the partition formed by the remaining points, $c(p') = 0 \mod s$. From this observation, one can prove by induction that $p \in \mathcal{C}_s$ using the formula

$$p = (q' \otimes p') \left(\mathrm{id}_\circ \otimes q' \otimes \mathrm{id}_*^{\otimes n} \right),$$

where once again id_* denotes suitable coloured versions of the identity partition and $q' \in \{\widetilde{q}, q_\bullet\}$.

Using this, one can conclude by induction exactly as in the proof of Lemma 7.8 □

We will now use this result to compute the representation theory of H_N^{s+}. As for U_N^+, we first define the abstract object indexing the equivalence classes of irreducible representations. Let M_s be the free monoid on $\mathbf{Z}_s$, that is to say the set of all words over $\{0, \cdots, s-1\}$. The monoid structure is given by the concatenation of words. However, the fact that $\mathbf{Z}_s$ is a group gives us another monoid structure as follows: for two words $w = w_1 \cdots w_k$ and $w' = w'_1 \cdots w'_\ell$, we set

$$w * w' = w_1 \cdots w_{k-1}(w_k + w'_1)w'_2 \cdots w'_\ell.$$

Eventually, the conjugation is given by

$$\overline{w_1 \cdots w_k} = (-w_k) \cdots (-w_1).$$

The following result is originally due to T. Banica and R. Vergnioux in [19].

Theorem 7.10 (Banica–Vergnioux) *For $N \geqslant 4$, the irreducible representations of H_N^{s+} can be indexed by the elements of M_s in such a way that $u^{\emptyset} = \varepsilon$, $u^0 = u$ and for any $w \in M_s$,*

$$\overline{u}^w = u^{\overline{w}}.$$

Moreover, for any $w, w' \in M_s$,

$$u^w \otimes u^{w'} = \bigoplus_{w=az, w'=\overline{z}b} \left(u^{ab} \oplus u^{a*b} \right).$$

Proof Let us define, for $1 \leqslant i \leqslant s-1$, π_i to be the projective partition in $\mathcal{C}_s(i, i)$ with one block and all points coloured in white, that is,

Let us also set for convenience $\pi_0 = \widetilde{q}$. We will consider the map $\Phi \colon M_s \to \mathrm{Irr}(H_N^{s+})$ given by

$$\Phi(w_1 \cdots w_k) = \left[u_{\pi_{w_1} \otimes \cdots \otimes \pi_{w_k}} \right].$$

1. We first prove that Φ is surjective. Let $p \in \mathcal{C}_s$ be a projective partition. As in the proof of Theorem 6.26, we decompose it as

$$p = (b_0^* b_0) \otimes p_1 \otimes (b_1^* b_1) \otimes p_2 \otimes \cdots \otimes p_k \otimes (b_k^* b_k),$$

where p_i is projective with $t(p_i) = 1$ and b_i lies on one line. Each block of b_i being a block of p, we have $b_i \in \mathcal{D}_s = \mathcal{C}_s$, hence (denoting by $n(p_i)$ the number of points of p_i) composing with

$$r = b_0^* \otimes \mathrm{id}_*^{\otimes n(p_1)/2} \otimes b_1^* \otimes \mathrm{id}_*^{\otimes n(p_2)/2} \otimes \cdots \otimes \mathrm{id}_*^{\otimes n(p_k)/2} \otimes b_k^* \in \mathcal{C}_s$$

(where id_* denote suitably coloured identity partitions) yields an equivalence between p and $p_1 \otimes \cdots \otimes p_k$. For a fixed $1 \leqslant i \leqslant k$ let r_i be defined as follows:

- If $c_u(p_i) \neq 0 \mod s$ and x is the unique representative of this class in $\{1, \cdots, s-1\}$, then r_i is the unique one-block partition in $\mathcal{C}_s(x, n(p_i)/2)$ with all upper points coloured in white and lower points coloured to match the upper points of p_i.
- If $c_u(p_i) = 0 \mod s$, then r_i is the unique one-block partition in $\mathcal{C}_s(2, n(p_i)/2)$ with upper points coloured with $\circ\bullet$ and lower points coloured to match the upper points of p_i.

Then, because $r_i \in \mathcal{C}_s$ for all i by construction,

$$r' = (p_1 \otimes \cdots \otimes p_k)(r_1 \otimes \cdots \otimes r_k) \in \mathcal{C}_s$$

is an equivalence between $p_1 \otimes \cdots \otimes p_k$ and $\pi_{s_1} \otimes \cdots \otimes \pi_{s_k}$. By transitivity of the equivalence relation, this proves that Φ is surjective by Theorem 4.10.

2. As for injectivity, let $w = w_1 \cdots w_k$ and $w' = w_1' \cdots w_n'$ be such that $\Phi(w) = \Phi(w')$, which is equivalent by Proposition 4.12 to

$$\pi_{w_1} \otimes \cdots \otimes \pi_{w_k} \sim \pi_{w_1'} \otimes \cdots \otimes \pi_{w_n'}.$$

Because two equivalent partitions have the same number of through-blocks, we must have $k = n$. Moreover, the equivalence is then implemented by the partition

$$r_{\pi_{w_1'}}^{\pi_{w_1}} \otimes \cdots \otimes r_{\pi_{w_k'}}^{\pi_{w_k}} \in \mathcal{C}_s,$$

where $r_{\pi_{w_i'}}^{\pi_{w_i}}$ is obtained by gluing the upper part of π_{w_i} to the lower part of $\pi_{w_i'}$, so that for any $1 \leqslant i \leqslant k$,

$$c_u(w_i) - c_d(w_i') = c_u(w_i) - c_u(w_i') = 0 \mod (s).$$

As a consequence, $w_i = w_i'$ for all $1 \leqslant i \leqslant k$ and $w = w'$.

3. We now have to compute the fusion rules and the reasoning is similar to the case of U_N^+ in Theorem 6.26. Let $w = w_1 \cdots w_k$ and $w' = w_1' \cdots w_n'$ be two words in M_s and consider $\Phi(w) \square^\ell \Phi(w')$ for some $\ell \leqslant \min(k, n)$. There is an interval in the middle obtained by gluing the upper blocks of π_{w_k} and $\pi_{w_1'}$, which has colour number $w_k + w_1'$. But this number must be a multiple of s by definition of $\mathcal{D}_s = \mathcal{C}_s$. Hence, $w_k = -w_1' \mod s$ and applying the same argument we see that there is a decomposition $w = az$, $w' = \overline{z}b$ such that

$$\Phi(w) \square^\ell \Phi(w') = \Phi(ab).$$

As for $\Phi(w) \,\square^\ell\, \Phi(w')$, we have similarly that $w = az$ and $w' = \overline{z}b$ with z of length $\ell - 1$. Moreover, the through-block obtained by gluing $\pi_{w_{k-\ell+1}}$ and $\pi_{w_\ell'}$ has upper colour number $w_{k-\ell+1} + w_\ell'$, hence the result.

4. Note eventually that for $u^w \otimes u^{w'}$ to contain the trivial representation $u^\emptyset$, one must have $w' = \overline{w}$. Hence, $\overline{u}^w = u^{\overline{w}}$. $\qquad\square$

Before going to the next section, let us comment on another example which discretely appeared without being named.

Exercise 7.3 We consider the category of non-crossing partitions $\mathcal{C}_\infty = \langle q_\circ \rangle = \mathcal{D}_\infty$. The corresponding unitary compact matrix quantum group is denoted by $H_N^{\infty+}$.

1. Prove that $\mathcal{O}(H_N^{\infty+}))$ is the quotient of $\mathcal{O}(U_N^+)$ by the relations
   - $(V_{ij}V_{ij}^*)^2 = V_{ij}V_{ij}^*$
   - $V_{ij}V_{ij}^* = V_{ij}^*V_{ij}$

   for all $1 \leqslant i, j \leqslant N$.
2. Describe the representation theory of $H_N^{\infty+}$.

Solution 1. We know that $\mathcal{O}(H_N^{\infty+})$ is the quotient of $\mathcal{O}(U_N^+)$ by the relations coming from the partition $q_\circ$. These were computed in Theorem 7.2 and are

$$V_{ij}V_{i'j} = 0 = V_{ij}V_{ij'}$$

for all $1 \leqslant i, j, i', j' \leqslant N$ with $i \neq i'$ and $j \neq j'$. Using the fact that V is unitary, we deduce from this that

$$V_{ij} V_{i'j}^* = V_{ij} \left(\sum_{k=1}^{N} V_{kj} V_{kj}^* \right) V_{i'j}^*$$
$$= 0$$

for $i \neq i'$. Similarly, we get $V_{ij} V_{ij'}^* = 0$ for $j \neq j'$. Therefore,

$$V_{ij} V_{ij}^* = V_{ij} \left(\sum_{k=1}^{N} V_{ik}^* V_{ik} \right) V_{ij}^*$$
$$= V_{ij} V_{ij}^* V_{ij} V_{ij}^*.$$

Moreover,

$$V_{ij} V_{ij}^* = \left(\sum_{k=1}^{N} V_{ik}^* V_{ik} \right) V_{ij} V_{ij}^*$$
$$= V_{ij}^* V_{ij} V_{ij} V_{ij}^*$$

and

$$V_{ij}^* V_{ij} = V_{ij}^* V_{ij} \left(\sum_{k=1}^{N} V_{ik} V_{ik}^* \right)$$
$$= V_{ij}^* V_{ij} V_{ij} V_{ij}^*,$$

hence the second relation.

To conclude we know have to prove that conversely, assuming that $V_{ij} V_{ij}^*$ is a projection and that V_{ij} commutes with its adjoint yields the relations corresponding to q_o. First, observe that, by unitarity,

$$\sum_{k=1}^{N} V_{ik} V_{ik}^* = 1$$

so that these orthogonal projections are pairwise orthogonal. Thus, setting $W = V_{ij} V_{ij}^* V_{ij} - V_{ij}$, we have

$$W^* W = (V_{ij}^* V_{ij})^3 - 2(V_{ij}^* V_{ij})^2 + V_{ij}^* V_{ij}$$
$$= 0$$

so that $h(W^* W)$ and by faithfulness of the Haar state h, $W = 0$. Therefore, for $i \neq i'$,

$$V_{ij} V_{i'j}^* = V_{ij} V_{ij}^* V_{ij} V_{i'j}^* V_{i'j} V_{i'j}^*$$
$$= 0.$$

As a consequence, for $i \neq i'$,

$$
\begin{aligned}
V_{ij}V_{i'j} &= V_{ij}V_{ij}^*V_{ij}V_{i'j}V_{i'j}^*V_{i'j} \\
&= V_{ij}V_{ij}^*V_{ij}V_{i'j}^*V_{i'j}V_{i'j} \\
&= 0.
\end{aligned}
$$

The same argument shows that $V_{ij}V_{ij'} = 0$ for $j \neq j'$, hence the proof is complete.

2. The representation theory of $H_N^{\infty+}$ is the same as that of H_N^{s+}, except that the group $\mathbf{Z}_s$ is replaced by $\mathbf{Z}$. More explicitly, let M_∞ be the set of all words over $\mathbf{Z}$. For two words $w = w_1 \cdots w_k$ and $w' = w_1' \cdots w_\ell'$, we set

$$
w * w' = w_1 \cdots w_{k-1}(w_k + w_1')w_2' \cdots w_\ell'.
$$

We also define a conjugation operation as follows:

$$
\overline{w_1 \cdots w_k} = (-w_k) \cdots (-w_1).
$$

The same proof as in Theorem 7.10 yields that the irreducible representations of $H_N^{\infty+}$ can be indexed by the elements of M_∞ in such a way that $u^\emptyset = \varepsilon$, $u^0 = u$ and, for any $w \in M_\infty$,

$$
\overline{u}^w = u^{\overline{w}}.
$$

Moreover, for any $w, w' \in M_\infty$,

$$
u^w \otimes u^{w'} = \bigoplus_{w=az,w'=\overline{z}b} \left(u^{ab} \oplus u^{a*b} \right). \qquad \square
$$

The abelianisation of $\mathcal{O}(H_N^{\infty+})$ produces the algebra of functions on the classical group H_N^∞ of all unitary monomial matrices. Note that H_N^∞ can also be recovered as the abelianisation of $\mathcal{O}(\widetilde{H}_N^+)$ (this follows from Exercises 6.7 and 6.3). This shows that the 'liberation procedure' with partitions is not unique: there can be several distinct categories of non-crossing partitions which become the same once crossings are added.

7.2 Quantum Automorphism Groups of Graphs

We have started our journey in the world of compact quantum groups by a problem from graph theory. Even though the connection explained in Chapter 1 is recent, the idea of using compact matrix quantum groups to study finite graphs had already been exploited before. In particular, it can shed a new light on quantum reflection groups, which strengthens the idea that they are the 'correct' generalisation of classical reflection groups. As usual, we will first review the basic features of classical automorphism groups of graphs, and this starts by making the notion of a graph precise.

Definition 7.11 A *finite oriented graph* X is the data of

- A finite set $V(X)$ called the *set of vertices*,
- A set of ordered pairs $E(X) \subset V(X) \times V(X)$ called the *set of edges*.

If we fix a numbering of the vertices so that $V(X) = \{1, \ldots, N\}$, the *adjacency matrix* of X is the matrix A_X with (i,j)-th coefficient 1 if $(i,j) \in E(X)$ and 0 otherwise.

Remark 7.12 In Chapter 1, we did not consider orientations on the edges of the graph so that our setting was more restricted. Indeed, unoriented graphs can be seen as a particular case of oriented graphs where $(i,j) \in E(X)$ if and only if $(j,i) \in E(X)$.

Let X be a finite oriented graph with N vertices, and let A_X be its adjacency matrix. An automorphism of X is a bijection of its vertex set which respects the edges, that is to say a bijection $f\colon V(X) \to V(X)$ such that $(f(x), f(x')) \in E(X)$ as soon as $(x, x') \in E(X)$. Seeing permutations of the vertices as matrices in $M_N(\mathbf{C})$, this means that we are considering those permutation matrices which commute with A_X. The key point is that this still makes sense if we start with quantum permutation matrices.

Proposition 7.13 *Let X be a finite oriented graph on N vertices with adjacency matrix A_X. Then, the quotient $\mathcal{O}(S_N^+)$ by the ideal $\mathcal{I}$ generated by the coefficients of*

$$A_X P - P A_X,$$

together with the image of P in this quotient, is an orthogonal compact matrix quantum group.

Proof We have to check that the ideal in the statement is stable under the involution and that the coproduct of S_N^+ factors through the quotient. But this was already done once and for all in the proof of Theorem 3.7! Indeed, consider the smallest[1] concrete rigid orthogonal C*-tensor category $\mathfrak{C}_X$ containing all the maps T_p for $p \in NC$ and the map $A_X \in \mathcal{L}(\mathbf{C}^N)$. Then, the quotient that

---

[1] That is to say the one with morphism spaces being the intersection of the morphism spaces of all the concrete rigid orthogonal C*-tensor categories satisfying the condition.

we are considering is the $*$-algebra of the compact quantum group associated to $\mathfrak{C}_X$ by Theorem 3.7.

Let us nevertheless be more down to earth to convince the reader. Using the notations of Theorem 3.7, the ideal $\mathcal{I}$ is generated by the elements

$$P_{A_X,i,j} = (A_X P - P A_X)_{ij} = \sum_{i \to k} P_{kj} - \sum_{k \to j} P_{ik},$$

where we write $i \to j$ if $(i,j) \in E(X)$. It was shown in Point (3) of the proof of Theorem 3.7 that

$$\Delta\left(P_{A_X,i,j}\right) = \sum_{k=1}^{N} P_{A_X,i,k} \otimes P_{kj} + P_{ik} \otimes P_{A_X,k,j}.$$

This readily implies

$$\Delta(\mathcal{I}) \subset \mathcal{I} \otimes \mathcal{O}(S_N^+) + \mathcal{O}(S_N^+) \otimes \mathcal{I},$$

hence the coproduct of S_N^+ factors through the quotient. Because $P^* = P$, it also commutes with A_X^*. Since, moreover, $P^*_{A_X,i,j} = P_{A_X^*,j,i}$, we have $\mathcal{I}^* = \mathcal{I}$ and the proof is complete. □

Definition 7.14 The orthogonal compact matrix quantum group defined by Proposition 7.13 is called the *quantum automorphism group of X* and is denoted by $\mathrm{Aut}^+(X)$.

Let us have a quick look at extremely basic examples.

Example 7.15 Let X be the graph on N vertices with no edge. Then, A_X is the identity so that $\mathcal{O}(\mathrm{Aut}^+(X))$ is the quotient of $\mathcal{O}(S_N^+)$ by the ideal $\{0\}$. In other words $\mathrm{Aut}^+(X) = S_N^+$.

Example 7.16 Let X be the *complete graph* on N vertices, meaning that there is an edge between any pair of distinct vertices. Its adjacency matrix is the matrix with all coefficients equal to 1. Therefore, $P A_X$ is the matrix where each row is constant equal to the sum of the corresponding row of P, while $A_X P$ is the matrix where each line is constant equal to the sum of the corresponding column of P. By definition of $\mathcal{O}(S_N^+)$, these two matrices coincide, so that once again $\mathrm{Aut}^+(X) = S_N^+$.

Our goal will be to prove that H_N^{s+} is, in fact, the quantum automorphism group of a graph. This may at first seem surprising, since it is defined as a genuinely *unitary* compact matrix quantum group. But the trick is that it will not be the quantum automorphism group of a graph on N vertices, but on sN

vertices. The precise graph can be found by looking at the classical reflection groups H_N^s. These are known to be the automorphism groups of a disjoint union[2] of N copies of an s-cycle oriented cyclically (see Corollary 7.24 for an a posteriori proof). More precisely, let C_s be the s-cycle graph oriented cyclically, which consists in s points with point i connected to point $i + 1$ modulo s. Then,

Theorem 7.17 (Banica–Vergnioux) *Let $s \geqslant 3$ and let $C_{s,N}$ be the disjoint union of N copies of the cyclically oriented s-cycle C_s. Then,*

$$H_N^{s+} \simeq \mathrm{Aut}^+(C_{s,N}).$$

Moreover, let $C'_{2,N}$ be the disjoint union of N copies of the unoriented *graph with two vertices connected by an edge. Then,*

$$H_N^+ \simeq \mathrm{Aut}^+(C'_{2,N}).$$

Remark 7.18 Even classically, it is known that the automorphism group of $C_{2,N}$ is not H_N^s but S_N. This comes from the fact that the 2-cycle graph has no non-trivial oriented automorphism, while it has a non-trivial one which reverses the orientation. One can also consider that cyclically orienting C_2 means giving the unique edge the two orientations, which is equivalent to considering it as an unoriented graph.

The reader may wonder about what happens for unoriented cycles. There are subtleties in that case which are slightly off-topic, but we will treat this as an exercise to give an idea of the tools used in computing a quantum automorphism group.

Exercise 7.4 Let D_N be the N-cycle with no orientation.

1. For $N \leqslant 3$, prove that $\mathrm{Aut}^+(D_N)$ is classical.
2. For $N = 4$, prove that $\mathrm{Aut}^+(D_4)$ is not classical.
3. For $N > 4$, we will prove that once again $\mathrm{Aut}^+(D_N)$ is classical.
   (a) Write down the adjacency matrix of D_N corresponding to a cyclic labelling of the vertices.
   (b) Let $\omega \in \mathbf{C}$ be an Nth root of unity and set $\xi^k = (1, \omega^k, \cdots, \omega^{k(N-1)}) \in \mathbf{C}^N$. Check that ξ^k is an eigenvector for A_{D_N} for all $1 \leqslant k \leqslant N$.

---

[2] The disjoint union of a family of graphs is the graph whose vertex set is the disjoint union of the vertex sets and similarly for edge sets.

(c) Prove that the eigenspaces of A_{D_N} are stable under ρ_u and that there exists $x, y \in \mathcal{O}(\mathrm{Aut}^+(D_N))$ such that

$$\rho_u(\xi^1) = x \otimes \xi^1 + y \otimes \xi^{-1},$$

where u denotes the fundamental representation of $\mathrm{Aut}^+(D_N)$.

(d) Prove that $\rho_u : \mathbf{C}^N \to \mathcal{O}(\mathrm{Aut}^+(D_N)) \otimes \mathbf{C}^N$ is an algebra homomorphism and that for all $k \neq N/2$,

$$\rho_u(\xi^k) = x^k \otimes \xi^k + y^k \otimes \xi^{-k},$$

Deduce that $xy = -yx$ and $xy^2 = 0 = yx^2$.

(e) Prove that ρ_u is a $*$-homomorphism and conclude that $xy = 0 = yx$.

(f) Prove that $\mathcal{O}(\mathrm{Aut}^+(D_N))$ is commutative.

Solution 1. For $N \leqslant 3$, the quantum permutation group is already classical, hence the result.

2. For $N = 4$, the adjacency matrix corresponding to a cyclic labelling of the vertices is

$$A_{D_4} = \begin{pmatrix} 0 & 1 & 0 & 1 \\ 1 & 0 & 1 & 0 \\ 0 & 1 & 0 & 1 \\ 1 & 0 & 1 & 0 \end{pmatrix}.$$

Let p and q be orthogonal projections in $\mathcal{L}(\mathbf{C}^2)$ which do not commute. Then,

$$\begin{pmatrix} p & 1-p & 0 & 0 \\ 1-p & p & 0 & 0 \\ 0 & 0 & q & 1-q \\ 0 & 0 & 1-q & q \end{pmatrix}$$

is a quantum permutation matrix which commutes with A_{D_4}. By universality, this implies the existence of a $*$-homomorphism $\mathcal{O}(\mathrm{Aut}^+(D_4)) \to \mathcal{L}(\mathbf{C}^2)$ sending u_{11} to p and u_{33} to q. Thus, these two coefficients do not commute in $\mathcal{O}(\mathrm{Aut}^+(D_4))$, concluding the proof.

3. (a) If the points of the unoriented cycle D_N are labelled by integers in cyclic order, then the corresponding adjacency matrix is the circulant matrix

$$A_{C_N} = \begin{pmatrix} 0 & 1 & \cdots & \cdots & 1 \\ 1 & 0 & 1 & \cdots & 0 \\ 0 & 1 & \ddots & \ddots & \vdots \\ 1 & 0 & 1 & \cdots & 0 \end{pmatrix}.$$

(b) If ξ_i is the i-th component of ξ^1, then

$$
\begin{aligned}
(A_{D_N}\xi)_i &= \omega^{i+1} + \omega^{N+i-1} \\
&= (\omega + \omega^{-1})\omega^i \\
&= (\omega + \omega^{-1})\xi_i
\end{aligned}
$$

hence ξ^1 is proper with eigenvalue $\omega + \omega^{-1}$. It follows that ξ^k is also proper with eigenvalue $\omega^k + \omega^{-k}$.

(c) Let η be an eigenvector for an eigenvalue λ. Then,

$$
((A_{D_N} - \lambda\,\mathrm{Id}) \otimes \mathrm{id}) \circ \rho_u(\eta) = \rho_u \circ (A_{D_N} - \lambda\,\mathrm{Id})(\eta) = 0,
$$

so that (by Lemma 1.21)

$$
\rho_u(\eta) \in \ker((A_{D_N} - \lambda\,\mathrm{Id}) \otimes \mathrm{id}) = \ker(A_{D_N} - \lambda\,\mathrm{Id}) \otimes \mathcal{O}(\mathrm{Aut}^+(D_N)).
$$

Now because the vectors $(\xi^k)_{1 \leqslant k \leqslant N}$ are linearly independent, they form a basis of eigenvectors. Therefore, the eigenvalues of A_{D_N} are the numbers $\omega^k + \omega^{-k}$ for $1 \leqslant i \leqslant N$. This set has repetitions and we have to distinguish according to the parity of N. If N is even, then the eigenspaces are $\mathbf{C}\xi^0$, $\mathbf{C}\xi^i + \mathbf{C}\xi^{-i}$ for $1 \leqslant i < N/2$ and $\mathbf{C}\xi^{N/2}$, while if N is odd, the eigenspaces are $\mathbf{C}\xi^0$ and $\mathbf{C}\xi^i + \mathbf{C}\xi^{-i}$ for $1 \leqslant i < N/2$. The last fact now follows since ξ^1 and ξ^{-1} span an eigenspace.

(d) If $(e_i)_{1 \leqslant i \leqslant N}$ denotes the canonical basis of $\mathbf{C}^N$, then

$$
\begin{aligned}
\rho_u(e_{i_1})\rho_u(e_{i_2}) &= \sum_{j_1,j_2=1}^{N} u_{i_1j_1}u_{i_2j_2} \otimes e_{j_1}e_{j_2} \\
&= \sum_{j_1,j_2=1}^{N} u_{i_1j_1}u_{i_2j_2} \otimes \delta_{j_1j_2}e_{j_1} \\
&= \sum_{j=1}^{N} u_{i_1j}u_{i_2j} \otimes e_j \\
&= \sum_{j=1}^{N} \delta_{i_1,i_2}u_{i_1j} \otimes e_j \\
&= \delta_{i_1i_2}\rho_u(e_{i_1}) \\
&= \rho_u(\delta_{i_1i_2}e_{i_1}) \\
&= \rho_u(e_{i_1}e_{i_2}),
\end{aligned}
$$

hence ρ_u is an algebra homomorphism. Let us prove the formula in the question by induction on k. For $k = 1$ this was done previously. If the formula holds for some k, then

$$
\begin{aligned}
\rho_u(\xi^{k+1}) &= \rho_u(\xi)\rho_u(\xi^k) \\
&= \left(x \otimes \xi^1 + y \otimes \xi^{-1}\right)\left(x^k \otimes \xi^k + y^k \otimes \xi^{-k}\right) \\
&= x^{k+1} \otimes \xi^{k+1} + y^{k+1} \otimes \xi^{-(k+1)} + xy^k \otimes \xi^{-k+1} + yx^k \otimes \xi^{k-1}.
\end{aligned}
$$

Because ξ^{k+1} and ξ^{k-1} span an eigenspace, the last two terms must vanish (note that $-k+1$ and $k-1$ cannot be equal to $k+1$ or $-k-1$), and this leaves us with the desired formula if $k + 1 \neq N/2$ (otherwise, $\xi^{k+1} = \xi^{-k+1}$). Moreover, for $k = 1$ the vanishing term is

$$
(xy + yx) \otimes \xi^0,
$$

yielding the first equality. For $k = 2$ (note that the preceding formula holds because $2 < N/2$ if $N \geqslant 5$), the vanishing term is

$$
xy^2 \otimes \xi^{-1} + yx^2 \otimes \xi^1,
$$

yielding the second equality.

(e) The elements $e_i \in \mathbf{C}^N$ are self-adjoint, it therefore suffices to check that their images also are. And indeed,

$$
\begin{aligned}
\rho_u(e_i)^* &= \left(\sum_{j=1}^{N} u_{ij} \otimes e_j\right) \\
&= \sum_{j=1}^{N} u_{ij}^* \otimes e_j \\
&= \sum_{j=1}^{N} u_{ij} \otimes e_j \\
&= \rho_u(e_i).
\end{aligned}
$$

It follows from this that

$$
\begin{aligned}
\rho_u(\xi^{-1}) &= \rho_u(\xi^{1*}) \\
&= \rho_u(\xi^1)^* \\
&= x^* \otimes \xi^{-1} + y^* \otimes \xi^1
\end{aligned}
$$

so that $x^{N-1} = y^*$ and $y^{N-1} = x^*$. We can now compute

$$(xy)(xy)^* = xyy^*x^*$$
$$= xyx^{N-1}y^{N-1}$$
$$= xyx^{N-2}(xy^2)y^{N-3}$$
$$= 0.$$

But then, $h((xy)^*(xy)) = 0$, hence $xy = 0$ by faithfulness of the Haar state h. We conclude by $yx = -xy = 0$.

(f) We have proven that x and y commute, and to conclude we must show that they generate $\mathcal{O}(\mathrm{Aut}^+(D_N))$. To do this observe that because

$$u_{ij} = (e_j^* \otimes \mathrm{id}) \circ \rho_u(e_i),$$

the linear span of these elements is the same as the linear span of the elements $(\xi^{j*} \otimes \mathrm{id}) \circ \rho_u(\xi^i)$. The latter are powers of x and y, hence the coefficients of u are polynomials in x and y. Because the coefficients of u generate $\mathcal{O}(\mathrm{Aut}^+(D_N))$, we are done. $\qquad\square$

Remark 7.19 When $N \neq 4$, the quantum automorphism group coincides by the previous result with the classical one, which is the dihedral group D_N. The computation of the quantum automorphism group of a square is not easy to do directly, but one may observe that the complement of a square is the union of two segments, hence these two graphs have the same quantum automorphism group (see the proof of Theorem 8.15 for more details on this). Therefore, it will follow from Theorem 7.17 that the quantum automorphism group of a square is H_2^+. As for disjoint unions of unoriented cyclic graphs, there is a general formula to compute them, given in Theorem 7.37.

The proof of Theorem 7.17, which is due to T. Banica and R. Vergnioux in [19, theorem 3.2] is not straightforward. It will require an alternate description of $\mathcal{O}(H_N^{s+})$.

7.2.1 Sudoku Matrices

We need to understand what the quantum automorphism group of $C_{s,N}$ can be. Let us number the vertices of $C_{s,N}$ as follows:

- For $1 \leqslant i \leqslant N - 1$ and $0 \leqslant \ell < s - 1$, vertex number $i + \ell N$ is connected to vertex number $i + (\ell + 1)N$,
- Vertex $i + (s - 1)N$ is connected to vertex number i.

The following representation of $C_{3,4}$ may help make this clearer:

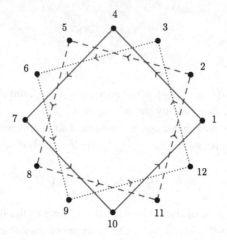

With these conventions, the adjacency matrix of $C_{s,N}$ is

$$A_{C_{s,N}} = \begin{pmatrix} 0 & I_N & 0 & \cdots & 0 \\ 0 & 0 & I_N & \cdots & 0 \\ \cdots & \cdots & \cdots & \cdots & \cdots \\ 0 & 0 & 0 & \cdots & I_N \\ I_N & 0 & 0 & \cdots & 0 \end{pmatrix} \in M_{sN}(\mathbf{C}).$$

The key point in what follows will be the following characterisation of commutation with $A_{C_{s,N}}$.

Exercise 7.5 Prove that $M \in M_{sN}(\mathbf{C})$ commutes with $A_{C_{s,N}}$ if and only if it is *block-circulant* in the sense that if we decompose the first N rows as s square matrices, then each row of M is obtained by applying the s-cycle $(1 \ldots s)$ to the previous row.

Solution Let us decompose M into s^2 blocks of $N \times N$ matrices and write $M = (B_{ij})_{1 \leqslant i,j \leqslant s}$. Then, the (i,j)-th block of $A_{C_{s,N}} M$ is

$$\sum_{k=1}^{s} (A_{C_{s,N}})_{ik} B_{kj} = \sum_{k=1}^{s} \delta_{k=i+1} B_{kj} = B_{(i-1)j},$$

where the indices are counted modulo s. Similarly, the (i,j)-th coefficient of $M A_{C_{s,N}}$ is

$$\sum_{k=1}^{s} B_{ik} (A_{C_{s,N}})_{kj} = \sum_{k=1}^{s} \delta_{k=j-1} B_{ik} = B_{i(j+1)},$$

counting once again the indices modulo s. As a consequence, M commutes with $A_{C_{s,N}}$ if and only if $B_{(i-1)j} = B_{i(j+1)}$ for all $1 \leqslant i, j \leqslant s$, which exactly means that the ith row is obtained by applying the s-cycle $(1 \ldots s)$ to the $(i-1)$-th row. In other words, this is equivalent to M being block-circulant. $\quad\square$

That observation naturally leads to the next definition.

Definition 7.20 A quantum permutation matrix $M \in M_{sN}(\mathcal{B}(H))$ of the form

$$
M = \begin{pmatrix}
M^{(0)} & M^{(1)} & \cdots & M^{(s-1)} \\
M^{(s-1)} & M^{(0)} & \cdots & M^{(s-2)} \\
\vdots & \vdots & \vdots & \vdots \\
M^{(1)} & M^{(2)} & \cdots & M^{(0)}
\end{pmatrix},
$$

for some matrices $M^{(0)}, \cdots, M^{(s-1)} \in M_N(\mathcal{B}(H))$, is called an (s, N)-sudoku matrix.

We will prove that $\mathcal{O}(H_N^{s+})$ can be characterised in terms of sudoku matrices, but this first requires establishing some properties of the coefficients of the fundamental representation of H_N^{s+}.

Lemma 7.21 *Let u be the fundamental representation of H_N^{s+}. Then, for any $1 \leqslant i, j, \leqslant N$,*

1. $u_{ij}^{s+\ell} = u_{ij}^{\ell}$,
2. $u_{ij}^{\ell*} = u_{ij}^{s-\ell}$.

Proof 1. Using the relations coming from $q_\circ$ as shown in the proof of Theorem 7.2 yields

$$
u_{ij}^{\ell} = \left(\sum_{k=1}^{N} u_{ik}^{s} \right) u_{ij}^{\ell} = u_{ij}^{s} u_{ij}^{\ell} = u_{ij}^{s+\ell}.
$$

2. First, notice that

$$
u_{ij} = u_{ij} \left(\sum_{k=1}^{N} u_{ik} u_{ik}^{*} \right) = u_{ij}^{2} u_{ij}^{*}
$$

which yields for any $1 \leqslant \ell < s$ by a straightforward induction

$$
u_{ij}^{\ell} = u_{ij}^{s} u_{ij}^{*(s-\ell)}.
$$

Another consequence of the first equality is that for any $k \neq j$,

$$
u_{ij} u_{ij'}^{*} = u_{ij}^{2} u_{ij}^{*} u_{ik}^{*} = u_{ij}^{2} \left(u_{ik} u_{ij} \right)^{*} = 0.
$$

so that

$$u_{ij}^* = \left(\sum_{k=1}^{N} u_{ik}^s \right) u_{ij}^* = u_{ij}^s u_{ij}^*.$$

We can now conclude that

$$u_{ij}^{*\ell} = \left(u_{ij}^s u_{ij}^{*(s-\ell)} \right)^* = \left(u_{ij}^{*(s-\ell)} \right)^* = u_{ij}^{s-\ell}. \qquad \square$$

Remark 7.22 The relations of the statement can also been recovered as the relations implemented by certain specific partitions of $\mathcal{C}_s$. Indeed, the first one corresponds to the one-block partition of $\mathcal{C}_s(s + \ell, \ell)$ with all points coloured in white while the second one corresponds to the rotation of p_s which belongs to $\mathcal{C}_s(s - \ell, \ell)$.

We can now establish the link between quantum reflection groups and sudoku matrices.

Proposition 7.23 *Let $\mathcal{A}$ be the universal $*$-algebra generated by the entries of a (s, N)-sudoku matrix and let ω be a primitive sth root of unity. Then, there exists a pair of mutually inverse $*$-homomorphisms $\Phi\colon \mathcal{O}(H_N^{s+}) \to \mathcal{A}$ and $\Psi\colon \mathcal{A} \to \mathcal{O}(H_N^{s+})$ such that*

$$\Phi(u_{ij}) = \sum_{p=0}^{s-1} \omega^{-p} M_{ij}^{(p)}$$

and

$$\Psi\left(M_{ij}^{(p)} \right) = \frac{1}{s} \sum_{r=0}^{s-1} \omega^{rp} u_{ij}^r$$

with the convention that $u_{ij}^0 = u_{ij}^s$.

Proof To show the existence of the $*$-homomorphism, we will use the universal properties of the $*$-algebras involved.

1. Let us start by checking that the elements $\Phi(u_{ij})$ satisfy the defining relations of $\mathcal{O}(H_N^{s+})$, that is, the relations corresponding to the partitions $q_\circ$ and p_s together with the relations making u and its conjugate unitary. Indeed,

$$\Phi(u_{ij})\Phi(u_{ij'})^* = \sum_{p,p'=0}^{s-1} \omega^{-p+p'} M_{ij}^{(p)} M_{ij'}^{(p')*}$$

$$= \sum_{p,p'=0}^{s-1} \omega^{-p+p'} \delta_{p,p'} \delta_{j,j'} M_{ij}^{(p)}$$

$$= \delta_{j,j'} \sum_{p}^{s-1} M_{ij}^{(p)}$$

so that the product vanishes if $j \neq j'$, giving the relations corresponding to $\widetilde{q}$, hence the relations corresponding to q_o are also satisfied. Note that we used here the fact that M is a quantum permutation matrix. Moreover, this computation also shows that

$$\sum_{k=1}^{N} \Phi(u_{ik}) \Phi(u_{jk}^*) = \sum_{k=1}^{N} \sum_{p,p'=0}^{s-1} \omega^{-p+p'} M_{ik}^{(p)} M_{jk}^{(p')*}$$

$$= \delta_{ij} \sum_{p=0}^{s-1} \sum_{k=1}^{N} M_{ik}^{(p)}$$

$$= \delta_{ij}$$

so that $\Phi(u)\Phi(u)^* = \text{Id}$. Similar computations work for $\Phi(u)^*\Phi(u)$ and $\overline{\Phi(u)}$, showing that $\Phi(u)$ and its conjugate are unitary. Eventually, because $M_{ij}^{(p)}$ and $M_{ij}^{(p')}$ are orthogonal to one another if $p \neq p'$,

$$\sum_{j=1}^{N} \Phi(u_{ij})^s = \sum_{j=1}^{N} \left(\sum_{p=0}^{s-1} \omega^p M_{ij}^{(p)} \right)^s$$

$$= \sum_{j=1}^{N} \sum_{p=0}^{s-1} \omega^{ps} M_{ij}^{(p)}$$

$$= \sum_{j=1}^{N} \sum_{p=0}^{s-1} M_{ij}^{(p)}$$

$$= 1$$

so that the relations corresponding to p_s are also satisfied. This shows the existence of the $*$-homomorphism Φ.

2. We now turn to Ψ. Let us first prove that the image of the coefficients of M are indeed orthogonal projections: using Lemma 7.21,

$$\Psi\left(M_{ij}^{(p)} \right)^2 = \frac{1}{s^2} \sum_{r,r'=0}^{s-1} \omega^{(r+r')p} u_{ij}^{r+r'}$$

$$= \frac{1}{s} \sum_{\ell=0}^{s-1} \omega^{\ell p} u_{ij}^{\ell}$$

$$= \Psi\left(M_{ij}^{(p)} \right)$$

and

$$\Psi\left(M_{ij}^{(p)}\right)^{*} = \frac{1}{s}\sum_{r=0}^{s-1}\omega^{-rp}u_{ij}^{r*}$$

$$= \frac{1}{s}\sum_{r=0}^{s-1}\omega^{(s-r)p}u_{ij}^{s-r}$$

$$= \Psi\left(M_{ij}^{(p)}\right).$$

The last condition is the sum on rows and columns. We have

$$\sum_{p=0}^{s-1}\sum_{j=1}^{N}\Psi\left(M_{ij}^{(p)}\right) = \frac{1}{s}\sum_{j=1}^{N}\sum_{p,r=0}^{s-1}\omega^{rp}u_{ij}^{r}$$

$$= \frac{1}{s}\sum_{j=1}^{N}\sum_{r=0}^{s-1}u_{ij}^{r}\sum_{p=0}^{s-1}\omega^{rp}$$

$$= \frac{1}{s}\sum_{j=1}^{N}\sum_{r=0}^{s-1}u_{ij}^{0}$$

$$= \sum_{j=1}^{N}u_{ij}^{s}$$

$$= 1,$$

and the same holds for the sum over rows by taking the transposed matrices.

3. To conclude, the only thing left is the check that Φ and Ψ are mutually inverse. Let us do it.

$$\Psi\circ\Phi(u_{ij}) = \sum_{p=0}^{s-1}\omega^{-p}\frac{1}{s}\sum_{r=0}^{s-1}\omega^{rp}u_{ij}^{r}$$

$$= \frac{1}{s}\sum_{r=0}^{s-1}u_{ij}^{r}\sum_{p=0}^{s-1}\omega^{(r-1)p}$$

$$= \frac{1}{s}\sum_{r=0}^{s-1}u_{ij}^{r}\delta_{r,1}s$$

$$= u_{ij}$$

and

$$\Phi \circ \Psi \left(M_{ij}^{(p)} \right) = \frac{1}{s} \sum_{r=0}^{s-1} \omega^{rp} \left(\sum_{\ell=0}^{s-1} \omega^{-\ell} M_{ij}^{\ell} \right)^r$$

$$= \frac{1}{s} \sum_{r=0}^{s-1} \sum_{\ell=0}^{s-1} \omega^{(p-\ell)r} M_{ij}^{\ell}$$

$$= \frac{1}{s} \sum_{\ell=0}^{s-1} M_{ij}^{\ell} \sum_{r=0}^{s-1} \sum_{\ell=0}^{s-1} \omega^{(p-\ell)r}$$

$$= M_{ij}^{(p)}. \qquad \square$$

We are now ready to prove that quantum reflection groups are quantum automorphism groups of graphs.

Proof of Theorem 7.17 Let us recall our numbering of the vertices of $C_{s,N}$:

- For $1 \leqslant i \leqslant N - 1$ and $0 \leqslant \ell < s - 1$, vertex number $i + \ell N$ is connected to vertex number $i + (\ell + 1)N$,
- Vertex $i + (s - 1)N$ is connected to vertex number i.

With these conventions, the adjacency matrix is

$$A_{C_{s,N}} = \begin{pmatrix} 0 & I_N & 0 & \cdots & 0 \\ 0 & 0 & I_N & \cdots & 0 \\ \cdots & \cdots & \cdots & \cdots & \cdots \\ 0 & 0 & 0 & \cdots & I_N \\ I_N & 0 & 0 & \cdots & 0 \end{pmatrix} \in M_{sN}(\mathbf{C}).$$

Note that for $s = 2$, this is not the adjacency matrix of $C_{2,N}$ but of $C'_{2,N}$, hence the following proof works for all cases. We will proceed in two steps.

1. It is a direct consequence of Exercise 7.5 that a quantum permutation matrix $P \in M_{Ns}(\mathcal{B}(H))$ commutes with $A_{C_{s,N}}$ if and only if it is an (s, N)-sudoku matrix. Thus, $\mathcal{O}(\mathrm{Aut}^{+}(C_{s,N}))$, which is the quotient of $\mathcal{O}(S_N^+)$ by the relations

$$P A_{C_{s,N}} = A_{C_{s,N}} P,$$

is also the universal $*$-algebra generated by the entries of an (s, N)-sudoku matrix.
2. All that is left to prove is that the isomorphism Φ of Proposition 7.23 behaves well with respect to the coproducts.

This first requires identifying the coproduct on $\mathcal{O}(\mathrm{Aut}^+(C_{s,N}))$ in the sudoku picture. Notice that if $i = i^* + pN$ and $j = j^* + p'N$ with $i^*, j^* \in \{1, \ldots, N\}$, then

$$M_{i,j} = M_{i^*,j^*}^{(p-p')}.$$

Using this equality, we see that for any $1 \leqslant i, j \leqslant N$ and $0 \leqslant p \leqslant s - 1$,

$$
\begin{aligned}
\Delta\left(M_{i,j}^{(p)}\right) &= \Delta(M_{i+pN,j}) \\
&= \sum_{k=1}^{sN} M_{i+pN,k} \otimes M_{k,j}, \\
&= \sum_{p'=0}^{s-1} \sum_{k=1}^{N} M_{i+pN,k+p'N} \otimes M_{k+p'N,j} \\
&= \sum_{p'=0}^{s-1} \sum_{k=1}^{N} M_{i^*,k}^{(p-p')} \otimes M_{k,j^*}^{(p')}.
\end{aligned}
$$

Note that all the exponents are taken modulo s. We are now left with an elementary computation,

$$
\begin{aligned}
\Delta \circ \Phi(u_{ij}) &= \sum_{p=0}^{s-1} \omega^{-p} \Delta\left(M_{ij}^{(p)}\right) \\
&= \sum_{p=0}^{s-1} \omega^{-p} \sum_{p'=0}^{s-1} \sum_{k=1}^{N} M_{i^*k}^{(p-p')} \otimes M_{kj^*}^{(p')},
\end{aligned}
$$

while

$$
\begin{aligned}
(\Phi \otimes \Phi) \circ \Delta(u_{ij}) &= \sum_{k=1}^{N} \Phi(u_{ik}) \otimes \Phi(u_{kj}) \\
&= \sum_{k=1}^{N} \sum_{p,p'=0}^{s-1} \omega^{-p-p'} M_{ik}^{(p)} \otimes M_{kj}^{(p')} \\
&= \sum_{k=1}^{N} \sum_{p',p''=0}^{s-1} \omega^{-p''} M_{i^*k}^{(p''-p')} \otimes M_{kj^*}^{(p')} \\
&= \Delta \circ \Phi(u_{ij}). \qquad \square
\end{aligned}
$$

Let us conclude by the a posteriori proof that H_N^s was indeed the automorphism group of $C_{s,N}$.

Corollary 7.24 *There is an isomorphism*

$$H_N^s \simeq \mathrm{Aut}(C_{s,N}).$$

Proof Simply consider the isomorphism induced by Φ between the maximal abelian quotients of the $*$-algebras of $\mathcal{O}(H_N^{s+})$ and $\mathcal{O}(\mathrm{Aut}^+(C_{s,N}))$. ☐

Before turning to the next section, we would like to come back to the question of deciding whether $\mathcal{O}(\mathrm{Aut}^+(X))$ is commutative or not by giving a nice criterion due to S. Schmidt in [65] as an exercise.

Exercise 7.6 Let X be a finite graph on N vertices and assume that there exists $\sigma_1, \sigma_2 \in \mathrm{Aut}(X)$ which are both non-trivial and have *disjoint supports* in the following sense: for any $x \in V(X)$, $\sigma_1(x) \neq x$ implies $\sigma_2(x) = x$ and $\sigma_2(x) \neq x$ implies $\sigma_1(x) = x$. Our goal is to prove that $\mathcal{O}(\mathrm{Aut}^+(X))$ is non-commutative.

We denote by n_i the order of σ_i. Let H be a Hilbert space and let $p_1, \cdots, p_{n_1}, q_1, \cdots q_{n_2}$ be orthogonal projections in H such that

$$\sum_{i=1}^{n_1} p_i = 1 = \sum_{j=1}^{n_2} q_j.$$

We denote by $\mathcal{A}$ the $*$-subalgebra of $\mathcal{B}(H)$ generated by these projections.

1. For $1 \leqslant i, j \leqslant N$, we set

$$v_{ij} = \sum_{k=1}^{n_1} \delta_{\sigma_1^k(i)j} p_k + \sum_{\ell=1}^{n_2} \delta_{\sigma_2^\ell(i)j} q_\ell - \delta_{ij} \in \mathcal{A}.$$

Prove that the matrix $v = (v_{ij})_{1 \leqslant i,j \leqslant N}$ is a quantum permutation matrix.
2. Prove that v commutes with A_X.
3. Conclude.

Solution 1. Even though it is clear on the definition that $v_{ij}^* = v_{ij}$, the fact that it is idempotent needs a proof. Observe that if $\sigma_1(i) \neq i$, then $\sigma_2(i) = i$ by assumption so that $\delta_{\sigma_2^\ell(i)j} = \delta_{ij}$ in that case. As a consequence, the sum over ℓ simplifies with the term $-\delta_{ij}$. Applying the same reasoning in the case $\sigma_2(i) \neq i$ leads to the following trichotomy:

$$v_{ij} = \begin{cases} \displaystyle\sum_{k=1}^{n_1} \delta_{\sigma_1^k(i)j} p_k & \text{if} & \sigma_1(i) \neq i \\ \displaystyle\sum_{\ell=1}^{n_2} \delta_{\sigma_2^\ell(i)j} q_\ell & \text{if} & \sigma_2(i) \neq i \\ \delta_{ij} & \text{otherwise} \end{cases}$$

and the conclusion follows.

We now have to prove that the sum on each row and column is the identity:

$$\sum_{i=1}^{N} v_{ij} = \sum_{i=1}^{N}\sum_{k=1}^{n_1} \delta_{\sigma_1^k(i)j} p_k + \sum_{i=1}^{N}\sum_{\ell=1}^{n_2} \delta_{\sigma_2^\ell(i)j} q_\ell - \sum_{i=1}^{N} \delta_{ij}$$

$$= \sum_{k=1}^{n_1} \left(\sum_{i=1}^{N} \delta_{\sigma_1^k(i)j} \right) p_k + \sum_{\ell=1}^{n_2} \left(\sum_{i=1}^{N} \delta_{\sigma_2^\ell(i)j} \right) q_\ell - 1$$

$$= \sum_{k=1}^{n_1} p_k + \sum_{\ell=1}^{n_2} q_\ell - 1$$

$$= 1 + 1 - 1$$

$$= 1.$$

For each ℓ and k, there exists exactly one i such that $\delta_{\sigma_1^k(i)j} \neq 0 \neq \delta_{\sigma_2^\ell(i)j}$. Thus,

$$\sum_{i=1}^{N} v_{ij} = \sum_{k=1}^{n_1} p_k + \sum_{\ell=1}^{n_2} q_\ell - 1 = 1.$$

The same argument works for the columns, concluding the proof.

2. We can write v as a sum of matrices in $M_N(\mathcal{A}) \simeq M_N(\mathbf{C}) \otimes \mathcal{A}$ in the following way:

$$v = \sum_{k=1}^{n_1} P_{\sigma_1^k} \otimes p_k + \sum_{\ell=1}^{n_2} P_{\sigma_2^\ell} \otimes q_\ell - \mathrm{Id}_{M_N(\mathbf{C})} \otimes 1_\mathcal{A}.$$

By definition of a graph automorphism, all the matrices $P_{\sigma_1^k}$ and $P_{\sigma_2^\ell}$ commute with the adjacency matrix A_X of X. Thus, v commutes with A_X.

3. By what precedes, there exists a surjective $*$-homomorphism from $\mathcal{O}(\mathrm{Aut}^+(X))$ to the $*$-subalgebra of $\mathcal{A}$ generated by the coefficients of v. To conclude, we therefore need to prove that this is non-commutative. Because one can obviously choose the projections so that $\mathcal{A}$ is non-commutative, we will simply prove that the coefficients of v generate $\mathcal{A}$. For that, let us fix $1 \leqslant k_0 \leqslant n_1$ and consider the set $E = \{i \mid \sigma_1(i) \neq i\}$. Because σ_1 is non-trivial, this is not the empty set. Moreover, we have

$$\prod_{i \in E} v_{i\sigma_1^{k_0}(i)} = \prod_{i \in I} \left(\sum_{k=1}^{n_1} \delta_{\sigma_1^k(i),\sigma_1^{k_0}(i)} p_k \right) = \sum_{k \in E_{k_0}} p_k$$

where

$$E_{k_0} = \bigcap_{i \in E} \{k \mid \sigma_1^k(i) = \sigma_1^{k_0}(i)\}.$$

It is clear on the definition that $k_0 \in E_{k_0}$. Moreover, if $k_1 \in E_{k_0}$, then $\sigma_1^{k_1-k_0}(i) = i$ for all $i \in E$. Since $\sigma_1(i) = i$ for all $i \notin E$, it follows that

$\sigma_1^{k_1-k_0} = \mathrm{id}$, which is impossible since the order of σ_1 is n_1. Therefore, we have proven that

$$p_{k_0} = \prod_{i \in E} v_{i \sigma_1^{k_0}(i)}.$$

The same argument works to prove that all the projections q_ℓ are in the algebra generated by the coefficients of v and this concludes the proof. $\square$

Remark 7.25 If the graph X is the square, then the reflections with respect to the diagonals are disjoint automorphisms of order 2. Thus, the previous result recovers the fact proven in Exercise 7.4 that the quantum automorphism group of the square has a non-commutative $*$-algebra.

Remark 7.26 The only assumption used on the projections $(p_i)_{1 \leqslant k \leqslant n_1}$ and $(q_j)_{1 \leqslant \ell \leqslant n_2}$ is that the sums equal the identity. Therefore, the proof, in fact, shows that there is a surjective $*$-homomorphism from $\mathcal{O}(\mathrm{Aut}^+(X))$ to the universal $*$-algebra generated by such projections, and the latter is easily seen to be isomorphic to $\mathcal{O}(\mathbf{Z}_{n_1} * \mathbf{Z}_{n_2})$.

7.2.2 Free Wreath Products

To conclude this chapter, we will give another description of quantum reflection groups, based on the classical notion of *wreath product*. Let us recall what this is.

Definition 7.27 Let G be a group and let N be an integer. The *(permutational) wreath product* of G by S_N is the semi-direct product

$$G^N \rtimes S_N,$$

where the action is given by

$$\sigma.(g_1, \cdots, g_N) = (g_{\sigma^{-1}(1)}, \cdots, g_{\sigma^{-1}(N)}).$$

It is denoted by $G \wr S_N$.

The reason why we give this definition is that there is an isomorphism (see Corollary 7.35 for an a posteriori proof)

$$H_N^s \cong \mathbf{Z}_s \wr S_N.$$

Our purpose is to prove a similar isomorphism involving H_N^{s+} on the left-hand side, and a new construction on the right-hand side, called the *free wreath product*. To explain the basic idea underlying the definition of that generalisation of the wreath product, note that the set underlying the classical wreath product is just the direct product $\mathbb{G}^N \times S_N$. To make this 'free', it is tempting to consider the $*$-algebra

$$\mathcal{O}(\mathbb{G})^{*N} * \mathcal{O}(S_N^+).$$

Now, the whole point of the construction of a semi-direct product is that the group law is not the product law. Similarly, we must here devise a coproduct which is not the canonical one for the free product. The issue is that the notion of semi-direct product is rather tricky at the level of compact quantum groups and does not work very well in this context. However, in our setting we know that the whole structure of a unitary compact matrix quantum group is encoded in its fundamental representation, so that we can try first to find a nice fundamental representation of $G \wr S_N$, and then to transport it to the free product.

Let us assume therefore that G is a unitary compact matrix group and let $\rho \colon G \to M_n(\mathbf{C})$ be a fundamental representation. Our first goal is to use ρ to realise $G \wr S_N$ as a group of matrices. For each $1 \leqslant i \leqslant N$ we define a representation ρ_i of G on

$$V = (\mathbf{C}^n)^{\oplus N} = \mathbf{C}^n \oplus \cdots \oplus \mathbf{C}^n$$

through the formula

$$\rho_i(g) = \mathrm{Id}_{\mathbf{C}^n}^{\oplus i-1} \oplus \rho(g) \oplus \mathrm{Id}_{\mathbf{C}^n}^{N-i}.$$

We also need a representation of S_N on V. Let us denote the canonical basis of the ith copy of $\mathbf{C}^n$ by $(e_p^{(i)})_{1 \leqslant p \leqslant n}$. Then we set, for $\sigma \in S_N$,

$$\pi(\sigma) \left(e_p^{(i)} \right) = e_p^{(\sigma(i))}.$$

We claim that these representations are enough to recover the wreath product.

Exercise 7.7 Prove that the subgroup H of $\mathrm{GL}(V)$ generated by $\rho_i(G)$ and $\pi(S_N)$ is isomorphic to $G \wr S_N$.

Solution First note that ρ_i is injective for all $1 \leqslant i \leqslant N$ as well as π and that $\rho_i(G)$ and $\rho_j(G)$ commute for $i \neq j$, so that H is generated by N commuting copies of G and a copy of S_N. Moreover, for any $1 \leqslant i \leqslant N$ and $1 \leqslant p \leqslant n$,

$$\pi(\sigma)\rho_i(g)\pi(\sigma^{-1}) \left(e_p^{(\sigma(i))} \right) = \pi(\sigma)\rho_i(g) \left(e_p^{(i)} \right)$$

$$= \pi(\sigma) \left(\sum_{q=1}^{n} \rho(g)_{qp} e_q^{(i)} \right)$$

$$= \sum_{q=1}^{n} \rho(g)_{qp} e_q^{(\sigma(i))}$$

$$= \rho_{\sigma(i)}(g) \left(e_p^{\sigma(i)} \right)$$

while

$$\pi(\sigma)\rho_i(g)\pi(\sigma^{-1}) \left(e_p^{(j)} \right)) = e_p^{(j)}$$

$$= \rho_{\sigma(i)}(g)(e_p^{(j)}),$$

if $j \neq \sigma(i)$ so that in the end

$$\pi(\sigma)\rho_i(g)\pi(\sigma^{-1}) = \rho_{\sigma(i)}(g).$$

In other words, the action of σ on G^N is implemented by conjugation in H (the ith component of $\sigma.g$ is g_j for j such that $\sigma^{-1}(j) = i$, hence $g_{\sigma(i)}$). In conclusion, H is the semi-direct product by that action. $\qquad\square$

The previous proposition can be restated by saying that the map

$$\tilde{\rho}: (g_1, \ldots, g_N, \sigma) \mapsto \pi(\sigma) \left(\prod_{i=1}^{N} \rho_i(g_i) \right)$$

is a fundamental representation,[3] so that the unitary compact matrix group

$$(\mathcal{O}(G^N \times S_N), \tilde{\rho})$$

is exactly $G \wr S_N$. Moreover, the coefficients of this representation are given by

$$\tilde{\rho}_{ip,jq}(g_1, \ldots, g_N, \sigma) = \delta_{\sigma(i)j} \rho(g_i)_{pq}$$

$$= \rho(g_i)_{pq} c_{ij}(\sigma),$$

where c_{ij} denotes the coefficient functions on S_N. We now know how to define a unitary compact matrix quantum group structure on

$$\mathcal{H} = \mathcal{O}(\mathbb{G})^{*N} * \mathcal{O}(S_N^+),$$

for a unitary compact matrix quantum group $(\mathbb{G}, u)$. Indeed, we should simply consider the fundamental representation w with coefficients

$$w_{ip,jq} = \nu_i(u_{pq}) P_{ij},$$

where $\nu_i: \mathcal{O}(\mathbb{G}) \to \mathcal{H}$ is the canonical inclusion into the ith copy of $\mathcal{O}(\mathbb{G})$. Let us check that this defines a unitary matrix.

---

[3] This is because it is injective, so that the group corresponding through Tannaka–Krein duality to the concrete rigid unitary C*-tensor category given by all tensor products of $\tilde{\rho}$ and its conjugate contains $G \wr S_N$. Since it is also contained in $G \wr S_N$ by definition, we have equality, hence any irreducible representation is contained in a tensor product of copies of $\tilde{\rho}$ and its conjugate.

Exercise 7.8 Prove that the matrix $w \in M_{nN}(\mathcal{H})$ is unitary, as well as the matrix $\overline{w}$.

Solution We simply compute

$$
\sum_{k=1}^{N} \sum_{r=1}^{n} w_{ip,kr} w_{jq,kr}^{*} = \sum_{k=1}^{N} \sum_{r=1}^{n} \nu_i(u_{pr}) P_{ik} P_{jk} \nu_j(u_{qr}^{*})
$$

$$
= \sum_{r=1}^{n} \nu_i(u_{pr}) \left(\sum_{k=1}^{N} P_{ik} P_{jk} \right) \nu_i(u_{qr}^{*})
$$

$$
= \delta_{ij} \sum_{r=1}^{n} \nu_i(u_{pr} u_{qr}^{*})
$$

$$
= \delta_{ij} \delta_{pq}
$$

so that $ww^{*} = \mathrm{id}$. Similar computations work for $w^{*}w$, as well as for $\overline{w}^{*}\overline{w}$ and $\overline{w}\,\overline{w}^{*}$. □

The problem is to check that there is a coproduct on the $*$-algebra $\mathcal{H}$ which turns this into a representation, and this is where things break down. Indeed, assume that there exists a $*$-homomorphism $\delta \colon \mathcal{H} \to \mathcal{H} \otimes \mathcal{H}$ such that

$$
\delta\left(\nu_i(u_{pq}) P_{ij}\right) = \sum_{k=1}^{N} \sum_{z=1}^{n} \nu_i(u_{pr}) P_{ik} \otimes \nu_k(u_{rq}) P_{kj}.
$$

Then, summing over j in both sides of the equation yields

$$
\delta\left(\nu_i(u_{pq})\right) = \sum_{k=1}^{N} \sum_{r=1}^{n} \nu_i(u_{pr}) P_{ik} \otimes \nu_k(u_{rq}).
$$

Assume, for instance, that u is orthogonal, that is, that all its coefficients are self-adjoint. Then,

$$
\sum_{k=1}^{N} \sum_{r=1}^{n} P_{ik} \nu_i(u_{pr}) \otimes \nu_k(u_{rq}) = \delta\left(\nu_i(u_{pq})\right)^{*}
$$

$$
= \delta\left(\nu_i(u_{pq})^{*}\right)
$$

$$
= \sum_{k=1}^{N} \sum_{r=1}^{n} \nu_i(u_{pr}) P_{ik} \otimes \nu_k(u_{rq})
$$

so that

$$
\sum_{k=1}^{N} \sum_{r=1}^{n} \left(P_{ik} \nu_i(u_{pr}) - \nu_i(u_{pr}) P_{ik}\right) \otimes \nu_k(u_{rq}) = 0.
$$

Such a relation does not hold in the $\mathcal{H}$ in general so that if we want to have a chance of turning w into a representation, we should add some relations to $\mathcal{H}$ making the above equality true.

Definition 7.28 Let $\mathbb{G}$ be a unitary compact matrix quantum group and let N be an integer. The *free wreath product algebra* is the quotient $\mathcal{A}$ of $\mathcal{O}(\mathbb{G})^{*N} * \mathcal{O}(S_N^+)$ by the relations

$$\nu_i(x)P_{ij} = P_{ij}\nu_i(x)$$

for all $1 \leqslant i, j \leqslant N$ and all $x \in \mathcal{O}(\mathbb{G})$.

Remark 7.29 In a sense, we are restoring here a symmetry which was broken when defining $w_{ip,jq}$ to be $\nu_i(u_{pq})P_{ij}$ while we could equally have chosen $P_{ij}\nu_i(u_{pq})$.

The fundamental result of J. Bichon in [23] is that these relations are enough to define a unitary compact matrix quantum group structure.

Theorem 7.30 (Bichon) *There exists a unique $*$-homomorphism $\Delta\colon \mathcal{A} \to \mathcal{A} \otimes \mathcal{A}$ such that for all $1 \leqslant p, q \leqslant n$ and $1 \leqslant i, j \leqslant N$,*

$$\Delta(w_{ip,jq}) = \sum_{k=1}^{N}\sum_{r=1}^{n} \nu_i(u_{pr})P_{ik} \otimes \nu_k(u_{rq})P_{kj}.$$

Proof We will proceed in two steps, first proving the existence of a suitable map from $\mathcal{H}$ to $\mathcal{A} \otimes \mathcal{A}$ and then checking that it passes to the quotient.

1. We start by defining the coproduct on $\nu_i(\mathcal{O}(\mathbb{G}))$. Let us consider the $*$-homomorphism

$$\Phi_{ik} = (\nu_i \otimes \nu_k) \circ \Delta\colon \nu_i(\mathbb{G}) \to \mathcal{H} \otimes \mathcal{H} \to \mathcal{A} \otimes \mathcal{A}$$

whose range is contained in $\nu_i(\mathcal{O}(\mathbb{G})) \otimes \nu_k(\mathcal{O}(\mathbb{G})) \subset \mathcal{A} \otimes \mathcal{A}$. On the latter subalgebra, the map

$$\Psi_{ik}\colon (x \otimes y) \mapsto xP_{ik} \otimes y = P_{ik}x \otimes y$$

is a $*$-homomorphism (note that it is well defined because it can be defined as a linear map on $\mathcal{F}(\nu_i(\mathcal{O}(\mathbb{G})) \times \nu_k(\mathcal{O}(\mathbb{G})))$ and can easily be checked to vanish on $\mathcal{I}(\nu_i(\mathcal{O}(\mathbb{G})), \nu_k(\mathcal{O}(\mathbb{G})))$). Indeed, because of the commutation relations defining $\mathcal{A}$,

$$(xP_{ij} \otimes y)(x'P_{ik} \otimes y') = (xP_{ik}x'P_{ik}) \otimes yy' = xx'P_{ik} \otimes yy'.$$

Let us now consider the linear map

$$\Phi_i = \sum_{k=1}^{N} \Psi_{ik} \circ \Phi_{ik}.$$

It is clear that $\Phi_i(x^*) = \Phi_i(x)^*$, and observing that for $k \neq k'$,

$$(\Psi_{ik} \circ \Phi_{ik}(x))\,(\Psi_{ik'} \circ \Phi_{ik'}(y)) = 0,$$

we see that Φ_i is multiplicative, hence defines a $*$-homomorphism from $\nu_i(\mathcal{O}(\mathbb{G}))$ to $\mathcal{A} \otimes \mathcal{A}$.

2. We still have to define the coproduct on $\mathcal{O}(S_N^+)$, but it suffices to consider the usual coproduct of S_N^+, seen as a $*$-homomorphism

$$\Delta \colon \mathcal{O}(S_N^+) \to \mathcal{H} \otimes \mathcal{H} \to \mathcal{A} \otimes \mathcal{A}.$$

3. By the universal property of the free product, there exists a unique $*$-homomorphism

$$\Phi \colon \mathcal{H} \to \mathcal{A} \otimes \mathcal{A}$$

restricting to Φ_i on $\nu_i(\mathcal{O}(\mathbb{G}))$ and to the usual coproduct on $\mathcal{O}(S_N^+)$. Let us check that it yields the correct formula for $w_{ip,jq}$:

$$\Phi(w_{ip,jq}) = \left(\sum_{k=1}^{N} \sum_{r=1}^{n} \nu_i(u_{pr}) P_{ik} \otimes \nu_k(u_{rq}) \right) \left(\sum_{k'=1}^{N} P_{ik'} \otimes P_{k'j} \right)$$

$$= \sum_{k,k'=1}^{N} \sum_{r=1}^{n} \nu_i(u_{pr}) P_{ik} P_{ik'} \otimes \nu_k(u_{rq}) P_{k'j}$$

$$= \sum_{k=1}^{N} \sum_{r=1}^{n} \nu_i(u_{pr}) P_{ik} \otimes \nu_k(u_{rq}) P_{kj}$$

$$= \sum_{k=1}^{N} \sum_{r=1}^{n} w_{ip,kr} \otimes w_{kr,jq}.$$

4. To conclude, we must prove that Φ vanishes on the commutators defining $\mathcal{A}$. This is a simple computation:

$$\Phi\left(\nu_i(u_{pq}) P_{ij}\right) = \sum_{k=1}^{N} \sum_{r=1}^{n} \nu_i(u_{pr}) P_{ik} \otimes \nu_k(u_{rq}) P_{kj}$$

$$= \sum_{k=1}^{N} \sum_{r=1}^{n} P_{ik} \nu_i(u_{pr}) \otimes P_{kj} \nu_k(u_{rq})$$

$$= \sum_{k=1}^{N} \sum_{r=1}^{n} P_{ik}\nu_i(u_{pr})P_{ik} \otimes P_{kj}\nu_k(u_{rq})$$

$$= \Phi(P_{ij})\,\Phi(\nu_i(u_{pq}))$$

$$= \Phi(P_{ij}\nu_i(u_{pq})). \qquad \square$$

Corollary 7.31 *The pair $(\mathcal{A}, w)$ is a unitary compact matrix quantum group. It is called the* (permutational) free wreath product *of $\mathbb{G}$ by S_N^+ and denoted by $\mathbb{G} \wr_* S_N^+$.*

Proof We already proved that w and $\overline{w}$ are unitary in Exercise 7.8 and Theorem 7.30 shows that the coproduct is well-defined. The only thing left to check is that the coefficients of w generate $\mathcal{A}$. For this, simply observe that

$$\sum_{j=1}^{N} w_{ip,jq} = \nu_i(u_{pq})$$

and

$$\sum_{q=1}^{n} w_{ip,jq}w_{ip,jq}^* = P_{ij},$$

hence the $*$-algebra generated by the coefficients of w contains both $\mathcal{O}(S_N^+)$ and $\mathcal{O}(\nu_i(\mathbb{G}))$ for all $1 \leqslant i \leqslant N$. In other words, it coincides with $\mathcal{A}$. $\qquad \square$

Remark 7.32 Recall that if G is a classical unitary compact matrix group, then the maximal abelian quotient of $\mathcal{O}(G)^{*N} * \mathcal{O}(S_N^+)$ is

$$\mathcal{O}(G)^{\times N} \times \mathcal{O}(S_N) = \mathcal{O}(G^N \times S_N) = \mathcal{O}(G \wr S_N),$$

and the same is, of course, true starting with $\mathcal{A}$. Moreover, the previous discussion shows that w is mapped to the fundamental representation of $G \wr S_N$ which was defined earlier, so that we recover the full group structure of the classical wreath product from the free wreath product.

Now that we have a quantum version of the wreath product construction, we may imagine that there is an isomorphism

$$H_N^{s+} \simeq \mathbf{Z}_s \wr_* S_N^+.$$

This is, in fact, true, but proving it first requires a suitable description of the unitary compact matrix quantum group $\mathbf{Z}_s$.

Lemma 7.33 *Let $\mathcal{B}_s$ be the quotient of $\mathcal{O}(S_s^+)$ by the relations*

$$P_{xy} = P_{x'y'}$$

for all $1 \leqslant x, y, x', y' \leqslant s$ such that $y - x = y' - x'$ mod s. Then, if u_{ij} denotes the image of P_{ij} in the quotient, $(\mathcal{B}_s, u)$ is a unitary compact matrix quantum group isomorphic to $\mathbf{Z}_s$.

Proof Let us consider the matrix

$$
B_s = \begin{pmatrix}
0 & 1 & 0 & \cdots & 0 \\
0 & 0 & 1 & \cdots & 0 \\
\cdots & \cdots & \cdots & \cdots & \cdots \\
1 & 0 & 0 & \cdots & 0.
\end{pmatrix}.
$$

It generates a subgroup of S_N isomorphic to $\mathbf{Z}_s$ and its coefficients satisfy the condition in the statement. As a consequence, if Q_{ij} is the (i, j)-th coefficient function on $M_s(\mathbf{C})$, then there is a surjective $*$-homomorphism

$$
\Phi \colon \mathcal{B}_s \to \mathcal{O}(\mathbf{Z}_s)
$$

sending u_{ij} to Q_{ij}. Moreover, the condition in the statement implies that $\mathcal{B}_s$ is spanned as a vector space by the elements $(P_{0,j})_{0 \leqslant j \leqslant s-1}$ so that it has dimension at most s. Because $\mathcal{O}(\mathbf{Z}_s)$ has dimension s, this forces Φ to be an isomorphism, proving both assertions. $\qquad\square$

We are now ready for the last important result of this chapter, which is originally due to T. Banica and R. Vergnioux in [10].

Theorem 7.34 (Banica–Vergnioux) *There is an isomorphism*

$$
H_N^{s+} \simeq \mathbf{Z}_s \wr_* S_N^+.
$$

Proof We must build an isomorphism $\Phi \colon \mathcal{O}(H_N^{s+}) \to \mathcal{O}(\mathbf{Z}_s \wr_* S_N^+)$ sending a fundamental representation of the first unitary compact matrix quantum group to a fundamental representation of the second one. The fundamental representation w of the right-hand side has size sN, so that it will be more practical to work with the sudoku characterisation of $\mathcal{O}(H_N^{s+})$ since it has the same size. Then, the natural guess is

$$
\Phi \left(M_{ij}^{(p)} \right) = w_{i0, jp}.
$$

To see that this is well defined, simply notice that if we set

$$
A_{ij}^{(p,q)} = w_{ip, jq},
$$

then the matrix $A = A_{ij}^{(p,q)}$ is a quantum permutation matrix such that $A^{(p,q)} = A^{(0,q-p)}$, hence an (s, N)-sudoku matrix. As a consequence, the preceding formula defines a surjective $*$-homomorphism $\Phi \colon \mathcal{O}(H_N^{s+}) \to \mathcal{O}(\mathbf{Z}_s \wr_* S_N^+)$.

Conversely, we have to prove that there exists a $*$-homomorphism sending $w_{ip, jq}$ to $A_{ij}^{(p,q)}$. We will do this using the universal property of the free product, hence we must first define the ∗-homomorphism at the level of $\nu_i(\mathcal{O}(\mathbb{G}))$ and $\mathcal{O}(S_N^+)$. But we have already seen in Corollary 7.31 how we can recover both factors from the fundamental representation w. Let us set, for a fixed $1 \leqslant i \leqslant N$,

$$N_i^{(p,q)} = \sum_{j=1}^{N} M_{ij}^{(q-p)}.$$

These are orthogonal projections satisfying

- $\sum_{p=0}^{s-1} N_i^{(p,q)} = \sum_{j=1}^{N} \sum_{p=0}^{s-1} M_{ij}^{(q-p)} = 1$ and similarly for the sum over q;
- $N_i^{(p,q)} = N_i^{(p',q')}$ if $q - p = q' - p' \mod s$.

By Lemma 7.33, we therefore have a ∗-homomorphism $\Psi_i \colon \nu_i(\mathcal{O}(\mathbf{Z}_s)) \to \mathcal{O}(H_N^{s+})$ sending u_{pq} to $N_i^{(p,q)}$.

In a similar fashion we set, for $1 \leqslant i, j \leqslant N$ and $0 \leqslant p \leqslant s-1$,

$$Q_{ij} = \sum_{q=0}^{s-1} A_{ij}^{(p,q)} A_{ij}^{(p,q)*} = \sum_{q=0}^{s-1} M_{ij}^{(q)}.$$

These are orthogonal projections, and

$$\sum_{i=1}^{N} Q_{ij} = \sum_{i=1}^{N} \sum_{q=0}^{s-1} M_{ij}^{(q)} = 1 = \sum_{j=1}^{N} Q_{ij}$$

so that they form a quantum permutation matrix. As a consequence, there exists a unique ∗-homomorphism $\Psi' \colon \mathcal{O}(S_N^+) \to \mathcal{O}(H_N^{s+})$ sending P_{ij} to Q_{ij}.

By the universal property of unital free products, there exists a unique ∗-homomorphism

$$\Psi \colon \mathcal{O}(\mathbb{G})^{*N} * \mathcal{O}(S_N^+) \to \mathcal{O}(H_N^{s+})$$

which coincides with Ψ_i on $\nu_i(\mathcal{O}(\mathbb{G}))$ and with Ψ' on $\mathcal{O}(S_N^+)$. Moreover, because

$$N_i^{(p,q)} Q_{ij} = \left(\sum_{k=1}^{N} M_{ik}^{(q-p)} \right) \left(\sum_{r=0}^{s-1} M_{ij}^{(r)} \right)$$

$$= \sum_{k=1}^{N} \sum_{r=0}^{s-1} M_{ik}^{(q-p)} M_{ij}^{(r)}$$

$$= M_{ij}^{(q-p)}$$

$$= \sum_{k=1}^{N} \sum_{r=0}^{s-1} M_{ij}^{(r)} M_{ik}^{(q-p)}$$

$$= Q_{ij} N_i^{(p,q)},$$

the $*$-homomorphism Ψ factors through $\mathcal{A}$. By construction, it is inverse to Φ, hence the result. $\qquad \square$

As a consequence, we can prove the classical isomorphism given as a motivation for this section.

Corollary 7.35 *There is an isomorphism*

$$H_N^s \sim \mathbf{Z}_s \wr S_N.$$

Proof This follows from the fact that taking the maximal abelian quotient of $\mathcal{O}(H_N^{s+})$ yields $\mathcal{O}(H_N^s)$ with its usual group structure because of their descriptions through categories of partitions, while we have already mentioned in Remark 7.32 that the maximal abelian quotient of $\mathcal{O}(G \wr_* S_N^+)$ is $\mathcal{O}(G \wr S_N)$ with its wreath product structure. $\qquad \square$

Remark 7.36 The reader might want to check that the proofs still work for $s = \infty$, yielding isomorphisms

$$H_N^{\infty+} \simeq \widehat{\mathbf{Z}} \wr_* S_N^+ \text{ and } H_N^{\infty} \simeq \widehat{\mathbf{Z}} \wr S_N.$$

Note that because the left-hand side in the free wreath product must be a compact quantum group, putting $\mathbf{Z}$ instead of $\widehat{\mathbf{Z}}$ would not make sense. This seems to contrast with the case of finite cyclic groups, but that phenomenon is just due to the hidden isomorphism between $\widehat{\mathbf{Z}}_s$ and $\mathbf{Z}_s$.

Let us conclude with a more general comment. A classical result asserts that if X is a connected oriented graph and X_N is the disjoint union of N copies of X, then

$$\mathrm{Aut}(X_N) = \mathrm{Aut}(X) \wr S_N.$$

This was, in fact, the original motivation behind the definition of the free wreath product by J. Bichon in [23], where he proved the following quantum analogue.

Theorem 7.37 (Bichon) *Let X be a connected oriented graph and let X_N be the disjoint union of N copies of X. Then,*

$$\mathrm{Aut}^+(X_N) = \mathrm{Aut}^+(X) \wr_* S_N^+.$$

8

Back to the Game

8.1 Perfect Quantum Strategies

In this section, we come back to the graph isomorphism game introduced in Chapter 1 and clarify the notion of a *perfect strategy* for that game, both in the classical and quantum contexts. We will in particular give a proof of Theorem 1.2, which served as a motivation for the introduction of compact matrix quantum groups. Quite unexpectedly, we will see that the proof partly relies on compact quantum group theory. From now on, we fix two unoriented graphs X and Y with the same number N of vertices. Because we do not care about orientation hereafter, we will write $x \sim y$ instead of $x \to y$ to indicate that $(x, y) \in E$.

8.1.1 What Is a Strategy?

8.1.1.1 Classical Setting

Classically, a strategy for the graph isomorphism game means that each player decides in advance which vertex they will answer depending on the input. This is quite naturally formalised through two functions

$$f_A, f_B \colon V(X) \to V(Y)$$

such that Alice answers $f_A(x)$ when given input x and Bob answers $f_B(x')$ when given input x'.

However, in order to improve their chances of winning, the players may introduce some randomness in their strategy. We will then have functions

$$p_A, p_B \colon V(X) \times V(Y) \to [0, 1]$$

such that for any $x \in V(X)$,

221

$$\sum_{y \in V(Y)} p_A(x, y) = 1 = \sum_{y \in V(Y)} p_B(x, y).$$

This has a simple probabilistic interpretation. For $x \in V(X)$, let $X_A(x)$ (respectively $X_B(x)$) be a random variable with values in $V(Y)$ and such that

$$\mathbb{P}(X_A(x) = y) = p_A(x, y) \quad \text{and} \quad \mathbb{P}(X_B(x') = y') = p_B(x', y').$$

Upon receiving their inputs x and x', Alice and Bob both perform a random experiment corresponding to the random variables $X_A(x)$ and $X_B(x')$. The outcome of their experiments are then the answers they give.

Of course, our first definition of a strategy is a particular case of that second one, given by

$$p_A(x, y) = \delta_{f_A(x), y} \quad \text{and} \quad p_B(x', y') = \delta_{f_B(x'), y'}.$$

This is equivalent to requiring all the random variables $X_A(x)$ and $X_B(x')$ to be deterministic, hence we will call these *deterministic strategies*.

Even though probabilistic strategies may seem more favourable, they are useless as long as one seeks a *perfect strategy*, that is to say one that wins at all times. Moreover, one can easily identify deterministic strategies which are perfect.

Exercise 8.1 Assume that there exists a perfect classical strategy for the graph isomorphism game. Prove that there exists a perfect deterministic strategy, which is moreover given by a graph isomorphism.

Solution Let p_A and p_B describe a perfect probabilistic strategy. Given $x, x' \in V(X)$, let $y, y' \in V(Y)$ be such that

$$p_A(x, y) \neq 0 \neq p_B(x', y').$$

Since there is a non-zero probability that Alice and Bob will answer y and y' respectively, and since the strategy is perfect, $\mathrm{rel}(x, x') = \mathrm{rel}(y, y')$. As a consequence, setting

$$p'_A(x, y) = 1 \text{ and } p'_A(x, z) = 0 \text{ for all } z \neq y,$$

and similarly for p_B, yields another perfect strategy. Iterating this process, we end up with a perfect deterministic strategy since the probability distributions p_A and p_B become deterministic.

Now, given such a perfect deterministic strategy, which is described by two functions f_A and f_B, first observe that by definition, $f_A(x) = f_B(x)$ for all x so that $f_A = f_B$. Moreover, if $x \neq x'$, then

$$f_A(x) \neq f_B(x') = f_A(x')$$

so that f_A is injective, hence bijective. Eventually, again by the definition of a perfect strategy, if x and x' are adjacent in X, if and only if $f_A(x)$ and $f_A(x')$ are adjacent in Y, hence f_A is a graph isomorphism. $\qquad\square$

8.1.1.2 Quantum Setting

We now turn to the quantum world. The reader may not be familiar with the mathematical formalism of quantum mechanics or with the basics of quantum information theory, and giving a detailed introduction to these topics is certainly beyond our scope (the reader may refer, for instance, to [74] for this). We will nevertheless try to motivate the definition of a quantum strategy by some quantum mechanical considerations.

The fundamental idea is that Alice and Bob will set up before the game starts a quantum mechanical system, and then bring part of it with them so that they can manipulate it. The quantum phenomenon known as *entanglement* will then enable them to modify the other one's system by manipulating theirs without any communication. The postulates of quantum mechanics assert that the states of a quantum system can be described by the vectors of a Hilbert space. Let, therefore, H be a Hilbert space fixed once and for all and let $\psi \in H$ represent the state of the system before the game starts.

Another postulate of quantum mechanics is that any measurable physical quantity of the system is given by a self-adjoint operator A on H, and that the possible outcomes of a measurement of A are the eigenvalues of A. Let us assume for the moment that H is finite-dimensional and write

$$\mathrm{Sp}(A) = \{\lambda_1, \dots, \lambda_n\}.$$

If Π_i denotes the orthogonal projection onto the eigenspace corresponding to λ_i, then these operators tell us 'how likely' the outcome λ_i is if the observable A is measured in the state ψ. This can be made concrete by considering the quantities

$$p_i = \langle \Pi_i(\psi), \psi \rangle \geqslant 0. \tag{8.1}$$

Since these numbers add up to 1, they form a probability distribution on the set of possible outcomes of the measurement. From a mathematical perspective, one can then forget about A and simply consider a family of orthogonal projections $(\Pi_i)_{i \in I}$ such that

$$\sum_{i=1}^{n} \Pi_i = \mathrm{Id}_H.$$

These describe a quantum measurement with possible outcomes $I = \{1, \ldots, n\}$, the probability of having outcome $i \in I$ being given by Equation (8.1).

Even though this is a correct and reasonable setting for quantum measurement, it turns out that we need to extend it slightly. Such an extension enables one, for instance, to consider statistical ensembles of quantum states instead of single ones, and is in a sense the most general notion of 'acting on a quantum mechanical system'.

Definition 8.1 A *Positive Operator-Valued Measure* (in short POVM) is a finite[1] family of positive semi-definite operators $\mathcal{E} = (E_i)_{i \in I}$ acting on a Hilbert space H such that

$$\sum_{i \in I} E_i = \mathrm{Id}_H \ .$$

Note that, given a state ψ, the previous probabilistic interpretation still makes sense. We will therefore consider that if we want to measure the 'observable' $\mathcal{E}$ in the state ψ, then we get outcome i with probability

$$p_i = \langle E_i(\psi), \psi \rangle \geqslant 0.$$

Let us go back to the strategy of Alice and Bob. Assume that they have prepared the state ψ and both have at their disposal, for each $x \in V(X)$, a POVM denoted respectively by

$$\mathcal{E}_x^A = (E_{xy})_{y \in V(Y)} \quad \& \quad \mathcal{E}_x^B = (F_{xy})_{y \in V(Y)}.$$

Upon receiving inputs x and x' respectively, each one wants to apply its corresponding POVM to ψ and send as an answer the outcome of the measurement. But here appears a crucial issue: in quantum mechanics, making a measure changes the system! More precisely, if Alice performs a measurement on a state ψ and the result is y, then the state of the system becomes $E_{xy}(\psi)$. Therefore, the probability of Bob having afterwards the outcome y' is not $\langle F_{x'y'}(\psi), \psi \rangle$ but $\langle F_{x'y'} E_{xy}(\psi), E_{xy}(\psi) \rangle$. Summing over all possible outcomes y of Alice's measurement eventually yields the probability for Bob to obtain y', namely

$$\sum_{y \in V(X)} \langle F_{x'y'} E_{xy}(\psi), E_{xy}(\psi) \rangle \langle E_{xy}(\psi), \psi \rangle.$$

---

[1] We restrict here the definition to finitely many operators for simplicity, since this is what we need. In general, one should consider a family of operators indexed by measureable subsets of a measure space, with natural conditions making it a 'measure' with operator values.

Since the operators E_{xy} and F_{xy} do not need to commute, the knowledge of the correlations of the inputs and outputs will depend on who measures first. But since Alice and Bob cannot communicate, they cannot agree on who measures first, ruining the mere possibility of building a strategy.

To overcome this problem, the players will need to choose their POVM wisely so that, even if they do not commute, the order of the measurements does not matter. Concretely, this means that for all x, x', y, y', we should have

$$\langle E_{xy} F_{x'y'}(\psi), \psi \rangle = \langle F_{x'y'} E_{xy}(\psi), \psi \rangle .$$

This can be seen as a condition on the linear map $\tau_\psi : \mathcal{B}(H) \to \mathbf{C}$ given by

$$\tau_\psi(T) = \langle T(\psi), \psi \rangle.$$

In that form, it says that the elements from $\mathcal{E}^A$ and $\mathcal{E}^B$ 'commute when τ_ψ is applied'.

Now that we have everything in hand, we can give the definition of a perfect quantum strategy for the graph isomorphism game.

Definition 8.2 A *perfect quantum strategy* for the graph isomorphism game is given by a Hilbert space H together with

- For each $x \in V(X)$, two POVMs

$$\mathcal{E}_x^A = \{E_{xy} \mid y \in V(Y)\} \quad \text{and} \quad \mathcal{E}_x^B = \{F_{xy} \mid y \in V(Y)\},$$

- A linear form τ on the $*$-algebra $\mathcal{A}$ generated by all the preceding operators, such that

1. τ is faithful, that is, $\tau(a) = 0$ implies $a = 0$ if a is a positive operator;
2. $\tau(E_{xy} F_{x'y'}) > 0$ if and only if $\mathrm{rel}(x, x') = \mathrm{rel}(y, y')$;
3. $\tau(E_{xy} F_{x'y'}) = \tau(E_{x'y'} F_{xy})$ for all $x, x' \in V(X)$ and $y, y' \in V(Y)$.

Let us make a crucial comment. The only thing which is important to us in this setting is the data of the *correlations* of the quantum strategy, that is to say the numbers

$$p(x, x', y, y') = \tau(E_{xy} F_{x'y'}).$$

In particular, if τ is not faithful, then we can make it so using the GNS construction (see the end of Section 5.1.2). This yields a new $*$-algebra and new measurement systems, but the correlations are the same. This means that removing the faithfulness condition in Definition 8.2 would not yield a more general notion. Since faithfulness is convenient, we choose to incorporate it directly.

As one may expect, quantum strategies include classical ones. Indeed, if we have a classical probabilistic strategy given by two functions p_A and p_B, let $H = \mathbf{C}$ and set

$$E_{xy}(1) = p_A(x,y).1 \quad \text{and} \quad F_{x'y'}(1) = p_B(x',y').1.$$

For fixed x, this defines two POVMs which commute. Moreover, the correlations associated to the vector

$$\psi = 1$$

factor as

$$\langle E_{xy}F_{x'y'}(\psi), \psi \rangle = p_A(x,y)p_B(x',y') = \langle F_{x'y'}E_{xy}(\psi), \psi \rangle.$$

8.1.2 Link with Quantum Permutation Matrices

The stage is now set for the proof of Theorem 1.2, namely that there exists a perfect quantum strategy if and only if there exists a quantum permutation matrix P such that $A_X P = P A_Y$. The proof proceeds in two steps of different nature. The first one is to see that a perfect quantum strategy must indeed have the form of a quantum permutation matrix.

Theorem 8.3 (Atserias–Mančinska–Roberson–Šamal–Severini–Varvitsiotis) *Let*

$$\left((\mathcal{E}^A_x)_{x \in V(X)}, (\mathcal{E}^B_x)_{x \in V(X)}), \tau\right)$$

be a perfect quantum strategy. Then,

1. $E_{xy} = F_{xy}$ *for all* $(x,y) \in V(X) \times V(Y)$;
2. τ *is a trace on* $\mathcal{A}$;
3. *The matrix* $P = (E_{xy})$ *is a quantum permutation matrix;*
4. $A_X P = P A_Y$.

Proof Let us prove each point one after the other.

1. We start by using the condition on τ to observe that

$$\tau(E_{xy}) = \sum_{y' \in V(Y)} \tau(E_{xy}F_{xy'}) = \tau(E_{xy}F_{xy})$$

so that, denoting by $E_{xy}^{1/2}$ the square root[2] of E_{xy},

---

[2] That is to say a positive operator T such that $T^2 = E_{xy}$. The existence and uniqueness of such an operator follows, for instance, from the spectral theorem applied to the self-adjoint operator E_{xy}.

$$\tau\left(E_{xy}(1 - F_{xy})\right) = \tau\left(E_{xy}^{1/2}(1 - F_{xy})E_{xy}^{1/2}\right) = 0.$$

The latter operator being positive, it is zero by faithfulness of τ, hence

$$E_{xy} = E_{xy}^{1/2}F_{xy}E_{xy}^{1/2}.$$

Now because E_{xy} is self-adjoint, it induces a bijection on the orthogonal complement of its kernel which is again a positive operator. Let us denote abusively by E_{xy}^{-1} the map which is 0 on $\ker(E_{xy})$ and coincides with the inverse of the restriction of E_{xy} on $\ker(E_{xy})^\perp$. Then for any $\xi \in \ker(E_{xy})^\perp$,

$$\begin{aligned}
\xi &= E_{xy}^{-1/2}E_{xy}E_{xy}^{-1/2}(\xi) \\
&= E_{xy}^{-1/2}E_{xy}^{1/2}F_{xy}E_{xy}^{1/2}E_{xy}^{-1/2}(\xi) \\
&= F_{xy}(\xi).
\end{aligned}$$

In other words,

$$(F_{xy})_{|\ker(E_{xy})^\perp} = \mathrm{Id}.$$

Note that since $\mathcal{E}_x^A$ and $\mathcal{E}_x^B$ play symmetric roles, we also have

$$F_{xy} = F_{xy}^{1/2}E_{xy}F_{xy}^{1/2}.$$

Let us now set

$$V = E_{xy}^{1/2}F_{xy}^{1/2}$$

and observe that $E_{xy} = VV^*$ and $F_{xy} = V^*V$. Then, the fact that F_{xy} restricts to the identity on the range of E_{xy} implies that

$$E_{xy}^2 = VV^*VV^* = VF_{xy}V^* = E_{xy}^{1/2}F_{xy}^2E_{xy}^{1/2} = E_{xy}.$$

In other words, E_{xy} is a projection, and the same holds for F_{xy} by symmetry. Moreover, the fact that F_{xy} dominates E_{xy} and its converse show that, in fact, $E_{xy} = F_{xy}$.

2. It follows from the previous point and the definition of a quantum strategy that

$$\tau(E_{xy}E_{x'y'}) = \tau(E_{xy}F_{x'y'}) = \tau(F_{x'y'}E_{xy}) = \tau(E_{x'y'}E_{xy}).$$

Thus, for any polynomials Q_1, Q_2 in the E_{xy}'s, we also have $\tau(Q_1Q_2) = \tau(Q_2Q_1)$, and these polynomials are exactly the elements of $\mathcal{A}$.

3. The fact that the sum over each row (i.e. for fixed x) is the identity is part of the definition of a quantum strategy. As for the columns, first note that,

because the strategy is perfect, Alice and Bob cannot give the same answer if they are given different inputs, so that for $x \neq x'$,

$$0 = \tau(E_{xy}F_{x'y}) = \tau(E_{xy}E_{x'y}) = \tau(E_{xy}E_{x'y}E_{xy}).$$

By faithfulness of τ, we infer that $E_{xy}E_{x'y}E_{xy} = 0$. Thus, E_{xy} and $E_{x'y}$ are orthogonal. As a consequence,

$$T_y = \sum_{x \in V(X)} E_{xy} \leqslant \mathrm{Id}_H$$

is a projection. But because

$$\sum_{y \in V(Y)} T_y = n \, \mathrm{Id}_H$$

each T_y must be exactly equal to Id_H. Indeed, if there exist y_0 such that $T_{y_0} < \mathrm{Id}_H$ and if $\xi \in \ker(T_{y_0})$, then

$$n\|\xi\|^2 = \left\langle \sum_{y \in V(Y)} T_y(\xi), \xi \right\rangle$$

$$= \sum_{y \in V(Y)} \|T_y(\xi)\|^2$$

$$= \sum_{y \in V(Y) \setminus \{y_0\}} \|T_y(\xi)\|^2$$

$$< (n-1)\|\xi\|^2,$$

a contradiction.

4. Let us fix $(x, y) \in V(X) \times V(Y)$. The corresponding coefficients of $A_X P$ and $P A_Y$ are respectively

$$L = \sum_{x' \in V(X)} \delta_{x \sim x'} E_{x'y} \quad \text{and} \quad R = \sum_{y' \in V(Y)} \delta_{y' \sim y} E_{xy'},$$

where $\sim$ means being adjacent in the corresponding graph. Note that both sides are projections, as sums of pairwise orthogonal projections. By assumption, if $x \sim x'$ but $y \nsim y'$, then $\tau(E_{xy}E_{x'y'}) = 0$ so that

$$E_{xy}E_{x'y'}E_{xy} = 0,$$

implying that the projections are orthogonal. Therefore,

$$L = \sum_{x' \in V(X)} \delta_{x \sim x'} E_{x'y}$$

$$= \sum_{x' \in V(X)} \sum_{y' \in V(Y)} \delta_{x \sim x'} E_{x'y} E_{xy'}$$

$$= \sum_{x' \in V(X)} \sum_{y' \in V(Y)} \delta_{x \sim x'} \delta_{y' \sim y} E_{x'y} E_{xy'}$$

$$= \sum_{x' \in V(X)} \sum_{y' \in V(Y)} \delta_{y' \sim y} E_{x'y} E_{xy'}$$

$$= \sum_{y' \in V(Y)} \delta_{y' \sim y} E_{xy'}$$

$$= R. \qquad \qquad \square$$

Starting now from a quantum permutation matrix, one can define two families of POVMs by setting

$$E_{xy} = P_{xy} = F_{xy}.$$

However, to upgrade it to a quantum strategy, one needs to produce a suitable tracial state τ on the $*$-algebra generated by the coefficients of a quantum permutation matrix. It turns out that there is no way to do this in general, and this is the reason why we have not claimed that quantum strategies corresponded to quantum permutation matrices. In fact, we could make things clearer, thanks to the following definition.

Definition 8.4 A quantum permutation matrix $P = (P_{ij})_{1 \leqslant i,j \leqslant N} \in M_N(\mathcal{B}(H))$ is said to be *tracial* if there exists a tracial state on the $*$-algebra $\mathcal{A} \subset \mathcal{B}(H)$ generated by its coefficients.

We then have

Corollary 8.5 *There exists a perfect quantum strategies for the graph isomorphism game if and only if there exists a tracial quantum permutation matrices P such that $A_X P = P A_Y$.*

Proof Setting $E_{xy} = P_{xy} = F_{xy}$ produces two families of POVMs and the traciality assumption yields the state τ on $\mathcal{A}$ which satisfies the commutation relation. Moreover, the condition $A_X P = P A_Y$ translates, for any $(x, y) \in V(X) \times V(Y)$, into

$$\sum_{a \sim x} P_{ay} = \sum_{y \sim b} P_{xb}.$$

Thus, if $x \sim x'$ but $y \not\sim y'$,

$$P_{xy} P_{x'y'} = P_{xy} \left(\sum_{x'' \sim x'} P_{x''y} \right) P_{x'y'}$$

$$= P_{xy} \left(\sum_{y \sim y''} P_{x'y''} \right) P_{x'y'}$$

$$= 0.$$

Taking the GNS representation if necessary, we may assume that τ is faithful, and this concludes the proof. □

What remains to be done in order to prove Theorem 1.2 as it is stated is therefore to make sure that as soon as there is a quantum permutation matrix intertwining the adjacency matrices, then there also exists a tracial one with the same property. The proof of this non-trivial fact uses, quite surprisingly, compact matrix quantum groups! It also requires some results on connectedness for graphs which we will detail first. Nevertheless, before delving into these one may ask whether there could automatically exist a trace on the $*$-algebra generated by the coefficients of a quantum permutation. Here is an exercise providing an explicit counter-example.

Exercise 8.2 Let A be a C*-algebra generated by two elements s_1, s_2 such that

$$s_1^* s_1 = 1 = s_2^* s_2$$

and

$$s_1 s_1^* + s_2 s_2^* = 1.$$

1. Prove that there is no tracial state on A.
2. Prove that $s_1^* s_2 = 0$.
3. We define, for $i = 1, 2$, the following elements of $M_2(A)$:

$$x_1 = \begin{pmatrix} 0 & s_1^* \\ s_1 & s_2 s_2^* \end{pmatrix} \quad ; \quad x_2 = \begin{pmatrix} 0 & s_2^* \\ s_2 & s_1 s_1^* \end{pmatrix}$$

$$x_3 = \begin{pmatrix} 1 & 0 \\ 0 & -1 \end{pmatrix} \quad ; \quad x_4 = \begin{pmatrix} 0 & 1 \\ 1 & 0 \end{pmatrix}$$

Prove that these are self-adjoint unitaries.
4. Prove that x_3 and x_4 generate $M_2(\mathbf{C})$ as a C*-algebra.
5. Prove that $M_2(A)$ is generated as a C*-algebra by x_1, x_2, x_3 and x_4.
6. We define a linear map $\Phi \colon M_2(A) \to A$ by

$$\Phi \left((z_{ij})_{1 \leqslant i,j \leqslant 2} \right) = \sum_{i,j=1}^{2} s_i z_{ij} s_j^*.$$

Prove that this is a $*$-isomorphism.

7. Build a quantum permutation matrix whose coefficients generate A.
8. By the way, does such a C*-algebra as A exist?

Solution 1. Let τ be a tracial state on A. Then,

$$
\begin{aligned}
1 &= \tau(1) \\
&= \tau(s_1 s_1^* + s_2 s_2^*) \\
&= \tau(s_1 s_1^*) + \tau(s_2 s_2^*) \\
&= \tau(s_1^* s_1) + \tau(s_2^* s_2) \\
&= 2,
\end{aligned}
$$

a contradiction.

2. We compute

$$
\begin{aligned}
(s_1^* s_2)(s_1^* s_2)^* &= s_1^* s_2 s_2^* s_1 \\
&= s_1^*(1 - s_1 s_1^*) s_1 \\
&= s_1^* s_1 - (s_1^* s_1)^2 \\
&= 1 - 1^2 \\
&= 0,
\end{aligned}
$$

from which the result follows.

3. The result is clear for x_3 and x_4, and the proofs for the first two ones are similar, hence we only deal with x_1. By definition, $x_1^* = x_1$ and moreover

$$
x_1^2 = \begin{pmatrix} s_1^* s_1 & s_1^* s_2 s_2^* \\ s_2 s_2^* s_1 & s_1 s_1^* + s_2 s_2^* \end{pmatrix} = \begin{pmatrix} 1 & 0 \\ 0 & 1 \end{pmatrix}.
$$

4. Let us denote by E_{ij} the matrix with a coefficient 1 in position (i, j) and 0 elsewhere. These form a basis of $M_2(\mathbf{C})$. First, we have $x_3^2 = \mathrm{Id}_{M_2(\mathbf{C})}$ so that

$$
\begin{aligned}
2E_{11} &= x_3 + x_3^2 \in M_2(\mathbf{C}), \\
-2E_{22} &= x_3 - x_3^2.
\end{aligned}
$$

Second,

$$
x_3 x_4 = \begin{pmatrix} 0 & 1 \\ -1 & 0 \end{pmatrix}
$$

so that

$$
\begin{aligned}
2E_{12} &= x_4 + x_3 x_4, \\
2E_{21} &= x_4 - x_3 x_4,
\end{aligned}
$$

concluding the proof.

5. Let us denote by B the sub-C*-algebra of $M_2(A)$ generated by x_1, x_2, x_3 and x_4. By the preceding question, $M_2(\mathbf{C}) \subset B$. Moreover, observe that for $i = 1, 2$,

$$\begin{pmatrix} 0 & 0 \\ 0 & 1 \end{pmatrix} x_i \begin{pmatrix} 1 & 0 \\ 0 & 0 \end{pmatrix} = \begin{pmatrix} 0 & 0 \\ s_i & 0 \end{pmatrix}.$$

Multiplying this last matrix with well-chosen elements of $M_2(\mathbf{C})$, one gets that $s_k E_{ij} \in B$ for all $1 \leqslant i, j, k \leqslant 2$, hence $B = M_2(A)$.

6. Given two elements $z = (z_{ij})_{1 \leqslant i,j \leqslant 2}$ and $t = (t_{k\ell})_{1 \leqslant k,\ell \leqslant 2}$ in $M_2(A)$, we compute

$$\begin{aligned} \Phi(z)\Phi(t) &= \sum_{i,j,k,\ell=1}^{2} s_i z_{ij} s_j^* s_k t_{k\ell} s_\ell^* \\ &= \sum_{i,j,\ell=1}^{2} s_i z_{ij} t_{j\ell} s_\ell^* \\ &= \sum_{i,\ell=1}^{2} s_i (zt)_{i\ell} s_\ell^* \\ &= \Phi(zt) \end{aligned}$$

so that Φ is multiplicative. Moreover,

$$\begin{aligned} \Phi(z)^* &= \sum_{i,j=1}^{2} s_j z_{ij}^* s_i^* \\ &= \sum_{i,j=1}^{n} s_i z_{ji}^* s_j^* \\ &= \Phi(z^*), \end{aligned}$$

proving that Φ is a $*$-homomorphism. Let us define a linear map $\Psi \colon A \to M_2(A)$ as follows:

$$\Psi(x) = \begin{pmatrix} s_1^* x s_1 & s_1^* x s_2 \\ s_2^* x s_1 & s_2^* x s_2 \end{pmatrix}.$$

On the one hand,

$$\begin{aligned} (\Psi \circ \Phi(z))_{k,\ell} &= s_k^* \Phi(z) s_\ell \\ &= \sum_{i,j=1}^{2} s_k^* s_i z_{ij} s_j^* s_\ell \\ &= z_{k\ell}, \end{aligned}$$

and on the other hand,

$$\Phi \circ \Psi(x) = \sum_{i,j=1}^{2} s_i \Psi(x)_{ij} s_j^*$$

$$= \sum_{i,j=1}^{2} s_i s_i^* x s_j s_j^*$$

$$= \left(\sum_{i=1}^{2} s_i s_i^* \right) x \left(\sum_{j=1}^{2} s_j s_j^* \right)$$

$$= x.$$

We have therefore proven that Φ is a $*$-isomorphism.

7. Let us set, for $1 \leqslant i \leqslant 4$,

$$p_i = \frac{\Phi(x_i) + 1}{2},$$

which are orthogonal projections generating A. We then define matrices $M_i \in M_2(A)$ as follows:

$$M_i = \begin{pmatrix} p_i & 1 - p_i \\ 1 - p_i & p_i \end{pmatrix}.$$

Let P be the block-diagonal matrix with diagonal blocks $(M_i)_{1 \leqslant i \leqslant 4}$. This is a quantum permutation matrix by construction, and its coefficients generate A.

8. Let us consider the Hilbert space

$$H = \ell^2(\mathbf{N}) \oplus \ell^2(\mathbf{N}).$$

It is separable and infinite-dimensional, hence there exists a unitary linear isomorphism

$$\Phi \colon H \to \ell^2(\mathbf{N}).$$

Consider now the inclusion maps $\widetilde{s}_1, \widetilde{s}_2 \colon \ell^2(\mathbf{N}) \to H$ respectively into the first and second summands. Then, setting

$$s_i = \Phi \circ \widetilde{s}_i \in B(\ell^2(\mathbf{N}))$$

for $i = 1, 2$ does the job. □

Remark 8.6 The C*-algebra A is usually known as the *Cuntz algebra* $\mathcal{O}_2$ and plays an important role in the study of the structure and classification of C*-algebras.

8.1.2.1 Connected Graphs and the Laplacian

Given a graph X, one may consider the equivalence relation obtained by taking the transitive closure of the adjacency relation $\sim$. The equivalence classes are then called the *connected components* of X, and X is said to be *connected* if it has only one connected component. Equivalently, this means that given any two vertices $x, y \in V(X)$, there exists an integer $n \in \mathbf{N}$ and vertices $x = z_0, \ldots, z_n = y$ such that $(z_i, z_{i+1}) \in E(X)$ for all $0 \leqslant i \leqslant n - 1$. In this subsection we will prove that if X is connected, then the existence of a perfect quantum strategy is only possible if Y is also connected. The proof is not very difficult once one has a good criterion for connectedness at hand. The one that we need here is based on the so-called Laplacian of a graph. For convenience, let us fix a numbering of the vertices of X so that we can write $V(X) = \{1, \ldots, N\}$.

Definition 8.7 Let X be a finite graph. For a vertex $i \in V(X)$, its *degree* d_i is defined as

$$d(i) = |\{j \in V(X) \mid (i, j) \in E(X)\}| .$$

The *Laplacian* of X is the matrix

$$L_X = D_X - A_X,$$

where D_X is the diagonal matrix with (i, i)-th coefficient d_i.

Consider the vector

$$\xi = \sum_{i=1}^{N} e_i.$$

The ith coordinate of $L_X(\xi)$ is

$$d_i - \sum_{j=1}^{N} (A_X)_{ij} = d_i - |\{j \in V(X) \mid (i, j) \in E(X)\}| = 0$$

so that $\xi \in \ker(L_X)$. The point is that in fact, the dimension of the kernel of L_X characterises connectedness.

Lemma 8.8 *The graph X is connected if and only if* $\ker(L_X)$ *has dimension one.*

Proof Let us consider an arbitrary vector

$$\zeta = \sum_{i=1}^{N} \lambda_i e_i \in \mathbf{C}^N.$$

Then,

$$
\langle L_X(\zeta), \zeta \rangle = \sum_{i,j=1}^{N} \lambda_i \bar{\lambda}_j \langle D_X(e_i), e_j \rangle - \sum_{i,j=1}^{N} \lambda_i \bar{\lambda}_j \langle A_X(e_i), e_j \rangle
$$

$$
= \sum_{i=1}^{N} d_i |\lambda_i|^2 - \sum_{i=1}^{N} \sum_{i \sim j} \lambda_i \bar{\lambda}_j
$$

$$
= \sum_{i,j=1}^{N} \delta_{i \sim j} |\lambda_i|^2 - \sum_{i,j=1}^{N} \delta_{i \sim j} \lambda_i \bar{\lambda}_j
$$

$$
= \sum_{i,j=1}^{N} \delta_{i \sim j} |\lambda_i - \lambda_j|^2.
$$

As a consequence, if $\zeta \in \ker(L_X)$, then $\lambda_i = \lambda_j$ as soon as i and j are connected. By transitivity, this implies that the coordinates of ζ are constant on connected components of X. In particular, if X is connected then $\lambda_i = \lambda_j$ for all $1 \leqslant i, j \leqslant N$ so that ζ is proportional to ξ and $\ker(L_X)$ is one-dimensional.

Conversely, assume that X is not connected, and let $S \subset V(X)$ be a connected component. Then,

$$
\xi_S = \sum_{i \in S} e_i \in \ker(L_X)
$$

and ξ_S is not colinear to ξ, so that $\ker(L_X)$ has dimension at least two. $\qquad \square$

Remark 8.9 The reader can easily show that $\dim(\ker(L_X))$ is, in fact, equal to the number of connected components of X.

Lemma 8.8 suggests to try to compare the Laplacians of X and Y under the assumption that there exists a quantum permutation matrix P such that $P A_X = A_Y P$. To see how one can do this, let us first note that because quantum permutation matrices are orthogonal, the intertwining property can be rewritten as

$$
P A_X {}^t P = A_Y
$$

and that more generally, for any polynomial $Q \in \mathbf{C}[X]$, we have $P Q(A_X)^t P = Q(A_Y)$, so that

$$
\Phi \colon T \mapsto P T^t P
$$

is an isomorphism between the subalgebras of $M_N(\mathbf{C})$ generated by A_X and A_Y respectively.[3] Unfortunately, the Laplacian does not belong to this algebra

---

[3] In particular, the adjacency matrices are isospectral.

in general. It is nevertheless possible to build it from the adjacency matrix using another operation.

Definition 8.10 The *Schur product* of two matrices A and B is the matrix $A \odot B$ with coefficients

$$(A \odot B)_{ij} = A_{ij}B_{ij}.$$

We can now connect conjugation by quantum permutation matrices with connectedness.

Proposition 8.11 *Assume that X and Y are* unoriented graphs *such that X is connected and that there exists a quantum permutation matrix P such that $PA_X = A_Y P$. Then, Y is connected.*

Proof Our aim is to prove that $PL_X = L_Y P$. Let us denote by $J_N \in M_N(\mathbf{C})$ the matrix with all coefficients equal to one and by $I_N \in M_N(\mathbf{C})$ the identity matrix. The proof is based on the following elementary computation,

$$D_X = (A_X J_N) \odot I_N$$

We will now use this to prove that $PD_X = D_Y P$. To start with, note that $PJ_N = J_N P$. Moreover, given matrices $A, A', B, B' \in M_N(\mathbf{C})$ such that $PA = A'P$ and $PB = B'P$, we have

$$
\begin{aligned}
(P(A \odot B))_{ij} &= \sum_{k=1}^{N} P_{ik} A_{kj} B_{kj} \\
&= \sum_{k,k'=1}^{N} P_{ik} P_{ik'} A_{kj} B_{k'j} \\
&= \left(\sum_{k=1}^{N} P_{ik} A_{kj} \right) \left(\sum_{k'=1}^{N} P_{ik'} B_{k'j} \right) \\
&= (PA)_{ij}(PB)_{ij} \\
&= (A'P)_{ij}(B'P)_{ij} \\
&= \left(\sum_{k=1}^{N} A'_{ik} P_{kj} \right) \left(\sum_{k'=1}^{N} B'_{ik'} P_{k'j} \right) \\
&= \sum_{k=1}^{N} A'_{ik} B'_{ik} P_{kj} \\
&= ((A' \odot B')P)_{ij}.
\end{aligned}
$$

As a consequence,

$$PD_X = P(A_X J_N) \odot I_N = (A_Y J_N) \odot I_N P = D_Y P$$

so that the claim is proven.

It follows that $PL_X = L_Y P$, hence for any polynomial $Q \in \mathbf{C}[X]$,

$$PQ(L_X)^t P = Q(L_Y).$$

Thus, $Q(L_X)$ vanishes if and only if $Q(L_Y)$ does. In particular, L_X and L_Y have the same characteristic polynomial. Since both are symmetric matrices, the multiplicity of their eigenvalues equals their multiplicity as roots of the characteristic polynomial. Applying this to the eigenvalue 0, we see that X is connected if and only if Y is connected. □

Remark 8.12 The idea of the preceding proof is that if there exists a perfect quantum strategy, then it implements an isomorphism between the so-called *coherent algebras* of the graphs X and Y. This was first noted and used in [53].

8.1.2.2 The Complete Characterisation

We now endeavour to prove that if there is a quantum permutation matrix intertwining the adjacency matrices, then there is a tracial one with the same property. Let us start with a small computation.

Lemma 8.13 Let X be a finite graph and let U be the fundamental representation of $\mathrm{Aut}^+(X)$. If $i \sim i'$ but $j \nsim j'$, then

$$U_{ij} U_{i'j'} = 0.$$

Proof Recall that U satisfies $A_X U = U A_X$, and that the (i, j')-th coefficient of this equality reads

$$\sum_{i \sim k} U_{kj'} = \sum_{k \sim j'} U_{ik}.$$

Multiplying both sides of the equality by U_{ij} on the left and on the right yields

$$\sum_{i \sim k} U_{ij} U_{kj'} U_{ij} = 0,$$

and since the operators in the sum are positive, it follows that for any $i \sim k$, $U_{ij} U_{kj'} U_{ij} = 0$. In particular, $U_{ij} U_{i'j'} U_{ij} = 0$ and because these are orthogonal projections, we conclude that $U_{ij} U_{i'j'} = 0$. □

The trick is now to consider the quantum automorphism group of the disjoint union of X and Y. There, we can separate in a strong sense the components coming from X and Y, provided that at least one of the graphs is connected.

Lemma 8.14 *Let X and Y be two finite graphs with X connected, and let U be the fundamental representation of $\mathrm{Aut}^+(X \amalg Y)$. If $x, x', x'' \in V(X)$ and $y \in V(Y)$, then*

$$U_{xx''}U_{x'y} = 0.$$

Proof Because X is connected, there is a finite path $x_0 = x, x_1, \cdots, x_n = x'$ connecting x to x'. Thus,

$$U_{xx''}U_{x'y} = U_{xx''}\left(\sum_{z\in V(X)\cup V(Y)} U_{x_1 z} \right) \cdots \left(\sum_{z\in V(X)\cup V(Y)} U_{x_{n-1} z} \right) U_{x'y}$$

$$= U_{xx''}\left(\sum_{z_1,\cdots z_n \in V(X)\cup V(Y)} U_{x_1 z_1} \cdots U_{x_{n-1} z_{n-1}} \right) U_{x'y}.$$

Let us consider one tuple $z_1, \cdots, z_{n-1}$. Because $x'' \in V(X)$ and $y \in V(Y)$ there exists an index $0 \leqslant m \leqslant n-1$ such that $z_m \in V(X)$ and $z_{m+1} \in V(Y)$ (with the convention that $z_n = y$). But then, $x_m \sim x_{m+1}$ and by definition of the disjoint union of graphs, $z_m \nsim z_{m+1}$. By Lemma 8.13, we therefore have

$$U_{x_m z_m}U_{x_{m+1}z_{m+1}} = 0$$

and the result follows. $\qquad\square$

We can now combine all our results to prove the desired characterisation of perfect quantum strategies. Let us first restate the result for clarity.

Theorem 8.15 (Lupini–Mančinska–Roberson) *Let X and Y be two finite unoriented graphs. Then, there exists a perfect quantum strategy for the graph isomorphism game if and only if there exists a quantum permutation matrix P such that*

$$A_X P = P A_Y.$$

Proof The only if part is Theorem 8.3. As for the if part, by Proposition 8.11, either X and Y are both connected, or they are both non-connected. In the second case, we claim that their complements[4] are both connected. Indeed, if $x, y \in V(X)$, then either $x \nsim y$, in which case they are connected in X^c,

---

[4] The *complement* of a graph X is the graph X^c with adjacency matrix $A_{X^c} = J_N - A_X$. It has the same vertex set as X, but two edges are connected if and only if they are not connected in X.

or $x \sim y$, but since X has at least two connected components, there exists $z \in V(X)$ such that $x \not\sim z$ and $z \not\sim y$. It then follows that z is connected to both x and y in X^c, proving the claim.

Since

$$A_{X^c}P = (J_N - A_X)P = P(J_N - A_Y) = PA_{Y^c},$$

there is a perfect quantum strategy between X and Y if and only if there is one between X^c and Y^c. Therefore, there is no loss of generality in assuming X and Y to be both connected.

The whole problem is to construct a quantum permutation matrix which intertwines A_X and A_Y but such that there exists a trace on the $*$-algebra generated by its coefficients. To do this, we will build such a quantum permutation matrix inside the quantum automorphism group of the disjoint union $X \amalg Y$, so that the Haar state will provide us with the desired trace. Let us start by considering the fundamental representation U of $\mathrm{Aut}^+(X \amalg Y)$. It can be written as a block-matrix

$$U = \begin{pmatrix} U_x & V_{xy} \\ V_{yx} & U_y \end{pmatrix},$$

and the idea is that U_x and U_y 'correspond' to the fundamental representations of $\mathrm{Aut}^+(X)$ and $\mathrm{Aut}^+(Y)$ respectively, while V_{xy} and V_{yx} should 'correspond' to quantum permutation matrices intertwining them. In order to make this precise, let us set, for $x \in V(X)$ and $y \in V(Y)$,

$$Q_x = \sum_{y' \in V(Y)} U_{xy'} \text{ and } Q_y = \sum_{x' \in V(X)} U_{x'y}.$$

These are orthogonal projections, and we claim that they are all equal. Indeed, for $x, x' \in V(X)$,

$$\begin{aligned} (1 - Q_x)Q_{x'} &= \left(\sum_{x'' \in V(X)} U_{xx''} \right) Q_{x'} \\ &= \sum_{x'' \in V(X)} \sum_{y \in V(Y)} U_{xx''} P_{x'y} \\ &= 0 \end{aligned}$$

by Lemma 8.14. This implies that $Q_x Q_{x'} = Q_{x'}$ and since x and x' play symmetric roles, we conclude that $Q_x = Q_{x'}$. The same reasoning shows that $Q_y = Q_{y'}$ for any $y, y' \in V(Y)$. Moreover,

$$NQ_x = \sum_{x' \in V(X)} Q_{x'}$$

$$= \sum_{x' \in V(X)} \sum_{y' \in V(Y)} U_{x'y}$$

$$= \sum_{y' \in V(Y)} Q_{y'}$$

$$= N Q_y$$

so that our claim is proven. Let us denote by Q that projection and observe that it is designed to cut down the U_x and U_y parts of U and only keep V_{xy} and V_{yx}

Before going further, we will prove that Q is non-zero. Let P be the quantum permutation matrix in the statement and let R be the block-matrix

$$R = \begin{pmatrix} 0 & P \\ {}^t P & 0 \end{pmatrix}.$$

This is a quantum permutation matrix again. Moreover, it commutes by construction with

$$A_{X \amalg Y} = \begin{pmatrix} A_X & 0 \\ 0 & A_Y \end{pmatrix}$$

so that by universality there is a surjective $*$-homomorphism

$$\pi \colon \mathcal{O}(\mathrm{Aut}^+(X \amalg Y)) \to \mathcal{B}(H)$$

sending U_{xy} to P_{xy}. In particular, Q is sent to Id_H by this map, hence it is non-zero.

Let us now consider the $*$-algebra $\mathcal{A} \subset \mathcal{O}(\mathrm{Aut}^+(X \amalg Y))$ generated by the elements U_{xy} for $x \in V(X)$ an $y \in V(Y)$. We have, for any $x, x' \in V(X)$ and $y, y' \in V(Y)$,

$$Q U_{xx'} = 0 = U_{xx'} Q,$$
$$Q U_{yy'} = 0 = U_{yy'} Q,$$
$$Q U_{xy} = U_{xy} = U_{xy} Q,$$
$$Q U_{yx} = U_{yx} = U_{yx} Q$$

so that Q is a central (i.e. it commutes with everything) projection in $\mathcal{O}(\mathrm{Aut}^+(X \amalg Y))$ and

$$\mathcal{A} = Q \mathcal{O}(\mathrm{Aut}^+(X \amalg Y)) Q.$$

If

$$\rho \colon \mathcal{O}(\mathrm{Aut}^+(X \amalg Y)) \to \mathcal{B}\left(L^2\left(\mathrm{Aut}^+(X \amalg Y)\right)\right)$$

denotes the GNS representation of the Haar state, then the same properties hold for $\rho(\mathcal{A})$. If K denotes the range of Q, then we can see $\rho(\mathcal{A})$ as a $*$-algebra of bounded operators on K. Moreover, $W = (\rho(U_{xy}))_{x \in V(X), y \in V(Y)}$ can be seen as a matrix in $M_N(\mathcal{B}(K))$ whose entries are projections and since

$$
\begin{aligned}
\sum_{y \in V(Y)} W_{xy} &= \sum_{y \in V(Y)} \rho(U_{xy}) \\
&= \rho(Q) \\
&= \mathrm{Id}_K \\
&= \sum_{x \in V(X)} W_{xy},
\end{aligned}
$$

this is a quantum permutation matrix. To conclude, simply observe that

- $A_X W = W A_Y$ because ρ is a $*$-homomorphism;
- The Haar state restricts to a faithful trace on $\mathcal{A}$ and ρ is injective, hence $\tau \colon \rho(x) \mapsto h(x)$ is a well-defined tracial state on $\rho(A)$. $\qquad \square$

8.2 Finite-Dimensional Strategies

In this final section, we will explain a recent result concerning the structure of the quantum permutation groups S_N^+ called *residual finiteness*. This result has several motivations. One of them is the theory of finite-dimensional approximation of operator algebras. This is not the topic of this text, but we suggest reading the book [27] for a comprehensive treatment of that fascinating subject. In particular, Theorem 8.21 implies that the von Neumann algebras $L^\infty(S_N^+)$ satisfy the so-called *Connes embedding conjecture*, an important problem in von Neumann algebra theory. Another motivation comes from the connection with quantum information theory through the graph isomorphism game, and this is the path that we will follow.

8.2.1 Residual Finite-Dimensionality

Let us get back to the graph isomorphism game described in Section 1.1. Recall that perfect quantum strategies are given by specific quantum permutation matrices in the sense of Definition 1.5, and that quantum permutation matrices are nothing but representations of the $*$-algebra $\mathcal{O}(S_N^+)$ on a Hilbert space H. From the physical point of view, the *finite-dimensional* ones, namely those for which H can be chosen to be finite-dimensional, are of peculiar interest. Indeed, for practical purposes it is important to know whether the strategy can be implemented with a finite number of degrees of freedom. Formalising this leads to the following general problem:

Question Given two finite graphs X and Y such that there exists a perfect quantum strategy for the corresponding isomorphism game, does there exist a perfect finite-dimensional quantum strategy?

This turns out to be a deep problem with connections to important conjectures in quantum information theory, quantum computation and operator algebras. We will not dig into this here but simply consider the following related question at the level of quantum permutation groups.

Question Does the $*$-algebra $\mathcal{O}(S_N^+)$ have many finite-dimensional representation?

The question is purposely vague because part of it is precisely to find a suitable meaning to the word 'many'. One reasonable notion of having many finite-dimensional representations is the possibility of separating the points of the $*$-algebra by them. This leads to the following notion.

Definition 8.16 A $*$-algebra A is said to be *residually finite-dimensional* if for any $x \in A$ such that $x \neq 0$, there exists a finite-dimensional $*$-algebra B and a $*$-homomorphism $\pi \colon A \to B$ such that $\pi(x) \neq 0$.

The preceding definition can be restated in several ways. For instance, any finite-dimensional $*$-algebra B embeds through left multiplication into $\mathcal{L}(B) \simeq M_{\dim(B)}(\mathbf{C})$, so that we can equivalently assume that the points of A are separated by finite-dimensional representations. Moreover, gathering all these maps shows the existence of a family of integers $(n_i)_{i \in I}$ such that there is an embedding of $*$-algebras[5]

$$A \hookrightarrow \prod_{i \in I} M_{n_i}(\mathbf{C}).$$

There is no general recipe for proving such a property for a $*$-algebra. However, when the algebra comes from a discrete group, then one may translate this into a property of the group.

Definition 8.17 A discrete group Γ is said to be *residually finite* if and only if there exists finite groups $(\Lambda_i)_{i \in I}$ such that there is a group embedding

$$\Gamma \hookrightarrow \prod_{i \in I} \Lambda_i.$$

---

[5] No assumption is made concerning the cardinality of the set I.

In other words, the points of Γ are separated by its finite quotients. There is at least one simple connection between this and residual finite-dimensionality.

Exercise 8.3 Prove that if Γ is a residually finite group then $\mathbf{C}[\Gamma]$ is residually finite-dimensional.

Solution If Γ is residually finite, then let
$$x = \sum_{g \in F} \lambda_g a_g \in \mathbf{C}[\Gamma]$$
be an arbitrary element, where F is some finite subset of Γ. If Λ_g is a finite quotient of Γ in which g is not trivial, then the image of x in $\mathbf{C}[\Lambda]$ is non-zero, where
$$\Lambda = \prod_{g \in F} \Lambda_g.$$
Since Λ is finite, $\mathbf{C}[\Lambda]$ is finite-dimensional and we have proven that $\mathbf{C}[\Gamma]$ is residually finite-dimensional. $\qquad\square$

Upon seeing this, it is natural to wonder whether the converse also holds. This is false in general, but true if the group Γ is, moreover, finitely generated. The complete proof of this fact is beyond our scope, but we can at least give an idea of why it holds.

Theorem 8.18 *Let Γ be a finitely generated discrete group. If $\mathbf{C}[\Gamma]$ is residually finite-dimensional, then Γ is residually finite.*

Sketch of proof The proof proceeds in two steps. The first one can be done using our setting of unitary compact matrix groups. We first prove that Γ embeds into a unitary compact matrix group G. To do this, let us consider the $*$-algebra A of all coefficient functions of finite-dimensional unitary representations of Γ. For such a coefficient function ρ_{ij}, define a function $\Delta(\rho_{ij})$ on $\Gamma \times \Gamma$ by
$$\Delta(\rho_{ij}): (g, h) \mapsto \rho_{ij}(gh) = \sum_{k=1}^{\dim(\rho)} \rho_{ik}(g) \otimes \rho_{kj}(h).$$
This uniquely defines a $*$-homomorphism
$$\Delta: A \to A \otimes A$$
so that if we can construct a fundamental representation, we will have a unitary compact matrix group. Let $S = \{g_1, \dots, g_N\}$ be a generating set for Γ and fix for each $1 \leqslant \ell \leqslant N$ a finite-dimensional representation π_ℓ of $\mathbf{C}[\Gamma]$ such that
$$\pi_\ell(a_{g_\ell} - a_e) \neq 0.$$

By definition, π_ℓ restricts to a unitary representation ρ_ℓ of Γ such that $\rho_\ell(g_\ell) \neq$ Id. As a consequence, setting

$$\rho = \bigoplus_{\ell=1}^{N} \rho_\ell,$$

the pair (A, ρ) satisfies the axioms of a unitary compact matrix quantum group. Since A is moreover commutative, there exists[6] a unitary compact matrix group G such that $A = \mathcal{O}(G)$. Recall that G can be recovered through the characters of A. For each $g \in \Gamma$, the evaluation map ev_g yields such a character. Moreover, two such characters are distinct because, as the points of $\mathbf{C}[\Gamma]$ are separated by finite-dimensional representation, the points of Γ are separated by finite-dimensional unitary representations. As a consequence, Γ embeds into G.

The second step is a theorem by A. Mal'cev from [56], stating that any finitely generated linear[7] group is residually finite. The interested reader can refer, for instance, to [27, theorem 6.4.13] for an elementary proof of this fact. Since we just proved that Γ embeds into a group of matrices, hence is linear, the proof is complete. □

Remark 8.19 The group G that we constructed out of Γ is known as its *Bohr compactification*. It is the 'largest' compactification of Γ in the sense that it is universal with respect to group homomorphisms with dense image into compact groups.

Based on this result, we can give the following definition.

Definition 8.20 A unitary compact matrix quantum group $\mathbb{G}$ is said to be *residually finite* if $\mathcal{O}(\mathbb{G})$ is a residually finite-dimensional *-algebra.

The main point of this restatement is that we have turned our algebraic problem into a quantum group problem, opening the door to the use of representation theory techniques, and in particular the combinatorics of partitions. We will illustrate this by answering Question 8.2.1.

Theorem 8.21 (Brannan–Chirvasitu–F.) *The quantum permutation groups S_N^+ are residually finite for all $N \in \mathbf{N}$.*

Before turning to the proof in the next section, let us show that the finite generation assumption in Theorem 8.18 is necessary.

---

[6] By a straightforward generalisation of Corollary 5.18.

[7] A discrete group is said to be *linear* if it is a subgroup of $GL_N(\mathbf{K})$ for some field $\mathbf{K}$ and some integer $N \in \mathbf{N}$.

Exercise 8.4 The purpose of this exercise is to show that the additive group $\mathbb{Q}$ of the rationals is not residually finite, while the $*$-algebra $\mathbf{C}[\mathbb{Q}]$ is residually finite-dimensional.

1. Let G be a finite group with neutral element e and let $\pi\colon \mathbb{Q} \to G$ be a group homomorphism. Show that $\pi(x) = e$ for all $x \in \mathbb{Q}$.
2. Let $x \in \mathbf{C}[\mathbb{Q}]$.
   (a) Prove that there is an element $y \in \mathbb{Q}$ such that $x \in \mathbf{C}[\langle y \rangle] \subset \mathbf{C}[\mathbb{Q}]$, where $\langle y \rangle$ denotes the subgroup generated by y.
   (b) Let $a \in \mathbf{C}[\mathbf{Z}]$, find a $*$-homomorphism $\varphi_a\colon \mathbf{C}[\mathbf{Z}] \to \mathbf{C}$ such that $\varphi_a(a) \neq 0$.
   (c) Deduce from this the existence of a $*$-homomorphism $\psi_x\colon \mathbf{C}[\mathbb{Q}] \to \mathbf{C}$ such that $\psi_x(x) \neq 0$ and conclude.

Solution 1. Let $x \in \mathbb{Q}$ and observe that by Lagrange's theorem, $\pi(x)^{|G|} = e$. Thus, setting $y = x/|G|$, we have

$$\pi(x) = \pi\left(\frac{x}{|G|}\right)^{|G|} = \pi(y)^{|G|} = e.$$

2. (a) By definition, x can be written as

$$x = \sum_{i=1}^{n} \lambda_i x_i$$

with $\lambda_i \in \mathbf{C}$ and $x_i = p_i/q_i \in \mathbb{Q}$ for $1 \leqslant i \leqslant n$. Let z be the least common multiple of q_i for $1 \leqslant i \leqslant n$ and set $y = z^{-1}$. Then, x_i is an integer multiple of y for all $1 \leqslant i \leqslant n$, hence the result.

   (b) Let us denote by a_k the basis vector of $\mathbf{C}[\mathbf{Z}]$ corresponding to the integer k and write

$$a = \sum_{k=1}^{n} \lambda_k a_k.$$

Let us pick $\theta \in \mathbf{R}$ such that $\omega = e^{i\theta}$ is not a root of the polynomial $\sum_k \lambda_k X^k$. Then, the map $\varphi_a\colon a_k \to \omega^k$ works.

   (c) Let $\theta \in \mathbf{R}$ be chosen as in the preceding question applied to $\mathbf{Z} = \langle y \rangle$. Then, the map $\psi_x\colon a_{p/q} \mapsto e^{ip\theta/q}$ works. $\qquad\square$

8.2.2 The Proof

8.2.2.1 Topological Generation

The strategy for the proof of Theorem 8.21 for S_N^+ is induction on N. This may seem strange at first, but it is rather natural if one thinks in terms of discrete groups. Indeed, let Γ be a finitely generated discrete group and let Γ_1 and Γ_2 be quotients of Γ. Assume that the two following properties hold:

- Both Γ_1 and Γ_2 are residually finite;
- Any element of Γ has a non-trivial image either in Γ_1 or in Γ_2.

Then, Γ is obviously residually finite.[8]

This observation suggests to reduce the problem to quotients which are easier to handle while being large enough to recover the original group. To express it at the level of compact matrix quantum groups, we should try to write the second condition in terms of the $*$-algebras $C[\Gamma_1]$, $C[\Gamma_2]$ and $C[\Gamma]$. First, note that the quotient map $\pi_i \colon \Gamma \to \Gamma_i$ for $i = 1, 2$ induces a surjective $*$-homomorphism $C[\Gamma] \to C[\Gamma_i]$ mapping the fundamental representation to the fundamental representation coefficientwise. Moreover, if $\Lambda = \ker(\pi_1) \cap \ker(\pi_2)$, then both π_1 and π_2 factor through $\Gamma \to \Gamma/\Lambda$, so that the $*$-algebra maps also factor through

$$C[\Gamma] \to C[\Gamma/\Lambda].$$

In other words, the second condition is equivalent to the impossibility of factoring both π_1 and π_2 at the same time.

The last step to obtain a tractable quantum version of the above strategy is to translate our observations in a 'compact' perspective. This will be done thanks to the key notion of *topological generation*. This idea was first introduced by A. Chirvasitu in [28] (though not under that name). To explain it, let us write $\mathbb{H} < \mathbb{G}$ if $\mathbb{G} = (\mathcal{O}(\mathbb{G}), u)$ and $\mathbb{H} = (\mathcal{O}(\mathbb{H}), v)$ are unitary compact matrix quantum groups with a surjective $*$-homomorphism

$$\pi \colon \mathcal{O}(\mathbb{G}) \to \mathcal{O}(\mathbb{H})$$

such that $\pi(u_{ij}) = v_{ij}$. We already mentionned in the beginning of Section 1.3.3 that this is a natural analogue of the inclusion of compact groups. Note that in such a situation, if both $\mathbb{G}$ and $\mathbb{H}$ are classical compact groups, then any character on $\mathcal{O}(\mathbb{H})$ yields a character on $\mathcal{O}(\mathbb{G})$ through composition with π, hence $\mathbb{H}$ embeds as a closed subgroup of $\mathbb{G}$, explaining the notation.

---

[8] We are saying here that residually 'residually finite' groups are residually finite, which is almost tautological.

Definition 8.22 Consider $\mathbb{G}_1, \mathbb{G}_2 < \mathbb{G}$ given by surjections π_1 and π_2. We say that $\mathbb{G}$ is *topologically generated* by $\mathbb{G}_1$ and $\mathbb{G}_2$ if there is no $\mathbb{H} < \mathbb{G}$ (except for $\mathbb{G}$ itself) given by a surjection π such that both π_1 and π_2 factor through π.

Remark 8.23 Assume that $G_1, G_2 < G$ are two closed compact subgroups, and let K be the closure of the subgroup of G generated by G_1 and G_2. Then, $G_1, G_2 < K < G$ so that π factors through the restriction map

$$\pi_K \colon \mathcal{O}(G) \to \mathcal{O}(K).$$

In other words, G is topologically generated by G_1 and G_2 if and only if it is generated by them as a topological group, hence the name.

The core result that we need is the following.

Proposition 8.24 *If $\mathbb{G}$ is topologically generated by two residually finite unitary compact matrix quantum subgroups $\mathbb{G}_1$ and $\mathbb{G}_2$, then $\mathbb{G}$ is residually finite.*

Proof Let $\mathcal{I}$ be the intersection of the kernels of all finite-dimensional $*$-representations of $\mathcal{O}(\mathbb{G})$, let $\mathcal{A} = \mathcal{O}(\mathbb{G})/\mathcal{I}$ be the corresponding quotient with canonical surjection

$$\pi \colon \mathcal{O}(\mathbb{G}) \to \mathcal{A}$$

and set $v_{ij} = \pi(u_{ij})$. Let $x \in \mathcal{I}$ and assume that $\pi_i(x) \neq 0$ for some $i \in \{1, 2\}$. Then, because $\mathbb{G}_i$ is residually finite, there exists a finite-dimensional representation ρ of $\mathcal{O}(\mathbb{G}_i)$ such that $\rho \circ \pi_i(x) \neq 0$. But $\rho \circ \pi_i$ is also a finite-dimensional representation of $\mathcal{O}(\mathbb{G})$, hence its kernel contains $\mathcal{I}$, a contradiction. As a consequence, both π_1 and π_2 factor through π. To conclude, we therefore simply have to show that $(\mathcal{A}, v)$ is a unitary compact matrix quantum group.

To do so, first note that $\mathcal{I}$ is an intersection of $*$-ideals, hence $\mathcal{A}$ is a $*$-algebra. Moreover, if $x \in \mathcal{I}$, and if ρ_1, ρ_2 are finite-dimensional representations of $\mathcal{O}(\mathbb{G})$, then

$$(\rho_1 \otimes \rho_2) \circ \Delta$$

is also a finite-dimensional representation, hence $(\rho_1 \otimes \rho_2) \circ \Delta(x) = 0$. By Lemma 1.21, this means that

$$\Delta(x) \subset \ker(\rho_1) \otimes \mathcal{O}(\mathbb{G}) + \mathcal{O}(\mathbb{G}) \otimes \ker(\rho_2).$$

In other words, writing

$$\Delta(x) = \sum_{i=1}^{n} x_i \otimes y_i$$

we have that for any $1 \leqslant i \leqslant n$, either $\rho_1(x_i) = 0$ or $\rho_2(y_i) = 0$. We claim that this implies

$$\Delta(x) \in \mathcal{I} \otimes \mathcal{O}(\mathbb{G}) + \mathcal{O}(\mathbb{G}) \otimes \mathcal{I}.$$

Indeed, if this was not the case, then there would exist $1 \leqslant i \leqslant n$ such that $x_i, y_i \notin \mathcal{I}$. By definition of $\mathcal{I}$, this means that there exists finite-dimensional representations ρ_1, ρ_2 such that $\rho_1(x_i) \neq 0$ and $\rho_2(y_i) \neq 0$, a contradiction.

It follows that there exists a *-homomorphism $\widetilde{\Delta}: \mathcal{A} \to \mathcal{A} \otimes \mathcal{A}$ uniquely determined by

$$\widetilde{\Delta} \circ \pi = (\pi \otimes \pi) \circ \Delta.$$

In particular,

$$\widetilde{\Delta}(v_{ij}) = \sum_{k=1}^{N} v_{ik} \otimes v_{kj}$$

and $(\mathcal{A}, v)$ is a unitary compact matrix quantum group. By the definition of topological generation, we must therefore have $\mathcal{A} = \mathcal{O}(\mathbb{G})$ and $v = u$, that is, $\mathcal{I} = \{0\}$. $\qquad\square$

In the sequel, we will need a criterion to prove topological generation which is better suited to the partition setting.

Proposition 8.25 *Let $\mathbb{G}$ be a unitary compact matrix quantum group and let $\mathbb{G}_1, \mathbb{G}_2 < \mathbb{G}$ be quantum subgroups. If for any word w over $\{\circ, \bullet\}$,*

$$\mathrm{Mor}_{\mathbb{G}}(u^w, \varepsilon) = \mathrm{Mor}_{\mathbb{G}_1}(u_1^w, \varepsilon) \cap \mathrm{Mor}_{\mathbb{G}_2}(u_2^w, \varepsilon),$$

then $\mathbb{G}$ is topologically generated by $\mathbb{G}_1$ and $\mathbb{G}_2$.

Proof First note that if $f: V^w \to \mathbf{C}$ is invariant under ρ_{u^w}, then it is also invariant under

$$(\mathrm{id} \otimes \pi_i) \circ \rho_{u^w} = \rho_{u_i^w}$$

for $i = 1, 2$ so that the left-hand side is always contained in the right-hand side.

If now we have a unitary compact matrix quantum group $(\mathbb{H}, v)$ and a surjection $\pi: \mathcal{O}(\mathbb{G}) \to \mathcal{O}(\mathbb{H})$ through which both π_1 and π_2 factor, then applying the previous remark yields

$$\mathrm{Mor}_{\mathbb{G}}(u^w, \varepsilon) \subset \mathrm{Mor}_{\mathbb{H}}(v^w, \varepsilon) \subset \mathrm{Mor}_{\mathbb{G}_1}(u_1^w, \varepsilon) \cap \mathrm{Mor}_{\mathbb{G}_2}(u_2^w, \varepsilon)$$

for all words w over $\{\circ, \bullet\}$. Therefore, the assumption forces $\mathrm{Mor}_{\mathbb{G}}(u^w, \varepsilon) \subset \mathrm{Mor}_{\mathbb{H}}(v^w, \varepsilon)$ for all w. Using Frobenius reciprocity, we deduce that for any words w and w',

$$\mathrm{Mor}_{\mathbb{G}}\left(u^w, u^{w'}\right) = \mathrm{Mor}_{\mathbb{H}}\left(v^w, v^{w'}\right),$$

hence $\mathbb{G} = \mathbb{H}$ by Theorem 3.7, and this finishes the proof. $\qquad\square$

There is another notion which is close to topological generation but in a sense more general. We will not need it afterwards, hence we simply study it through an exercise.

Exercise 8.5 Let $(\mathbb{G}, u)$ be an orthogonal compact matrix quantum group and let $\mathcal{A}$ be a $*$-algebra. A $*$-homomorphism $\pi \colon \mathcal{O}(\mathbb{G}) \to \mathcal{A}$ is said to be *inner faithful* if there is no quantum subgroup $\mathbb{H}$ of $\mathbb{G}$ (except for $\mathbb{G}$ itself) through which π factors.

1. Prove that if $\mathcal{A}$ is residually finite-dimensional and π is inner faithful, then $\mathbb{G}$ is residually finite.
2. Let $\mathbb{G}_1, \mathbb{G}_2 < \mathbb{G}$. Prove that they topologically generate $\mathbb{G}$ if and only if the map

$$\pi = (\pi_1 \otimes \pi_2) \circ \Delta : \mathcal{O}(\mathbb{G}) \to \mathcal{O}(\mathbb{G}_1) \otimes \mathcal{O}(\mathbb{G}_2)$$

   is inner faithful.

Solution 1. The proof is the same as that of Proposition 8.24. Consider the intersection $\mathcal{I}$ of the kernels of all finite-dimensional $*$-representations. Then, $\mathcal{I} \subset \ker(\pi)$ because $\mathcal{A}$ is residually finite-dimensional, hence π factors through $\mathcal{O}(\mathbb{G})/\mathcal{I}$ which defines a quantum subgroup. By inner faithfulness, this implies $\mathcal{I} = \{0\}$.

2. Let $\mathbb{H} < \mathbb{G}$ be a quantum subgroup given by a surjection $\pi' \colon \mathcal{O}(\mathbb{G}) \to \mathcal{O}(\mathbb{H})$ and assume that there are factorisations $\pi_1 = \pi'_1 \circ \pi'$ and $\pi_2 = \pi'_2 \circ \pi'$. Then,

$$\pi = (\pi'_1 \otimes \pi'_2) \circ (\pi' \otimes \pi') \circ \Delta$$
$$= (\pi'_1 \otimes \pi'_2) \circ \Delta \circ \pi'.$$

hence π factors through $\mathbb{H}$. Therefore, if π is inner faithful then $\mathbb{G}_1$ and $\mathbb{G}_2$ topologically generate $\mathbb{G}$. Conversely, if $\mathbb{G}_1$ and $\mathbb{G}_2$ topologically generate $\mathbb{G}$, let $\mathbb{H} < \mathbb{G}$ be such that π factors through π'. Denoting by ε_1 to counit of $\mathbb{G}_1$ (see Corollary 2.18), we have

$$(\varepsilon_1 \otimes \mathrm{id}) \circ \pi = (\varepsilon_1 \circ \pi_1 \otimes \pi_2) \circ \Delta$$
$$= (\varepsilon_{\mathbb{G}} \otimes \pi_2) \circ \Delta$$
$$= \pi_2 \circ (\varepsilon_{\mathbb{G}} \otimes \mathrm{id}) \circ \Delta$$
$$= \pi_2.$$

Therefore,

$$\pi_2 = [(\varepsilon_1 \otimes \mathrm{id}) \circ \pi''] \circ \pi'$$

and π_2 factors through $\mathbb{H}$. Similarly, π_1 factors through $\mathbb{H}$ so that by the topological generation assumption, $\mathbb{H} = \mathbb{G}$. $\qquad\square$

8.2.2.2 At Least Six Points

The following result [29, theorem 3.3 and theorem 3.12] is the key tool to prove residual finite-dimensionality. In this statement, we understand the inclusion $S_{N-1}^+ < S_N^+$ through the surjection

$$\pi_1 \colon \mathcal{O}(S_N^+) \to \mathcal{O}(S_{N-1}^+)$$

sending u_{11} to 1.

Theorem 8.26 (Brannan–Chirvasitu–F.) *For any* $N \geqslant 6$*, the quantum permutation group* S_N^+ *is topologically generated by* S_{N-1}^+ *and* S_N*.*

Proof of Theorem 8.26 Set $V = \mathbf{C}^N$ and let P_N, P_{N-1} and P_N^{class} denote the fundamental representations of S_N^+, S_{N-1}^+ and S_N respectively. We will say that a map is S_N^+ (respectively S_{N-1}^+, respectively S_N)-invariant if it commutes with appropriate tensor powers of P_N (respectively P_{N-1}, respectively P_N^{class}). By Proposition 8.25, it is enough to prove that if

$$f \colon V^{\otimes k} \to \mathbf{C}$$

is a linear map which is both S_{N-1}^+-invariant and S_N-invariant, then f is S_N^+-invariant.

Let us therefore consider such a map f and, for $1 \leqslant i \leqslant N$, let $V_i = e_i^{\perp}$. Because f is S_{N-1}^+ invariant, its restriction to $V_1^{\otimes k}$ is a linear combination of partition maps: there exist complex numbers $(\lambda_p)_{p \in NC(k)}$ such that

$$f_{|V_1^{\otimes k}} = \sum_{p \in NC(k)} \lambda_p f_p.$$

Let us set

$$\tilde{f} = f - \sum_{p \in NC(k)} \lambda_p f_p.$$

This is still invariant under S_{N-1}^+ and S_N and vanishes on $V_1^{\otimes k}$. Our task is to show that it vanishes on the whole of $V^{\otimes k}$.

For this purpose, let us set $V_i' = \mathbf{C}e_i$, so that

$$V^{\otimes k} = \bigoplus_{\epsilon_1, \cdots, \epsilon_k} V_1^{\epsilon_1} \otimes \cdots \otimes V_1^{\epsilon_k}.$$

where ϵ is either prime or nothing. Let us consider one of these summands where V_i appears ℓ times and denote it by W. Since S_{N-1}^+ acts trivially on V_1', there exists a linear S_{N-1}^+-equivariant isomorphism

$$\Phi: W \to V_1^{\otimes \ell}.$$

As a consequence, there exist complex numbers $(\mu_p)_{p \in NC(\ell)}$ such that

$$\widetilde{f} \circ \Phi^{-1} = \sum_{p \in NC(\ell)} \mu_p f_p.$$

The idea now is to use the linear independence of the partition maps to conclude that $\mu_p = 0$ for all p, hence that $\widetilde{f} \circ \Phi^{-1} = 0$.

To do this, set $V_{1,N} = V_1 \cap V_N$ and observe that

$$\Phi^{-1}\left(V_{1,N}^{\otimes \ell}\right) \subset V_N^{\otimes k}.$$

Now, by S_N-invariance, we can exchange e_1 and e_N without changing the value of $\widetilde{f}$, hence it vanishes on $V_N^{\otimes k}$. Thus, $\widetilde{f} \circ \Phi^{-1}$ vanishes on $V_{1,N}^{\otimes \ell}$. Since $N \geqslant 6$, $\dim(V_{1,N}) \geqslant 4$ so that non-crossing partition maps on $V_{1,N}^{\otimes \ell}$ are linearly independent by Theorem 4.1. This forces $\mu_p = 0$ for all p, hence $\widetilde{f} = 0$. $\qquad\square$

8.2.2.3 At Most Five Points

So far, Theorem 8.26 is useless for an inductive proof since we do not know the base case: is $\mathcal{O}(S_5^+)$ residually finite-dimensional? Before addressing this question, let us comment on the case $N = 4$. It was shown by B. Collins and T. Banica in [13, theorem 4.1] that there is an embedding

$$\pi: C(S_4^+) \hookrightarrow C(SU(2), M_4(\mathbf{C})).$$

As a consequence, the *-representations $\pi_g: x \mapsto \pi(x)(g)$ for all $g \in SU(2)$ separate the points, that is, S_4^+ is residually finite-dimensional.

Only $N = 5$ remains, and this is indeed a non-trivial matter which can be solved using the classification of subfactors.[9] It was shown by T. Banica in [8, theorem 7.10] that S_5^+ enjoys a much stronger property: the canonical map

$$\pi_{ab}: \mathcal{O}(S_5^+) \to \mathcal{O}(S_5)$$

does not factor through any compact matrix quantum group. The idea of the proof is that by [5] and [67], any quantum subgroup of S_5^+ yields a subfactor planar algebra at index 5, with extra properties if it contains S_5. Moreover,

---

[9] These are objects from the theory of von Neumann algebras which have some similarities and connections with quantum groups; see, for instance, [46].

this correspondence is one-to-one. Now one has to look at the complete list of subfactor planar algebras at index 5 satisfying the extra properties (see, for instance, the survey [45]) and check that none of the corresponding quantum groups contains S_5. We can now complete our proof.

Proof of Theorem 8.21 First, we can extend the statement of Theorem 8.26 to $N = 5$. Indeed, if S_5^+ was not topologically generated by S_5 and S_4^+, then there would be a compact matrix quantum group $S_5 < \mathbb{H} < S_5^+$ which is not classical because $S_4^+ < \mathbb{H}$, contradicting the aforementioned result. Thus, S_4^+ and S_5 topologically generate S_5^+. We can now conclude by induction, starting from the fact that S_4^+ is residually finite.   □

Remark 8.27 Let us mention the following interesting problem: is there a compact matrix quantum group through which the quotient map

$$\pi_{\mathrm{ab}} \colon \mathcal{O}(S_N^+) \to \mathcal{O}(S_N)$$

factors? If not, then the inclusion $S_N < S_N^+$ is said to be *maximal*. For $N \leqslant 3$, maximality trivially follows from the equality of the two quantum groups. For $N = 4$, it can be proven by checking the list of all quantum subgroups of S_4^+ given, for instance, in [10] and we already mentioned the proof of T. Banica in [8, theorem 7.10] for $N = 5$. This is all that is known to the day of this writing.

8.2.3 Further Examples

To conclude this chapter, we will prove that our other friends U_N^+, O_N^+ and H_N^{s+} also are residually finite. There exist several ways of proving this and we choose some which involve different techniques in each case. We will therefore treat them separately.

8.2.3.1 Quantum Unitary Groups

We start with the quantum unitary group U_N^+. In that case, one can prove a topological generation result in a way paralleling the proof for S_N^+. This, however, requires the use of an algebraic result which is interesting in its own right. To make the proofs clear, we first isolate an elementary property of analytic functions.

Lemma 8.28 *Let $f \colon \mathbf{C}^n \to \mathbf{C}$ be an analytic map which vanishes on $\mathbf{R}^n$. Then, $f = 0$.*

Proof We will prove the result by induction on n. For $n = 1$, the result holds because a non-zero analytic function has isolated zeros. Assume now that the result holds for some $n \geqslant 1$ and consider $f \colon \mathbf{C}^{n+1} \to \mathbf{C}$. For $x_1, \ldots, x_n \in \mathbf{R}$, we define an analytic function on $\mathbf{C}$ by

$$f_{x_1, \ldots, x_n} \colon z \mapsto f(z, x_1, \ldots, x_n).$$

Because $f_{x_1, \ldots, x_n}$ vanishes on the real line by assumption, it vanishes everywhere. If now $z \in \mathbf{C}$ is fixed, we have just proven that the analytic function

$$f_z \colon (z_1, \ldots, z_n) \in \mathbf{C}^n \mapsto f(z, z_1, \ldots, z_n)$$

vanishes on $\mathbf{R}^n$, hence by the induction hypothesis it vanishes on all of $\mathbf{C}^n$, concluding the proof. $\qquad \Box$

We can now prove the technical result that we need, linking the unitary group U_N with the general linear complex group $\mathrm{GL}_N(\mathbf{C})$.

Proposition 8.29 *Let* $P \in \mathcal{O}(M_N(\mathbf{C}))$ *be such that* $P(M) = 0$ *for all* $M \in U_N$. *Then,* $P(M) = 0$ *for all* $M \in \mathrm{GL}_N(\mathbf{C})$.

Proof Let us denote by $\mathfrak{u}_N \subset M_N(\mathbf{C})$ the set of matrices such that $M + M^* = 0$. It is well known that the exponential map

$$\exp \colon M_N(\mathbf{C}) \to \mathrm{GL}_N(\mathbf{C})$$

sends $\mathfrak{u}_N$ surjectively onto U_N. Therefore, the map

$$\varphi = P \circ \exp \colon M_N(C) \to \mathbf{C},$$

which is analytic by construction, vanishes on $\mathfrak{u}_N$. The key observation is that as a *real* vector space, we have a decomposition $M_N(\mathbf{C}) = \mathfrak{u}_N \oplus i\mathfrak{u}_N$. Let $A_1, \ldots, A_{N^2}$ be a real basis of $\mathfrak{u}_N$ and set

$$\widetilde{\varphi}(z_1, \ldots, z_{N^2}) = \varphi \left(\sum_{j=1}^{N^2} z_j A_j \right).$$

This is an analytic map on $\mathbf{C}^{N^2}$ which vanishes on $\mathbf{R}^{N^2}$. By Lemma 8.28, $\widetilde{\varphi} = 0$, hence also $\varphi = 0$. This means that P vanishes on the range of exp, which is precisely $\mathrm{GL}_N(\mathbf{C})$. $\qquad \Box$

Remark 8.30 Proposition 8.29 can be rephrased in algebraic terms by saying that U_N is dense in $\mathrm{GL}_N(\mathbf{C})$ for the Zariski topology.

We will not use this result directly in that form, but rather restate it in terms of invariants.

Corollary 8.31 *Let $f \colon (\mathbf{C}^N)^{\otimes k} \to \mathbf{C}$ be a polynomial function which is invariant under the action of U_N. Then, it is invariant under the action of $\mathrm{GL}_N(\mathbf{C})$.*

Proof Let $\xi \in (\mathbf{C}^N)^{\otimes k}$ and set

$$P_\xi \colon M \in M_N(\mathbf{C}) \mapsto f(M\xi).$$

This is a polynomial function which is by assumption constant on U_N. If λ_ξ is that constant value, then $P_\xi - \lambda_\xi$ is a polynomial vanishing on U_N, hence vanishing on $\mathrm{GL}_N(\mathbf{C})$ by Proposition 8.29. As a consequence, the value of $f(M\xi)$ is the same for all $M \in \mathrm{GL}_N(\mathbf{C})$. This being true for any ξ, the proof is complete. $\square$

We are now ready to prove a topological generation result for quantum unitary groups. That proof is due to A. Chirvasitu in [29, proposition 2.3].

Theorem 8.32 (Chirvasitu) *For any $N \geqslant 3$, U_N^+ is topologically generated by U_{N-1}^+ and U_N.*

Proof Let us set $V = \mathbf{C}^N$, $W = \mathrm{Vect}(e_1, \ldots, e_{N-1})$ and let W' be any supplement of W in V. Let us fix a tensor product of k copies of V and $\overline{V}$

$$V^\epsilon = V^{\epsilon_1} \otimes \cdots \otimes V^{\epsilon_k},$$

where $\epsilon = (\epsilon_1, \ldots, \epsilon_k)$ is a list of elements of $\{\circ, \bullet\}$, $V^\circ = V$ and $V^\bullet = \overline{V}$, and let f be a linear map on that tensor product which is invariant under the action of U_{N-1}^+ and U_N. Writing $V = W \oplus W'$, we can split the tensor product as a sum of tensors involving W, W' and their conjugates,

$$V^\epsilon = \sum_{\eta_i \in \{\emptyset, \prime\}} W^{\eta_i, \epsilon_i}.$$

Among these there is one summand, denoted by $W_\emptyset$, which corresponds to $\eta_i = \emptyset$ for all $1 \leqslant i \leqslant k$ and therefore only involves W and its conjugate. The restriction of f to that subspace is by definition U_{N-1}^+-invariant, hence can be written as a linear combination of non-crossing partitions,

$$f_{|W_\emptyset} = \sum_{p \in \mathcal{U}(k)} \lambda_p f_p.$$

Thus, setting

$$\tilde{f} = f - \sum_{p \in \mathcal{U}(k)} \lambda_p f_p,$$

we now have a map which is still U_{N-1}^+-invariant and U_N-invariant and such that $\widetilde{f}_{|W_\emptyset} = 0$.

The key argument now is that all this (including the coefficients λ_p) does not depend on the precise choice of W'. Let us set, for $t \in \mathbf{R}$,

$$W_t' = \mathrm{Span}\{e_1 + te_N\}.$$

Then, W_t' is a supplement of W. Moreover, for $s, t \neq 0$, there exists an invertible matrix $M_{t,s} \in \mathrm{GL}_N(\mathbf{C})$ such that $M_{t,s}(W_t') = W_s'$ while $M_{t,s|W} = \mathrm{Id}_W$. But by Corollary 8.31, $\widetilde{f}$ is $\mathrm{GL}_N(\mathbf{C})$-invariant because it is U_N-invariant. Let us restate this in terms of vectors. Let $\xi \in V^\epsilon$ and decompose it on the tensor product basis coming from the decomposition $V = W \oplus W_1'$. Let now $\xi(t)$ be the vector obtained by taking the same coordinates but in the tensor product basis coming from $V = W \oplus W_t'$. Then,

$$\widetilde{f}(\xi_t) = \widetilde{f}(M_{1,t}\xi_1) = \widetilde{f}(\xi_1).$$

We can now conlude using the continuity of $\widetilde{f}$. Keeping the previous notations, let t go to 0. Then ξ_t converges to a vector $\xi_0 \in W^\epsilon$. Because $\widetilde{f}(\xi_t)$ is constant, it follows by continuity that

$$\widetilde{f}(\xi_1) = \widetilde{f}(\xi_0) = 0$$

and the proof is complete. □

We can now prove residual finiteness.

Corollary 8.33 *The compact matrix quantum group U_N^+ is residually finite for all integers N.*

Proof For $N = 1$, $U_1^+ = U_1 = \mathbb{T}$ is residually finite because it is a classical compact group, hence its algebra of regular functions is separated by evaluations at elements of the group, which are one-dimensional representations. For $N = 2$, it was shown in [4, lemma 7] that $\mathcal{O}(U_2^+)$ embeds into $\mathcal{O}(SU(2)) * \mathcal{O}(\mathbb{T})$ (the embedding is in fact a free complexification, and the proof is extremely close to that of Theorem 6.34 using the fact that the representation theory of $SU(2)$ is the same as that of O_2^+) so that the result can be derived from the fact that a free product of residually finite-dimensional $*$-algebras is again residually finite-dimensional by [28, proposition 2.7]. Now, the result follows by induction on N using Theorem 8.32 and Proposition 8.24. □

8.2.3.2 Quantum Orthogonal Group

A proof of the same result for O_N^+ may be doable along the same line, even though one has to be careful that one cannot obtain GL_N-invariance out of O_N-invariance. In any case, one can also deduce the result from O_N^+ directly from the result for U_N^+. The proof makes use of a natural analogue of the projective quotients of matrix groups.

Definition 8.34 Let $(\mathbb{G}, u)$ be a unitary compact matrix quantum group. Then, the $*$-algebra $\mathcal{O}(P\mathbb{G})$ generated by the coefficients of $u \otimes \overline{u}$, together with the representation $u \otimes \overline{u}$, is a unitary compact matrix quantum group called the *projective version of* $\mathbb{G}$.

Remark 8.35 One could, of course, also consider the representation $\overline{u} \otimes u$ in the definition. The antipode $S \colon \mathcal{O}(\mathbb{G}) \to \mathcal{O}(\mathbb{G})$ then provides a $*$-anti-isomomorphism between the $*$-subalgebras generated by the coefficients of $u \otimes \overline{u}$ and $\overline{u} \otimes u$, so that the corresponding compact matrix quantum groups are anti-isomorphic. Being anti-isomorphic is almost like being isomorphic, and this does not make any difference in the sequel.

For the sake of clarity, we will prove two separate lemmata before establishing the residual finiteness of O_N^+.

Lemma 8.36 *The canonical surjection $\pi \colon \mathcal{O}(U_N^+) \to \mathcal{O}(O_N^+)$ restricts to an isomorphism between $\mathcal{O}(PU_N^+)$ and $\mathcal{O}(PO_N^+)$.*

Proof Consider the $*$-algebra $A = \mathcal{O}(O_N^+) * \mathcal{O}(\mathbb{T})$ and set $v_{ij} = tu_{ij}$. The matrices $v = (v_{ij})_{1 \leqslant i,j \leqslant N}$ and $\overline{v}$ are easily seen to be unitary, hence there is by the universal property of $\mathcal{O}(U_N^+)$ a surjective $*$-homomorphism

$$\psi \colon \mathcal{O}(U_N^+) \to A$$

sending V_{ij} to v_{ij}. But then, $\psi(V_{ij}^* V_{k\ell}) = u_{ij} u_{k\ell} \in \mathcal{O}(PO_N^+)$ so that ψ restricts to a surjective $*$-homomorphism $\mathcal{O}(PU_N^+) \to \mathcal{O}(PO_N^+)$ which coincides with the restriction of π. To conclude, recall that by Theorem 6.34, ψ is an isomorphism on its image, hence injective. The same, therefore, holds for the restriction of π and the proof is complete. $\qquad\square$

Remark 8.37 The previous statement is rather surprising, since in the classical case PU_N and PO_N are not isomorphic (consider the case $N = 2$). This confirms in particular that the tensor complexification $\overline{O}_N$ of O_N (see 6.4.3) is not isomorphic to U_N. Indeed, the preceding proof adapts verbatim to show that $P\overline{O}_N = PO_N$.

The second lemma is a useful characterisation of when a surjective homomorphism between compact quantum groups is an isomorphism. Recall that if $\pi\colon \mathcal{O}(\mathbb{G}) \to \mathcal{O}(\mathbb{H})$ is a $*$-homomorphism such that

$$(\pi \otimes \pi) \circ \Delta_{\mathbb{G}} = \Delta_{\mathbb{H}} \circ \pi,$$

then for any representation $v = (v_{ij})_{1 \leqslant i,j \leqslant n}$ of $\mathbb{G}$, the matrix $\widehat{\pi}(v) = (\pi(v_{ij}))_{1 \leqslant i,j \leqslant n}$ defines a representation of $\mathbb{H}$. Moreover, if $T \in \mathrm{Mor}_{\mathbb{G}}(v, w)$, then applying π to all the coefficents of the equation $Tv = wT$ yields $T\widehat{\pi}(v) = \widehat{\pi}(w)T$, so that $T \in \mathrm{Mor}_{\mathbb{H}}(\widehat{\pi}(v), \widehat{\pi}(w))$. In particular, if $\widehat{\pi}(v)$ is irreducible, then so is v.

Lemma 8.38 *Let $\pi\colon \mathcal{O}(\mathbb{G}) \to \mathcal{O}(\mathbb{H})$ be a surjective $*$-homomorphism. Then, the following are equivalent:*

1. *For any irreducible representation v of $\mathbb{G}$, $\widehat{\pi}(v)$ does not contain the trivial representation;*
2. *π is an isomorphism.*

Proof Assume that the first point holds and let v be an irreducible representation of $\mathbb{G}$. Since $\widehat{\pi}(v)$ does not contain the trivial representation, $\widehat{h}_{\mathbb{H}}(\widehat{\pi}(v)) = 0$, so that in particular

$$h_{\mathbb{H}}(\pi(v_{ij})) = 0 = h_{\mathbb{G}}(v_{ij})$$

for all $1 \leqslant i,j \leqslant \dim(v)$. Because coefficients of irreducible representations span $\mathcal{O}(\mathbb{G})$, it follows that $h_{\mathbb{G}} = h_{\mathbb{H}} \circ \pi$. Then, as in the proof of Theorem 6.34, the faithfulness of the Haar states yields the conclusion.

Conversely, if π is an isomorphism and v is a finite-dimensional representation such that $\widehat{v} \simeq \varepsilon_{\mathbb{H}} \oplus w$, then

$$v = \widehat{\pi^{-1}}(v) \simeq \varepsilon_{\mathbb{G}} \oplus \widehat{\pi^{-1}}(w),$$

so that v is not irreducible. $\qquad\qquad\qquad\qquad\qquad\qquad\qquad\qquad\square$

We are now ready for the proof, which is a special case of [28, theorem 3.6], with a proof simplified using ideas from [29, theorem 3.3].

Theorem 8.39 (Chirvasitu) *The compact matrix quantum groups O_N^+ are residually finite for all N.*

Proof The result is trivial for $N = 1$ since in that case $O_1^+ = O_1 = \mathbf{Z}_2$ is finite. Assume, therefore, that $N \geqslant 2$ and let $\mathcal{I}$ be the intersection of the kernels of all the finite-dimensional representations of $\mathcal{O}(O_N^+)$. We have shown in the proof of Proposition 8.24 that $\mathcal{O}(\mathbb{G}) = \mathcal{O}(O_N^+)/\mathcal{I}$ together with the image of the fundamental representation U is an orthogonal compact matrix quantum group. If we can prove that the canonical surjection π associated to $\mathcal{I}$ is injective, then the conclusion will follow. Note that we already know that π is injective on $\mathcal{O}(PO_N^+)$ because that algebra is residually finite-dimensional. Indeed, it is isomorphic to $\mathcal{O}(PU_N^+)$ and the latter is a subalgebra of the residually finite-dimensional algebra $\mathcal{O}(U_N^+)$.

Let us consider the group $\Gamma = \{\pm \operatorname{Id}_N\} \subset \operatorname{GL}_N(\mathbf{C})$. There is a surjective $*$-homomorphism

$$s \colon \mathcal{O}(O_N) \to \mathcal{O}(\Gamma)$$

given by the inclusion of the compact groups, and we will be interested in the map

$$s' = s \circ \pi_{\mathrm{ab}} \colon \mathcal{O}(O_N^+) \to \mathcal{O}(\Gamma).$$

Let us consider the $*$-subalgebra

$$H = \left\{ x \in \mathcal{O}(O_N^+) \mid (s' \otimes \operatorname{id}) \circ \Delta(x) = 1 \otimes x \right\} \subset \mathcal{O}(O_N^+).$$

Observe that $\mathcal{O}(\Gamma)$ is generated by one function c which satisfies $c^2 = 1$: this is just the function associating to a matrix in Γ the common value of its diagonal coefficients. Therefore, $s'(U_{ij}) = \delta_{ij} c$ and

$$(s' \otimes \operatorname{id}) \circ \Delta(U_{ij}) = \sum_{k=1}^{N} s'(U_{ik}) \otimes U_{kj} = c \otimes U_{ij}.$$

This implies that $U_{ij} U_{k\ell} \in H$ for all $1 \leqslant i, j, k, \ell \leqslant N$ so that $\mathcal{O}(PO_N^+) \subset H$ and we claim that the converse inclusion also holds. Indeed, for any integer k,

$$(s' \otimes \operatorname{id}) \circ \Delta \left(U_{i_1 \cdots i_{2k+1}, j_1 \cdots j_{2k+1}}^{\otimes(2k+1)} \right) = c \otimes U_{i_1 \cdots i_{2k+1}, j_1 \cdots j_{2k+1}}^{\otimes(2k+1)}$$

and since any coefficient of the irreducible representation u^{2k+1} is a linear combination of coefficients of $U^{\otimes(2k+1)}$, we conclude that

$$(s' \otimes \operatorname{id}) \circ \Delta(u_{ij}^{2k+1}) = c \otimes u_{ij}^{2k+1}.$$

Now, an arbitrary element of H can be written as a linear combination of coefficients of irreducible representations so that by taking linear combinations, the only non-zero coefficients are those corresponding to even representations. In other words, $H \subset \mathcal{O}(PO_N^+)$, proving our claim.

Note that $\pi_{\mathrm{ab}} \colon \mathcal{O}(O_N^+) \to \mathcal{O}(O_N)$ factors through $\mathcal{O}(\mathbb{G})$ by construction because if $\pi_{\mathrm{ab}}(x) \neq 0$, then evaluating that function at a point of O_N where it does not vanish yields a one-dimensional representation of $\mathcal{O}(O_N^+)$ whose kernel does not contain x. This implies that s' factors as $s'' \circ \pi$. Setting

$$K = \{x \in \mathcal{O}(\mathbb{G}) \mid (s'' \otimes \mathrm{id}) \circ \Delta(x) = 1 \otimes x\},$$

we have as before $\mathcal{O}(P\mathbb{G}) \subset K \subset \pi(\mathcal{O}(PO_N^+))$. But π is injective on $\mathcal{O}(PO_N^+)$, hence $K = \mathcal{O}(P\mathbb{G})$.

We are now close to the result. Indeed, if π is not injective on $\mathcal{O}(O_N^+)$, then by Lemma 8.38 there exists an irreducible representation v^n whose image under $\widehat{\pi}$ contains the trivial representation. If n is even, then v^n is a representation of PO_N^+ so that this cannot happen by injectivity. Thus, $n = 2k + 1$. But then,

$$
\begin{aligned}
(s' \otimes \mathrm{id}) \circ \Delta \circ \pi(v_{ij}^{2k+1}) &= (s' \circ \pi_{\mathcal{I}} \otimes \pi) \circ \Delta(v_{ij}^{2k+1}) \\
&= (s \otimes \pi) \circ \Delta(v_{ij}^{2k+1}) \\
&= c \otimes \pi(v_{ij}^{2k+1}).
\end{aligned}
$$

By linear independence, this implies that all the coefficients of the irreducible subrepresentations of $\widehat{\pi}(v^{2k+1})$ satisfy $(s'' \otimes \mathrm{id}) \circ \Delta(x) = c \otimes x$, but this is not true for the unit 1, which is the only coefficient of the trivial subrepresentation. We therefore have a contradiction and the proof is complete. $\qquad\square$

Remark 8.40 The previous argument may be better understood if one thinks in terms of would-be 'discrete quantum groups' dual to the compact ones. Then, $\widehat{\Gamma}$ is a quotient of $\widehat{O}_N^+$ and the kernel is $\widehat{PO}_N^+$. The latter therefore has finite index inside $\widehat{O}_N^+$, which enables to 'induce' residual finiteness.

Remark 8.41 There is also a 'compact' interpretation of the construction in the proof of Theorem 8.39. If one thinks of Γ as a quantum subgroup of O_N^+, then, it is in a sense 'normal' and the corresponding quotient is exactly PO_N^+. In other words, O_N^+ is a degree-two cover of PO_N^+.

8.2.3.3 Quantum Reflection Groups

We conclude with H_N^+, and the proof will be of a different nature. Indeed, we will rely on a trick making use of the free wreath product decomposition given in Section 7.2.2. This requires a slight extension of the unital free product construction for $*$-algebras which we now briefly introduce.

Definition 8.42 Let $\mathcal{A}_1$, $\mathcal{A}_2$ and $\mathcal{B}$ be $*$-algebras together with $*$-homomorphisms

$$\pi_i \colon \mathcal{A}_i \to \mathcal{B}$$

for $i = 1, 2$. Then, the quotient of the unital free product $\mathcal{A}_1 * \mathcal{A}_2$ by the $*$-ideal generated by the elements $\pi_1(x) - \pi_2(x)$ for all $x \in \mathcal{B}$ is called the *unital free product of $\mathcal{A}_1$ and $\mathcal{A}_2$ amalgamated over* $\mathcal{B}$ and denoted by $\mathcal{A}_1 *_{\mathcal{B}} \mathcal{A}_2$.

Remark 8.43 The reader is invited to state and prove the universal property of amalgamated free products, as well as to check that the construction carries over to the setting of compact quantum groups, where it extends the notion of amalgamated free product for discrete groups (see, for instance, [27, definition E.8] for the definition).

Theorem 8.44 *Let s, N be integers. Then, the compact matrix quantum group H_N^{s+} is residually finite.*

Proof It was proven in Theorem 7.34 that $H_N^+ = \mathbf{Z}_s \wr_* S_N^+$ and this can be restated in the following way: consider the sequence of $*$-algebras $\mathcal{A}_k$ where $\mathcal{A}_0 = \mathcal{O}(S_N^+)$ and for any $k \geqslant 0$,

$$\mathcal{A}_{k+1} = \mathbf{C}[\mathbf{Z}_s] * \mathcal{A}_k / \langle [\mathbf{C}[\mathbf{Z}_s], u_{k+1j}] \mid 1 \leqslant j \leqslant N \rangle.$$

Then, $\mathcal{A}_N = \mathcal{O}(H_N^{s+})$. All that we need is therefore a stability result for residual finite-dimensionality. Let us first notice that $\langle u_{k+1} \mid 1 \leqslant j \leqslant N \rangle \simeq \mathbf{C}^N$ so that

$$\mathcal{A}_{k+1} \simeq \left(\mathbf{C}[\mathbf{Z}_s] \otimes \mathbf{C}^N \right) *_{\mathbf{C}^N} \mathcal{A}_k \hookrightarrow \left(\mathbf{C}[\mathbf{Z}_s] \otimes \mathcal{A}_k \right) *_{\mathbf{C}^N} \left(\mathbf{C}[\mathbf{Z}_s] \otimes \mathcal{A}_k \right).$$

It therefore suffices to show that $\mathcal{A} *_{\mathcal{B}} \mathcal{A}$ is residually finite-dimensional as soon as $\mathcal{A}$ is and $\mathcal{B}$ is finite-dimensional. We will not give the proof but simply explain the two steps.

1. Using a free product decomposition of a Hilbert space on which $\mathcal{A}$ acts faithfully, it is easy to see that any element has non-zero image into a similar free product with $\mathcal{A}$ replaced by a finite-dimensional quotient. It is therefore sufficient to do it for finite-dimensional $\mathcal{A}$.
2. The result is then a particular case of [1, theorem 4.2], where the strategy is to embed $\mathcal{A}$ into a matrix algebra in a trace-preserving way. One can then use [26, theorem 2.3] to conclude. $\square$

The case of $H_N^{\infty+}$ can be treated by resorting to a topological generation trick. This requires us, nevertheless, to use a variant of our topological generation criterion.

Exercise 8.6 Prove that a unitary compact matrix quantum group $(\mathbb{G}, u)$ is topologically generated by $(\mathbb{H}_i)_{i \in I}$ if, for any irreducible representation v of $\mathbb{G}$, there exists at least one $i \in I$ such that $\widehat{\pi}_i(v)$ is irreducible.

Solution The condition in the statement implies that for any irreducible representation v, there exists $i_0 \in I$ such that $\mathrm{Mor}_{\mathbb{H}_{i_0}}(v, v) = \mathbb{C}$, so that in particular

$$\bigcap_{i \in I} \mathrm{Mor}_{\mathbb{H}_i}(v, \varepsilon) \subset \mathrm{Mor}_{\mathbb{H}_{i_0}}(v, \varepsilon) = \{0\} = \mathrm{Mor}_{\mathbb{G}}(v, \varepsilon).$$

In fact, the same result holds for $v^{\oplus n}$, $n \in \mathbb{N}$. Assume now v splits as a direct sum

$$v = \bigoplus_{\ell=1}^{m} v_\ell^{\oplus n_\ell}$$

with v_ℓ irreducible and not equivalent to $v_{\ell'}$ for $\ell \neq \ell'$ as representations of $\mathbb{G}$. Then, if none of the summands is the trivial representation, by the beginning of the proof, there exists $i_1, \ldots, i_m \in I$ such that

$$\mathrm{Mor}_{\mathbb{G}}(v, \varepsilon) = \bigoplus_{\ell=1}^{m} \mathrm{Mor}_{\mathbb{G}}(v_\ell^{\oplus n_\ell}, \varepsilon)$$

$$= \bigoplus_{\ell=1}^{m} \mathrm{Mor}_{\mathbb{H}_{i_\ell}}(v_\ell^{\oplus n_\ell}, \varepsilon)$$

$$\supset \bigoplus_{\ell=1}^{m} \bigcap_{i \in I} \mathrm{Mor}_{\mathbb{H}_i}(v_\ell^{\oplus n_\ell}, \varepsilon)$$

$$= \bigcap_{i \in I} \mathrm{Mor}_{\mathbb{H}_i}(v, \varepsilon).$$

The equality being trivial satisfied if $v = \varepsilon^{\oplus n}$, we conclude that for any finite-dimensional representation v,

$$\bigcap_{i \in I} \mathrm{Mor}_{\mathbb{H}_i}(v, \varepsilon) = \mathrm{Mor}_{\mathbb{G}}(v, \varepsilon).$$

Taking $v = u^w$ and applying Proposition 8.25 then yields the result. $\square$

We can now deal with our final example.

Exercise 8.7 For any integer $s \in \mathbb{N}$, we denote by $\pi_s \colon \mathcal{O}(H_N^{\infty+}) \to \mathcal{O}(H_N^{s+})$ the surjective $*$-homomorphism given by the inclusion of categories of partitions $\mathcal{C}_\infty \subset \mathcal{C}_s$.

1. Prove that for any word $w = w_1 \cdots w_n$ over $\mathbf{Z}$, there exists $s(w) \in \mathbf{N}$ such that for all $s \geqslant s(w)$,

$$\widehat{\pi}_s(u^w) = u^{[w_1]_s \cdots [w_n]_s},$$

where $[\cdot]_s$ denotes the class modulo s.

2. Conclude that $H_N^{\infty+}$ is residually finite.

Solution 1. We will prove the result by induction on n. For $n = 1$, the result holds with $s(w) = 1$. Assume now that the result holds for all words of length n and consider $w = w_1 \cdots w_{n+1}$. Then,

$$u^{w_1 \cdots w_n} \otimes u^{w_{n+1}} = u^w \oplus \delta_{w_n = -w_{n+1}} u^{w_1 \cdots w_{n-1}}.$$

- If $w_{n+1} = -w_n$, then for $s \geqslant s(w_1 \cdots w_n)$,

$$\begin{aligned}
\widehat{\pi}_s(u^{w_1 \cdots w_n}) \otimes \widehat{\pi}_s(u^{w_{n+1}}) &= u^{[w_1]_s \cdots [w_n]_s} \otimes u^{[w_{n+1}]_s} \\
&= u^{[w_1]_s \cdots [w_{n+1}]_s} \oplus u^{[w_1]_s \cdots [w_{n-1}]_s} \\
&= u^{[w_1]_s \cdots [w_{n+1}]_s} \oplus \widehat{\pi}_s\left(u^{w_1 \cdots w_{n-1}}\right)
\end{aligned}$$

and the result follows.

- Otherwise, set $s(w) = \max(s(w_1 \cdots w_n), w_n + w_{n+1} + 1)$. Then, for any $s \geqslant s(w)$, $[w_n]_s \neq -[w_{n+1}]_s$ so that

$$\begin{aligned}
\widehat{\pi}_s(u^w) &= \widehat{\pi}_s(u^{w_1 \cdots w_n}) \otimes \widehat{\pi}_s(u^{w_{n+1}}) \\
&= u^{[w_1]_s \cdots [w_n]_s)} \otimes u^{[w_{n+1}]_s} \\
&= u^{[w_1]_s \cdots [w_{n+1}]_s}.
\end{aligned}$$

2. A consequence of the previous question is that any irreducible representation of $H_N^{\infty+}$ remains irreducible in some H_N^{s+}. By Exercise 8.6, this implies that $H_N^{\infty+}$ is topologically generated by the quantum subgroups $(H_N^{s+})_{s \in \mathbf{N}}$. We can therefore conclude by Proposition 8.24 and Theorem 8.44. $\qquad\square$

Remark 8.45 Once the representation theory of free wreath products is known (this was computed in [52]), the same strategy as in the proof of Theorem 8.44 works to prove that if Γ is a finite group, then $\widehat{\Gamma} \wr_* S_N^+$ is residually finite-dimensional. An argument similar to that of Exercise 8.7 then shows that if Γ is residually finite, then $\widehat{\Gamma} \wr_* S_N^+$ is topologically generated[10] by $\widehat{\Lambda} \wr_* S_N^+$ for all finite quotients Λ of Γ; we see that the free wreath product is residually finite as soon as Γ is.

---

[10] We are using here topological generation by a possibly infinite family of quantum groups instead of just two like in Definition 8.22, but everything works the same.

Remark 8.46 Using the fact (used in the proof of Theorem 8.44) that a free product of residually finite compact matrix quantum groups is again residually finite, it is straightforward to see that the free complexification of a residually finite compact matrix quantum group is residually finite. Hence, $\widetilde{H}_N^+$ is residually finite.

Appendix A

Two Theorems on Complex Matrix Algebras

For the sake of completeness, we give detailed proofs of two results from the representation theory of algebras which were used in the text. Since we only need them for algebras of matrices over the field of complex numbers, we will give the proofs in that restricted setting, allowing us to simplify some of the arguments.

A subalgebra $A \subset M_n(\mathbf{C})$ is said to be *irreducible* if there is no proper subspace in $\mathbf{C}^n$ which is globally fixed by all the elements of A, except for $\{0\}$. The nature of irreducible matrix algebras is elucidated by the following celebrated result of W. Burnside.

Theorem A.1 (Burnside's Theorem) *Let* $A \subset M_n(\mathbf{C})$ *be an irreducible subalgebra. Then,* $A = M_n(\mathbf{C})$.

Proof The proof proceeds in two steps. First, we will prove that A contains a rank 1 matrix. Then, we will deduce from this that it contains all rank-one matrices, hence all matrices. We will do this following [43]. Before we start, note that by irreducibility, for any non-zero $x \in \mathbf{C}^n$,

$$A.x := \{T(x) \mid T \in A\} = \mathbf{C}^n.$$

Indeed, $A.x \neq \{0\}$ is globally fixed by all the elements of A by construction.

For the first part, we will proceed by contradiction. Let $T \in A$ be a matrix with minimal rank and assume that $d = \mathrm{rk}(T) \geqslant 2$. Then, there exists $x_1, x_2 \in \mathbf{C}^n$ such that the vectors $T(x_1)$ and $T(x_2)$ are linearly independent. Let us choose $S \in A$ such that $ST(x_1) = x_2$ (such an S exists because $A.T(x_1) = \mathbf{C}^n$). Then, $T(x_1)$ and

$$TST(x_1) = T(x_2)$$

264

are linearly independent, so that the operator $TST - \lambda T$ is non-zero for all $\lambda \in \mathbf{C}$. However, notice that TS is a linear operator on the range of T, hence it has an eigenvalue: there exists $\mu \in \mathbf{C}$ such that $TS - \mu \operatorname{Id}$ is not invertible. Then,

$$TST - \mu T = (TS - \mu \operatorname{Id})T \in A$$

has rank strictly between 0 and d, contradicting minimality. In conclusion, there is a rank-one matrix in A.

The rank-one matrix obtained in the previous paragraph can be written as the operator

$$T_{\phi,y} \colon x \mapsto \phi(x)y$$

for some linear form $\phi \in (\mathbf{C}^n)^*$ and a vector $y \in \mathbf{C}^n$. Because $A.y = \mathbf{C}^n$, we also have $T_{\phi,z} \in A$ for all $z \in \mathbf{C}^n$. Moreover, A acts on $(\mathbf{C}^n)^*$ by

$$(S, \psi) \mapsto \psi \circ S.$$

Let $V \subset (\mathbf{C}^n)^*$ be a subspace of dimension $0 < m < n$ which is globally fixed by all the elements of A. Then, using for instance a basis, we see that

$$\cap_{\psi \in V} \ker(\psi)$$

equals the intersection of m subspaces of dimension $n - 1$, hence has dimension at most $n - m > 0$. If now y_1 is in that intersection and y_2 is not, then there is an $S \in A$ such that $S(y_1) = y_2$, so that, for any $\psi \in V$,

$$\psi \circ S(y_1) \neq 0.$$

Thus, $\psi \circ S \notin V$, contradicting stability. Therefore, A acts irreducibly on $(\mathbf{C}^n)^*$, hence

$$\{\phi \circ S \mid S \in A\} = (\mathbf{C}^n)^*.$$

Putting things together, for any $\psi \in (\mathbf{C}^n)^*$ and $z \in \mathbf{C}^n$, there exists $S_1, S_2 \in A$ such that

$$S_1 \circ T \circ S_2 \colon x \mapsto \psi(x)z.$$

As a consequence, A contains all rank-one matrices, hence equals $M_n(\mathbf{C})$. $\square$

Remark A.2 Note that the argument only requires the existence of an eigenvalue for the matrix ST, hence works for any algebraically closed field.

Our second result concerns the double commutant of a matrix $*$-algebra. Recall that, given a subalgebra $A \subset M_n(\mathbf{C})$, its *commutant* A' is by definition

$$A' = \{T \in M_n(\mathbf{C}) \mid AT = TA\}$$

and its *double commutant* is $A'' = (A')'$.

Theorem A.3 (Double Commutant Theorem) *Let $A \subset M_n(\mathbf{C})$ be a subalgebra which is stable under taking adjoints. Then, $A'' = A$.*

Proof By definition $A \subset A''$. Moreover, if $V \subset \mathbf{C}^n$ is stable under the action of A, then by stability under taking adjoints, its orthogonal complement also is stable. As a consequence, there is an orthogonal decomposition

$$\mathbf{C}^n = V_1 \oplus \cdots \oplus V_m$$

into irreducible subspaces. Each of these subspaces yields an irreducible representation of A, and some of them may be equivalent. Let $\{i_1, \cdots, i_k\}$ be such that any subspace is equivalent to V_{i_j} for some j, and let us denote by n_j the number of such subspaces. By Schur's Lemma, A is then block diagonal with k blocks corresponding to the restrictions to each $V_{i_j}^{n_j}$.

Let us consider one of those diagonal blocks, which has size $n_j \times \dim(V_{i_j})$, and assume for the moment that this is all of $\mathbf{C}^n$. The restriction of A to any of the summands is $M_{\dim(V_{i_j})}(\mathbf{C})$ by Theorem A.1. Moreover, given two summands V, V' of V_{i_j}, the fact that they are equivalent as representations of A means that there is a linear isomorphism $\varphi \colon V \to V'$ such that for all $T \in A$,

$$\varphi \circ T_{|V} = T_{|V'} \circ \varphi.$$

Fixing such an isomorphism φ_k from the first to the kth summand for all $k \leqslant n_j$ yields an invertible matrix $\Phi \in \mathrm{GL}_n(\mathbf{C})$ such that $\Phi^{-1} A \Phi$ consists in all matrices of the form $\mathrm{diag}(T, \cdots, T)$ for $T \in M_{\dim(V_{i_j})}(\mathbf{C})$.

We now investigate the commutant of $\Phi^{-1} A \Phi$, which is nothing but the self-intertwiners of $V_{i_j}^{n_j}$. Such an intertwiner is given by a block $n_j \times n_j$ matrix, where the (a, b)th block is an intertwiner from the ath copy of V_{i_j} to the bth one. By Schur's Lemma again, each block must be a scalar matrix so that A' is $M_{n_j}(\mathbf{C})$ seen as block scalar matrices in $M_{n_j \times \dim(V_{i_j})}(\mathbf{C})$. As a consequence, the bicommutant consists in all $M_{\dim(V_{i_j})}(\mathbf{C})$-block diagonal matrices with constant diagonal, that is, A again.

We now return to the general case. A final application of Schur's Lemma shows that A is block diagonal, with blocks of the previous form, and the result follows. $\qquad\square$

By way of conclusion, let us isolate an important structure result that we proved along the way.

Corollary A.4 *Let $A \subset M_N(\mathbf{C})$ be a subalgebra stable under taking adjoints. Then, A is isomorphic to a direct sum of matrix algebras.*

Appendix B
Classical Compact Matrix Groups

On several occasions in the main text, we have used results from the classical theory of compact matrix groups to motivate important definitions. In this appendix, we will show how one can concretely recover the whole theory of continuous representations of compact matrix groups from what has been done in the general quantum setting. The challenge is, of course, to extract informations on representations which are continuous out of what has been done on *polynomial* representations.

Definition B.1 Let $G \subset M_N(\mathbf{C})$ be a compact matrix group. A finite-dimensional representation $\rho \colon G \to M_n(\mathbf{C})$ is said to be polynomial if, for any $1 \leqslant i, j \leqslant n$, the function

$$g \mapsto \rho(g)_{ij}$$

is polynomial in the coefficient functions of G.

As we saw in Chapter 5, non-commutative geometry teaches us that to recover topology from an algebraic perspective, one needs to build a C*-algebra. Moreover, that C*-algebra should have properties paralleling those of the *-algebra $\mathcal{O}(G)$ so that we can transport our results without further work. The correct way to do this is through a universal construction which was briefly alluded to in Section 5.2.1. We will here give a bit more detail about it.

Definition B.2 Let G be a unitary compact matrix group. The *enveloping C*-algebra* of $\mathcal{O}(G)$ is the completion of $\mathcal{O}(G)$ with respect to the norm $\| \cdot \|_{\max}$ defined as

$$\| \cdot \|_{\max} = \sup \left\{ \|\pi(x)\| \mid \pi \colon \mathcal{O}(G) \to \mathcal{B}(H) \ *\text{-homomorphism} \right\}.$$

It is denoted by $C^*(\mathcal{O}(G))$.

Let us check that this enables us to recover the topological structure of G.

Proposition B.3 *Let G be a unitary compact matrix group. Then, there is an isomorphism of C^*-algebras*

$$C^*(\mathcal{O}(G)) \simeq C(G)$$

which is the identity on $\mathcal{O}(G)$.

Proof Because $\mathcal{O}(G)$ is commutative, so is $C^*(\mathcal{O}(G))$ so that by Theorem 5.2, it must be of the form $C(X)$ for some compact topological space X. We know, moreover, that X can be identified with the set of $*$-homomorphism from $C^*(\mathcal{O}(G))$ to $\mathbf{C}$. Any such $*$-homomorphism restricts to a $*$-homomorphism on $\mathcal{O}(G)$. Conversely, if

$$\varphi \colon \mathcal{O}(G) \to \mathbf{C} \simeq M_1(\mathbf{C})$$

is a $*$-homomorphism, then by definition $\|\varphi(x)\| \leqslant \|x\|_{\max}$ so that it extends at least to a bounded linear map on $C(G)$. Because that map is multiplicative and involutive on the dense $*$-subalgebra $\mathcal{O}(\mathbb{G})$, so is it on $C^*(\mathcal{O}(G))$.

Let us consider again such a φ and set

$$(M_{ij})_{1 \leqslant i,j \leqslant N} = (\varphi(u_{ij}))_{1 \leqslant i,j \leqslant N} \in M_N(\mathbf{C})$$

so that $\varphi = \mathrm{ev}_M$. The same argument as in Exercise 1.10 and Corollary 5.18 shows that if we write $\mathcal{O}(G) = \mathcal{O}(O_N)/\mathcal{I}$, then G exactly consists in matrices on which all the polynomials in $\mathcal{I}$ vanish. Because φ is a $*$-homomorphism, we have, for any $P \in I$,

$$P(M) = P(\varphi(u_{ij})) = \varphi(P(u)) = 0.$$

As a consequence, $M \in G$ and the proof is complete. $\qquad\square$

Using universality again, we see that the $*$-homomorphism

$$\Delta \colon \mathcal{O}(G) \to \mathcal{O}(G) \otimes \mathcal{O}(G) \simeq \mathcal{O}(G \times G) \subset C(G \times G)$$

extends to $C(G)$ and satisfies $\Delta(f)(g,h) = f(gh)$ for any $f \in C(G)$, so that we also recover the group law of G. The first fundamental result that we can prove with this is the existence of the Haar measure.

Theorem B.4 *Let G be a unitary compact matrix group. Then, G has a unique Borel probability measure which is invariant by both left and right translation. Moreover, that measure has full support.*

Proof Let $h \colon \mathcal{O}(G) \to \mathbf{C}$ be the Haar state. The GNS construction ensures that h is of the form

$$h(x) = \langle \pi(x)\xi, \xi \rangle$$

for some $*$-representation π, which extends to $C^*(\mathcal{O}(G)) = C(G)$ because it is bounded for the norm $\|\cdot\|_{\max}$. Thus, h extends to $C(G)$, and by the Riesz–Markov–Kakutani representation theorem, there exists a Borel probability measure μ on G such that for any $f \in C(G)$,

$$h(f) = \int_G f(g)\mathrm{d}\mu(g).$$

Moreover, we know that for any $f \in C(G)$ and $g \in G$,

$$\int_G f(gg')\mathrm{d}\mu(g') = (h \otimes \mathrm{id}) \circ \Delta(f)$$
$$= h(f).1$$
$$= \int_G f(g)\mathrm{d}\mu(g),$$

and this implies that μ is invariant by left translations. Invariance by right translations is, of course, proven similarly.

As for the support, assume that there is an open set $O \subset G$ with measure 0. If $g \in O$, then, for any $g' \in G$, $g' \in g'g^{-1}O$ which is open so that $(gO)_{g \in G}$ is an open covering of G. By compactness, we can extract a finite covering, and since each of these open sets has measure 0, if follows that G itself has measure 0, a contradiction. $\qquad\square$

We can now turn to the main subject of this appendix, that is to say, continuous representations. Note that, because $\mathcal{O}(G)$ is dense in $C(G)$, it follows from Theorem 5.15 that $L^2(G)$ has an orthonormal basis consisting in coefficients of polynomial (in particular continuous) finite-dimensional representations. This is a version of the celebrated Peter–Weyl Theorem.

One is easily convinced from this that all continuous finite-dimensional representations should be equivalent to polynomial ones, hence polynomial themselves, as claimed in Section 2.1.

Proposition B.5 *Let G be a unitary compact matrix group. Then, any continuous finite-dimensional representation is polynomial.*

Proof We will first prove the result assuming that the representation is unitary. For a finite-dimensional unitary representation $\rho\colon G \to M_n(\mathbf{C})$, recall that the map

$$\widehat{h}(\rho)\colon \xi \in \mathbf{C}^n \mapsto \int_G \rho(g)\xi \mathrm{d}\mu(g)$$

is the orthogonal projection onto the subspace of fixed vectors of ρ. Moreover, the proof of Schur's Lemma (Lemma 2.14) still works with continuous representations. As a consequence, if ρ is irreducible and not polynomial, then, for any polynomial representation $\rho'\colon G \to M_m(\mathbf{C})$, the same argument as in the proof of Theorem 5.15 (note that $\bar{\rho}$ is continuous as soon as ρ is) shows that

$$h(\rho_{ij}\rho'_{k\ell}) = 0$$

for all $1 \leqslant i, j \leqslant n$ and $1 \leqslant k, \ell \leqslant m$. But because coefficients of polynomial representations are dense in $L^2(G)$, this forces $\rho = 0$. If ρ is not irreducible, it suffices to show that it splits into a sum of irreducible continuous representations to conclude. But this follows from the same argument as in the proof of point 1 of Theorem 2.15.

We now turn to the case of an arbitrary finite-dimensional representation ρ acting on a vector space V. Let us pick an arbitrary inner product $\langle \cdot, \cdot \rangle$ on V and define a bilinear form on V through the formula (which makes sense because continuous functions on a compact space are integrable)

$$(\xi \mid \eta)_\rho = \int_G \langle \rho(g)\xi, \rho(g)\eta \rangle.$$

Then, $(\cdot \mid \cdot)_\rho$ is an inner product and ρ is a unitary representation on the corresponding Hilbert space. In other words, any finite-dimensional continuous representation is equivalent to a unitary one, and the proof is complete. $\qquad \square$

Appendix C
General Compact Quantum Groups

As we explained at the beginning of this text, we have made the choice to work in a restricted setting, that of compact *matrix* quantum groups. Even though this was perfectly justified by several reasons, we would like to provide the reader with a few elements concerning the more general setting of compact quantum groups. Or to put it differently, we would like to try to answer the question: what is the quantum version of arbitrary compact groups?

C.1 CQG-Algebras

Let us try to generalise the notion of a compact group using our algebraic setting. One idea for that is based on a specific property of compact groups: any continuous representation of a compact group splits as a sum of irreducible representations. In particular, the so-called *regular representation* π_{reg} on $L^2(G)$ given by

$$\pi_{\text{reg}}(g)(f) \colon h \mapsto f(gh)$$

decomposes as a direct sum of irreducible ones. Because π_{reg} is injective, we deduce the following crucial fact.

Proposition C.1 *Let G be a compact group. Then, for any $g \in G \setminus \{e\}$, there exists a finite-dimensional unitary representation ρ such that $\rho(g) \neq \text{Id}$.*

In other words, the $*$-algebra $\mathcal{O}(G) \subset C(G)$ of coefficients of finite-dimensional unitary representations separate the points of G. Thus, by the Stone–Weierstrass theorem, it is dense. A reasonable idea is therefore to give a non-commutative version of $\mathcal{O}(G)$. For that, observe that if ρ is a continuous finite-dimensional unitary representation of G, then the subalgebra of $\mathcal{O}(G)$

271

generated by the coefficients of ρ equals $\mathcal{O}(\rho(G))$. Therefore, $\mathcal{O}(G)$ is in a sense *locally a compact matrix quantum group algebra*. This leads to the following notion.

Definition C.2 A *CQG-algebra of Kac type* is a $*$-algebra $\mathcal{A}$ such that there exists a family $(u^{(i)})_{i \in I}$ of unitary matrices $u^{(i)} \in M_{N_i}(\mathcal{O}(\mathbb{G}))$ satisfying three conditions:

1. If $\mathcal{A}_i$ denotes the $*$-subalgebra generated by the coefficients of $u^{(i)}$, then $\mathcal{A}$ is generated by $(\mathcal{A}_i)_{i \in I}$.
2. For all $i \in I$, $(\mathcal{A}_i, u^{(i)})$ is a unitary compact matrix quantum group with coproduct Δ_i.
3. For all $i, j \in I$, $\Delta_{i|\mathcal{A}_i \cap \mathcal{A}_j} = \Delta_{j|\mathcal{A}_i \cap \mathcal{A}_j}$.

Remark C.3 One could restate the definition in a more abstract way by saying that a CQG-algebra of Kac type is an *inductive limit* of algebras of unitary compact quantum matrix groups, a notion which we will not define here. This parallels the fact that any compact group is a *projective limit* of unitary compact matrix groups.

By the preceding discussion, for any compact group G, $\mathcal{O}(G)$ is a commutative CQG-algebra of Kac type from which one can conversely recover G. Our goal now will be to give an alternate description of CQG-algebras of Kac type in the general case.

Note that a direct consequence of the axioms is that there is a unique $*$-homomorphism

$$\Delta \colon \mathcal{A} \to \mathcal{A} \otimes \mathcal{A}$$

such that for all $i \in \mathcal{A}$, $\Delta_{|\mathcal{A}_i} = \Delta_i$. As a consequence, a CQG-algebra of Kac type is a special case of a *bialgebra*. We will not explain what this means, because $\mathcal{A}$, in fact, satisfies more properties which will be explained in Proposition C.5 hereafter. Before that, however, we need to introduce some vocabulary generalising the notions introduced in the matrix case. Let us say that a matrix $v = (v_{ij})_{1 \leqslant i,j \leqslant N} \in M_N(\mathcal{A})$ is a *finite-dimensional representation* of $\mathcal{A}$ if, for all $1 \leqslant i, j \leqslant N$,

$$\Delta(v_{ij}) = \sum_{k=1}^{N} v_{ik} \otimes v_{kj}.$$

The definitions of an intertwiner, a subrepresentation, a unitary representation and an irreducible representation are the same as in Definition 2.10 for the matrix case. This yields a particularly convenient generating set for a CQG-algebra of Kac type.

Lemma C.4 *Let $\mathcal{A}$ be a CQG-algebra of Kac type and let v be an irreducible unitary finite-dimensional representation. Then, $\bar{v}$ is also an irreducible unitary representation. Moreover, $\mathcal{A}$ is spanned by coefficients of unitary finite-dimensional representations and coefficients of inequivalent unitary representations are linearly independent.*

Proof By assumption, the coefficients v_{ij} are polynomials in the coefficients of $(u^{(i)})_{i \in I}$. In particular, there exists a finite subset $F \subset I$ such that the coefficients of v are linear combinations of the coefficients of tensor powers of $u_{(F)} = \bigoplus_{i \in F} u^{(i)}$. Now, the $*$-subalgebra $\mathcal{A}_F$ spanned by the coefficients of $u^{(F)}$ is nothing but $\langle \mathcal{A}_i \mid i \in F \rangle$ and, together with $u^{(F)}$, this is a unitary compact matrix quantum group. Since v is a unitary representation of it, the first assertion follows from Corollary 5.13.

As for the second one, observe that for each $i \in I$, $u^{(i)}$ splits as a sum of irreducible representations of $(\mathcal{A}_i, u^{(i)})$, which are therefore also irreducible representations of $\mathcal{A}$. This proves that these coefficients span $\mathcal{A}$. Given now a finite family of pairwise inequivalent irreducible representations, there is a finite subset $F \subset I$ such that they are all representations of $(\mathcal{A}_F, u^{(F)})$, and applying Theorem 2.15 yields linear independence. $\square$

We can now show that CQG-algebras of Kac type are specific instances of Hopf $*$-algebras.

Proposition C.5 *Let $\mathcal{A}$ be a CQG-algebra of Kac type. Then $\mathcal{A}$ is a Hopf $*$-algebra in the sense that there exists $*$-homomorphisms $\Delta \colon \mathcal{A} \to \mathcal{A} \otimes \mathcal{A}$ and $\varepsilon \colon \mathcal{A} \to \mathbf{C}$ as well as a $*$-anti-homomorphism $S \colon \mathcal{A} \to \mathcal{A}$ satisfying*

1. $(\Delta \otimes \mathrm{id}) \circ \Delta = (\mathrm{id} \otimes \Delta) \circ \Delta$;
2. $m \circ (S \otimes \mathrm{id}) \circ \Delta = \varepsilon.1 = m \circ (\mathrm{id} \otimes S) \circ \Delta$;
3. $(\varepsilon \otimes \mathrm{id}) \circ \Delta = \mathrm{id} = (\mathrm{id} \otimes \varepsilon) \circ \Delta.$

Proof We have already mentionned the existence of the coproduct Δ, and the first property is satisfied because it is satisfied by Δ_i for all $i \in I$. Using Lemma C.4, let us set, for any irreducible unitary representation v of $\mathcal{A}$,

$$\varepsilon(v_{ij}) = \delta_{ij} \quad \& \quad S(v_{ij}) = v_{ji}^*.$$

These are linear maps which restrict on each $\mathcal{A}_i$ to a $*$-homomorphism and a $*$-anti-homomorphism respectively, satisfying the equations in the statement.

(It is enough to check them on coefficients of the fundamental representation, see also the proof of Proposition 2.8 and Corollary 2.18.) The proof is therefore complete. □

We will now give two characterisations of CQG-algebras of Kac type, including one in terms of the existence of a Haar state in the following sense.

Definition C.6 An *integral* on a Hopf ∗-algebra $\mathcal{A}$ is a *faithful* state $h\colon \mathcal{A} \to$ C such that for any $x \in \mathcal{A}$,

$$(h \otimes \mathrm{id}) \circ \Delta(x) = h(x).1 = (\mathrm{id} \otimes h) \circ \Delta(x). \tag{C.1}$$

Remark C.7 If an integral exists on a given Hopf ∗-algebra, then it is unique. Indeed, if both h and h' satisfy Equation (C.1), then for any $x \in \mathcal{A}$,

$$\begin{aligned} h(x) &= h(x)h'(1) \\ &= (h \otimes h') \circ \Delta(x) \\ &= h'(x)h(1) \\ &= h'(x). \end{aligned}$$

The other characterisation will be in terms of representations.

Definition C.8 A Hopf ∗-algebra is said to be *co-semisimple* if any finite-dimensional representation is equivalent to a direct sum of irreducible ones. It is, moreover, said to be *unimodular*,[1] if the conjugate of any irreducible finite-dimensional unitary representation is again unitary.

We are now ready for the main result of this section.

Theorem C.9 *Let $\mathcal{A}$ be a ∗-algebra. Then the following are equivalent:*

1. *$\mathcal{A}$ is a CQG-algebra of Kac type;*
2. *$\mathcal{A}$ is a Hopf ∗-algebra with a tracial integral;*
3. *$\mathcal{A}$ is a co-semisimple unimodular Hopf ∗-algebra.*

Proof Assuming that $\mathcal{A}$ is a CQG-algebra of Kac type, we can define a linear map $h\colon \mathcal{A} \to$ C by setting $h(v_{ij}) = 0$ for all non-trivial irreducible representation v and $h(1) = 1$. On each $\mathcal{A}_i$, this coincides with the Haar state, hence the invariance property is satisfied, as well as traciality.

---

[1] This is a non-standard terminology but it fits well with the problem of unimodularity for compact quantum groups that we want to highlight; see Section C.2.

Assume now that $\mathcal{A}$ is a Hopf $*$-algebra with a tracial integral h. For clarity we will split the proof into several steps.

(a) We will first prove unitarisability, that is, that any finite-dimensional representation of $\mathcal{A}$ is equivalent to a unitary one. Let $v \in M_N(\mathcal{A})$ be a finite-dimensional representation, let $\langle \cdot, \cdot \rangle$ be the canonical inner product on $\mathbf{C}^N$ and define another inner product in the following way on the canonical basis $(e_i)_{1 \leqslant i \leqslant N}$:

$$\langle e_i, e_j \rangle_v = h \left(\sum_{k=1}^{N} v_{ik} v_{jk}^* \right).$$

We have to check that this is indeed an inner product. For an arbitrary vector

$$\xi = \sum_i \lambda_i e_i,$$

we have

$$\langle \xi, \xi \rangle_v = \sum_{i,j=1}^{N} \lambda_i \bar{\lambda}_j \langle e_i, e_j \rangle_v$$

$$= h \left(\sum_{i,j,k=1}^{N} \lambda_i \bar{\lambda}_j v_{ik} v_{jk}^* \right)$$

$$= \sum_{k=1}^{N} h \left(\left(\sum_{i=1}^{N} \lambda_i v_{ik} \right) \left(\sum_{i=1}^{N} v_{ik} \right)^* \right)$$

$$\geqslant 0.$$

If, moreover, the sum vanishes, then for each k the corresponding term must be 0 so that, by faithfulness of h,

$$\sum_i \lambda_i v_{ik} = 0.$$

Multiplying by v_{kj} on the right and summing over k then yields $\lambda_j = 0$ for all $1 \leqslant j \leqslant N$, hence $\xi = 0$.

(b) It is now easy to see that the corepresentation ρ_v associated to v is unitary with respect to that inner product in the following sense: if we set $(a \otimes x \mid b \otimes y)_v = \langle x, y \rangle_v ab^*$, then

$$(\rho_v(e_i), \rho_v(e_j))_v = \sum_{k,k'=1}^{N} h \left(\sum_{\ell,\ell'=1}^{N} v_{k\ell} v_{k'\ell'}^* \right) v_{ik} v_{jk'}^*$$

$$= \sum_{\ell,\ell'} (\mathrm{id} \otimes h) \left(\Delta(v_{i\ell} v_{j\ell}^*) \right)$$

$$= \langle e_i, e_j \rangle_v.$$

As a consequence, if $(f_i)_{1 \leqslant i \leqslant N}$ is an orthonormal basis with respect to $\langle \cdot, \cdot \rangle_v$ and if $(\widetilde{v}_{ij})_{1 \leqslant i,j \leqslant N}$ is the matrix of v in that basis, then

$$\sum_{k=1}^{N} \widetilde{v}_{ik} \widetilde{v}_{jk}^* = \sum_{k=1}^{N} \sum_{k'=1}^{N} \delta_{kk'} \widetilde{v}_{ik} \widetilde{v}_{jk}^*$$

$$= \sum_{k=1}^{N} \sum_{k'=1}^{N} \widetilde{v}_{ik} \widetilde{v}_{jk}^* \langle f_k, f_{k'} \rangle_v$$

$$= (\rho_v(f_i), \rho_v(f_j))_v$$

$$= \langle f_i, f_j \rangle_v$$

$$= \delta_{ij}$$

so that $\widetilde{v}$ is a unitary representation equivalent to v.

(c) Let now v be a finite-dimensional representation, which can therefore be assumed to be unitary. The same argument as in the proof of the first point of Theorem 2.15 shows that v splits as a direct sum of irreducible sub-representations, hence $\mathcal{A}$ is co-semisimple. What is left now is to prove that, given a unitary representation v, its conjugate $\overline{v}$ is again unitary. Recall that if v acts on $\mathbf{C}^N$, then $\overline{v}$ acts on the conjugate Hilbert space, which is endowed with the canonical inner product

$$\langle \overline{e}_i, \overline{e}_j \rangle = \langle e_j, e_i \rangle.$$

We know by the previous point that it is unitary with respect to the inner product

$$\langle \overline{e}_i, \overline{e}_j \rangle_{\overline{v}} = \sum_{k=1}^{N} h(\overline{v}_{ik} \overline{v}_{jk}^*)$$

$$= \sum_{k=1}^{N} h(v_{ik}^* v_{jk})$$

$$= \sum_{k=1}^{N} h(v_{jk} v_{ik}^*)$$

$$= \delta_{ij}$$

$$= \langle \overline{e}_i, \overline{e}_j \rangle,$$

where we used the traciality of h when going from the second to the third line. In other words, $\bar{v}$ is unitary for the canonical inner product, that is, it is a unitary matrix.

Assuming now $\mathcal{A}$ to be a unimodular co-semisimple Hopf $*$-algebra, we will again proceed stepwise.

(a) The reader may simply copy and paste the proof of Theorem 5.15 here.
(b) We will now prove that the coefficients of finite-dimensional representations span $\mathcal{A}$. This is, in fact, more or less the same as a standard result known as the *fundamental theorem of coalgebras*. Let $x \in \mathcal{A}$ and write

$$\Delta(x) = \sum_{i=1}^{N} y_i \otimes z_i,$$

where $(y_i)_{1 \leqslant i \leqslant N}$ and $(z_i)_{1 \leqslant i \leqslant N}$ are linearly independent families. Because

$$(\Delta \otimes \mathrm{id}) \circ \Delta = (\mathrm{id} \otimes \Delta) \circ \Delta,$$

we have

$$\sum_{i=1}^{N} y_i \otimes \Delta(z_i) = \sum_{i,j=1}^{N'} y_i \otimes z'_{ij} \otimes z''_{ij} = \sum_{i=1}^{N} \Delta(y_i) \otimes z_i$$

so that, by linear independence, we can express the elements $(z''_{ij})_{1 \leqslant i,j \leqslant N'}$ as linear combinations of the elements $(z_i)_{1 \leqslant i \leqslant N}$. This means that, up to re-indexing, there exist elements $(v_{ij})_{1 \leqslant i,j \leqslant N}$ such that

$$(\Delta \otimes \mathrm{id}) \circ \Delta(x) = (\mathrm{id} \otimes \Delta) \circ \Delta(x) = \sum_{i,j=1}^{N} y_i \otimes v_{ij} \otimes z_j.$$

The properties of the counit ε readily imply that

$$x = \sum_{i,j=1}^{N} \varepsilon(y_i)\varepsilon(z_j)v_{ij}$$

so that x is in the linear span of the coefficients of v. Hence, proving that v is a representation will prove the claim. And indeed, denoting by X the sum of triple tensor products just given, we have

$$\sum_{i,j=1}^{N} y_i \otimes \Delta(v_{ij}) \otimes z_j = (\mathrm{id} \otimes \Delta \otimes \mathrm{id})(X)$$

$$= (\mathrm{id} \otimes \mathrm{id} \otimes \Delta)(X)$$

$$= \sum_{i,j=1}^{N} y_i \otimes v_{ij} \otimes \Delta(z_j)$$

$$= \sum_{i,j,k=1}^{N} y_i \otimes v_{ij} \otimes v_{jk} \otimes z_k.$$

By linear independence, this means that, for all $1 \leqslant i \leqslant N$,

$$\sum_{j=1}^{N} \Delta(v_{ij}) \otimes z_j = \sum_{j,k=1}^{N} v_{ij} \otimes v_{jk} \otimes z_k,$$

and again, by linear independence, this means that, for all $1 \leqslant k \leqslant N$,

$$\Delta(v_{ik}) = \sum_{j=1}^{N} v_{ij} \otimes v_{jk}.$$

(c) Summing up, given $x \in \mathcal{A}$, there is a Hopf $*$-subalgebra $\mathcal{B} \subset \mathcal{A}$ containing x, which is generated by the coefficients of a finite-dimensional represen- tation. Because $\mathcal{A}$ has a tracial integral, by the first part of the proof any finite-dimensional representation is equivalent to a unitary one. We may therefore assume that the representation is unitary and has a unitary con- jugate. In other words, $\mathcal{B}$ can be chosen to be the $*$-algebra of a unitary compact matrix quantum group. This proves that $\mathcal{A}$ is a CQG-algebra of Kac type. □

C.2 Unimodularity

The restriction to matrix quantum groups is not the only one that was made in this text, and the statement of Theorem C.9 is designed to suggest a nat- ural generalisation of the setting. Indeed, why restrict to Hopf $*$-algebras with tracial Haar state? Or to co-semisimple $*$-algebras which are unimodular? It turns out that this is not necessary. More precisely, we have the following result.

Theorem C.10 *Let $\mathcal{A}$ be a Hopf $*$-algebra. The following are equivalent:*

1. *$\mathcal{A}$ is generated by the coefficients of its finite-dimensional unitary represen- tations;*
2. *$\mathcal{A}$ has an integral;*
3. *$\mathcal{A}$ is co-semisimple.*

Proof The implication from 1 to 2 is one of the main results of [34]. The main point is to prove that any finite-dimensional representation is equivalent to a unitary one. In particular, if v is unitary, then its conjugate $\bar{v}$ need not be unitary, but there needs to exist an invertible matrix F such that $F\bar{v}F^{-1}$ is unitary. One can then go back to the beginning of the book and change the definition of a unitary compact matrix quantum group, replacing the assumption that $\bar{u}$ is unitary by the assumption that $F\bar{u}F^{-1}$ is for some $F \in \mathrm{GL}_N(\mathbf{C})$. It turns out that all the theorems still hold in that setting, including the positivity and faithfulness of the Haar state.

The implication from 2 to 3 can be proven exactly as in the proof of Theorem C.9. As for the implication from 3 to 1, invoking [36], any co-semisimple Hopf $*$-algebra has an integral, hence all its finite-dimensional representations are unitarisable. Combining this with the fact that coefficients of irreducible representations span $\mathcal{A}$ concludes the proof. $\qquad\square$

Remark C.11 This result can be translated as follows: the Haar integral is a trace if and only if the conjugate of any unitary representation is again unitary.

In that case, $\mathcal{A}$ is called a *CQG-algebra*. This means that all along the text, we should have been writing 'compact matrix quantum group of Kac type' instead of 'compact matrix quantum group'. We hope that the reader will forgive our choice of practicality in dropping the Kac type qualification.

Nevertheless, there is one last question which should be asked: why does it make sense to remove the Kac type assumption since we already have a quantum generalisation of all classical compact groups and (as the reader is invited to check) all duals of discrete groups? To answer this, we will conclude with a natural example where unimodularity fails.

Definition C.12 For $q \in [-1, 1] \setminus \{0\}$, let $\mathcal{A}_q$ be the universal $*$-algebra generated by two elements α and γ such that the matrix

$$u = \begin{pmatrix} \alpha & -q\gamma^* \\ \gamma & \alpha^* \end{pmatrix} \in M_2(\mathcal{A}_q)$$

is unitary.

Remark C.13 Computing u^*u and uu^* shows that the matrix u is unitary if and only if the following relations are satisfied:

$$\alpha^*\alpha + \gamma^*\gamma = 1,$$
$$\alpha\alpha^* + q^2\gamma\gamma^* = 1,$$
$$\gamma\gamma^* - \gamma^*\gamma = 0,$$

$$\alpha\gamma - \gamma\alpha = 0,$$
$$\alpha\gamma^* - q\gamma^*\alpha = 0.$$

In order to have a compact quantum group, we must be able to define a coproduct turning u into a representation, and this is easily done.

Lemma C.14 *There exists a unique $*$-homomorphism $\Delta\colon \mathcal{A}_q \to \mathcal{A}_q \otimes \mathcal{A}_q$ such that u is a representation.*

Proof One has to check that $a = \alpha \otimes \alpha - q\gamma^* \otimes \gamma$ and $c = \gamma \otimes \alpha + \alpha^* \otimes \gamma$ satisfy the defining relations of $\mathcal{A}_q$ so that we can conclude by the universal property. The reader may refer to [69, proposition 6.2.3] for details concerning the computations.   □

As a consequence, $(\mathcal{A}_q, u)$ is close to a unitary compact matrix quantum group. Indeed, it is even one in the very specific case where $q = 1$. It follows from the defining relations that $\mathcal{A}_1$ is commutative, and one easily checks that it is, in fact, isomorphic to the algebra $\mathcal{O}(SU(2))$ of regular functions on the compact group $SU(2)$. Moreover, Δ then corresponds to the usual matrix product. For that reason, the notation $\mathcal{A}_q = \mathcal{O}(SU_q(2))$ is often used.

The only property that we have not checked is that $\overline{u}$ is again unitary. But that is precisely where things break down.

Proposition C.15 *The matrix $\overline{u} \in M_2(\mathcal{A}_q)$ is not unitary for $q \in\,]-1, 1[\setminus\{0\}$.*

Proof Let us set

$$F = \begin{pmatrix} 0 & 1 \\ -q^{-1} & 0 \end{pmatrix}.$$

A straightforward computation shows that $F\overline{u}F^{-1} = u$, so that

$$\overline{u}^*\overline{u} = F^{-1*}u^*F^*FuF^{-1}.$$

If this were the identity, then we would have $u^*F^*Fu = F^*F$, that is, $F^*Fu = uF^*F$. Because

$$F^*F = \begin{pmatrix} q^{-2} & 0 \\ 0 & 1 \end{pmatrix},$$

this is equivalent to $\gamma = q^2\gamma$. For $q \neq \pm1$, this would force $\gamma = 0$, a contradiction.[2]   □

---

[2] It is not completely obvious that the defining relations of $\mathcal{A}_q$ can be satisfied with $\gamma \neq 0$. This can be proven by finding explicit operators on a Hilbert space satisfying these; see, for instance, [69, proposition 6.2.5].

We therefore have an object which resembles ours while having an extra subtlety in its representation theory. In fact, we can see more of the general principle in this example: for an arbitrary compact quantum group, the conjugate of a unitary representation is not necessarily unitary, but it is always *unitarisable* in the sense that it is equivalent to a unitary one.

Note that for $q = -1$, the conjugate of u is unitary so that we have a unitary compact matrix quantum group. It can be shown that this is indeed isomorphic to O_2^+. (See, for instance, [4, proposition 7].)

Remark C.16 There is an explicit formula for the Haar state on $\mathcal{O}(SU_q(2))$ (see, for instance, [69, theorem 6.2.17]), on which it can be checked that it fails to be tracial for $q \neq \pm 1$.

References

[1] S. Armstrong, K. Dykema, R. Exel and H. Li. On embeddings of full amalgamated free product C*-algebras. *Proc. Amer. Math. Soc.*, 132(7):2019–30, 2004.

[2] A. Atserias, L. Mančinska, D. Roberson et al. Quantum and non-signalling graph isomorphisms. *J. Comb. Theory, B*, 136:289–328, 2019.

[3] T. Banica. Théorie des représentations du groupe quantique compact libre $O(n)$. *C. R. Acad. Sci. Paris Sér. I Math.*, 322(3):241–4, 1996.

[4] T. Banica. Le groupe quantique compact libre $U(n)$. *Comm. Math. Phys.*, 190(1):143–72, 1997.

[5] T. Banica. Representations of compact quantum groups and subfactors. *J. Reine Angew. Math.*, 509:167–98, 1999.

[6] T. Banica. Symmetries of a generic coaction. *Math. Ann.*, 314(4):763–80, 1999.

[7] T. Banica. Quantum automorphism groups of small metric spaces. *Pacific J. Math.*, 219(1):27–51, 2005.

[8] T. Banica. Homogeneous quantum groups and their easiness level. *Kyoto J. Math.*, 61(1):1–30, 2021.

[9] T. Banica, S. T. Belinschi, M. Capitaine and B. Collins. Free Bessel laws. *Canad. J. Math.*, 63(1):3–37, 2011.

[10] T. Banica and J. Bichon. Quantum groups acting on 4 points. *J. Reine Angew. Math.*, 626:75–114, 2009.

[11] T. Banica, J. Bichon and B. Collins. The hyperoctahedral quantum group. *J. Ramanujan Math. Soc.*, 22:345–84, 2007.

[12] T. Banica and B. Collins. Integration over compact quantum groups. *Publ. Res. Inst. Math. Sci.*, 43(2):277–302, 2007.

[13] T. Banica and B. Collins. Integration over the Pauli quantum group. *J. Geom. and Phys.*, 58(8):942–61, 2008.

[14] T. Banica, B. Collins and P. Zinn-Justin. Spectral analysis of the free orthogonal matrix. *Int. Math. Res. Not.*, 2009(17):3286–309, 2009.

[15] T. Banica and S. Curran. Decomposition results for Gram matrix determinants. *J. Math. Phys.*, 51(11):113503, 2010.

[16] T. Banica, S. Curran and R. Speicher. Stochastic aspects of easy quantum groups. *Prob. Theory Related Fields*, 149(3–4):435–62, 2011.

282

[17] T. Banica, S. Curran and R. Speicher. De Finetti theorems for easy quantum groups. *Ann. Probab.*, 40(1):401–35, 2012.

[18] T. Banica and R. Speicher. Liberation of orthogonal Lie groups. *Adv. Math.*, 222(4):1461–501, 2009.

[19] T. Banica and R. Vergnioux. Fusion rules for quantum reflection groups. *J. Noncommut. Geom.*, 3(3):327–59, 2009.

[20] T. Banica and R. Vergnioux. Growth estimates for discrete quantum groups. *Infin. Dimens. Anal. Quantum Probab. Relat. Top.*, 12(02):321–40, 2009.

[21] H. Bercovici and V. Pata. Stable laws and domains of attraction in free probability theory. *Ann. Math.*, 149(3):1023–60, 1999.

[22] J. Bichon. Quantum automorphism groups of finite graphs. *Proc. Amer. Math. Soc.*, 131(3):665–73, 2003.

[23] J. Bichon. Free wreath product by the quantum permutation group. *Algebr. Represent. Theory*, 7(4):343–62, 2004.

[24] B. Blackadar. *Operator algebras*, volume 122 of Encyclopædia of Mathematical Sciences. Springer, 2006.

[25] R. Brauer. On algebras which are connected with the semisimple continuous groups. *Ann. of Math.*, 38(4):857–72, 1937.

[26] N. Brown and K. Dykema. Popa algebras in free group factors. *J. Reine Angew. Math.*, 573:157–80, 2004.

[27] N. P. Brown and N. Ozawa. *C*-algebras and finite-dimensional approximation*, volume 88 of Graduate Studies in Mathematics. American Mathematical Society, 2008.

[28] A. Chirvasitu. Residually finite quantum group algebras. *J. Funct. Anal.*, 268(11):3508–33, 2015.

[29] A. Chirvasitu. Topological generation results for free unitary and orthogonal groups. *Internat. J. Math.*, 31(1):2050003, 2020.

[30] A. Connes. *Noncommutative geometry*. Academic Press, 1994.

[31] J. Conway. *A course in functional analysis*, volume 96 of Graduate Texts in Mathematics. Springer, 1994.

[32] J. Conway. *A course in operator theory*. Graduate Studies in Mathematics. American Mathematical Society, 2000.

[33] P. Di Francesco, O. Golinelli and E. Guitter. Meanders and the Temperley–Lieb algebra. *Comm. Math. Phys.*, 186(1):1–59, 1997.

[34] M. S. Dijkhuizen and T. H. Koornwinder. CQG algebras: A direct algebraic approach to compact quantum groups. *Lett. Math. Phys.*, 32:315–30, 1994.

[35] V. G. Drinfeld. Quantum groups. *Proc. Int. Congr. Math.*, 1(2):798–820, 1987.

[36] E. G. Effros and Z.-J. Ruan. Discrete quantum groups I. The Haar measure. *International J. Math.*, 5(5):681–723, 1994.

[37] P. Etingof, S. Gelaki, D. Nikshych and V. Ostrik. *Tensor categories*, volume 205 of Mathematical Surveys and Monographs. American Mathematical Society, 2015.

[38] A. Freslon. Fusion (semi)rings arising from quantum groups. *J. Algebra*, 417:161–97, 2014.

[39] A. Freslon. On the partition approach to Schur–Weyl duality and free quantum groups – with an appendix by A. Chirvasitu. *Transform. Groups*, 22(3):705–51, 2017.

[40] A. Freslon. On two-coloured noncrossing partition quantum groups. *Trans. Amer. Math. Soc.*, 372(6):4471–508, 2019.

[41] A. Freslon and M. Weber. On the representation theory of partition (easy) quantum groups. *J. Reine Angew. Math.*, 720:155–97, 2016.

[42] D. Gromada. Free quantum analogue of Coxeter group D_4. *J. Algebra*, 604:577–613, 2022.

[43] I. Halperin and P. Rosenthal. Burnside's theorem on algebras of matrices. *Amer. Math. Monthly*, 87(10):810, 1980.

[44] M. Jimbo. A q-difference analogue of $U(\mathfrak{g})$ and the Yang–Baxter equation. *Lett. Math. Phys.*, 10(1):63–9, 1985.

[45] V. Jones, S. Morrison and N. Snyder. The classification of subfactors of index at most 5. *Bull. Amer. Math. Soc.*, 51(2):277–327, 2014.

[46] V. Jones and V. Sunder. *Introduction to subfactors*, volume 234 of London Mathematical Society Lecture Note Series. Cambridge University Press, 1997.

[47] V. F. R. Jones. The Potts model and the symmetric group. In H. Araki, Y. Kawahigashi and H. Kosaki, eds., *Subfactors: Proceedings of the Taniguchi Symposium on Operator Algebras (Kyuzeso, 1993)*, pages 259–67, World Scientific Publishing 1993.

[48] C. Kassel. *Quantum groups*, volume 155 of Graduate Texts in Mathematics. Springer-Verlag, 1995.

[49] V. Kodiyalam and V. Sunder. Temperley–Lieb and non-crossing partition planar algebras. *Contemp. Math.*, 456:61–72, 2008.

[50] C. Köstler and R. Speicher. A noncommutative de Finetti theorem: invariance under quantum permutations is equivalent to freeness with amalgamation. *Comm. Math. Phys.*, 291(2):473–90, 2009.

[51] J. Kustermans and S. Vaes. Locally compact quantum groups. *Ann. Sci. Éc. Norm. Sup.*, 33(6):837–934, 2000.

[52] F. Lemeux. Haagerup property for quantum reflection groups. *Proc. Amer. Math. Soc.*, 143(5):2017–31, 2015.

[53] M. Lupini, L. Mančinska and D. Roberson. Nonlocal games and quantum permutation groups. *J. Funct. Anal.*, 279(5):108592, 2020.

[54] A. Maes and A. Van Daele. Notes on compact quantum groups. *arXiv:9803122*, 1998.

[55] S. Malacarne. Woronowicz's Tannaka–Krein duality and free orthogonal quantum groups. *Math. Scand.*, 122(1):151–60, 2018.

[56] A. Malcev. On isomorphic matrix representations of infinite groups. *Mat. Sb.*, 50(3):405–22, 1940.

[57] A. Mang and M. Weber. Categories of two-colored pair partitions, Part I: Categories indexed by cyclic groups. *Ramanujan J.*, 53:181–208, 2020.

[58] A. Mang and M. Weber. Categories of two-colored pair partitions, Part II: Categories indexed by semigroups. *J. Combin. Theory Ser. A*, 180:105509, 2021.

[59] P. Martin. Temperley–Lieb algebras for non-planar statistical mechanics – the partition algebra construction. *J. Knot Theory Ramifications*, 3(1):51–82, 1994.

[60] S. Neshveyev and L. Tuset. *Compact quantum groups and their representation categories*, volume 20 of Cours Spécialisés. Société Mathématique de France, 2013.

[61] A. Nica and R. Speicher. *Lectures on the combinatorics of free probability*, volume 335 of Lecture Note Series. London Mathematical Society, 2006.

[62] D. Radford. *Hopf algebras*, volume 49 of Series on Knots and Everything. World Scientific, 2011.

[63] S. Raum. Isomorphisms and fusion rules of orthogonal free quantum groups and their free complexifications. *Proc. Amer. Math. Soc.*, 140(9):3207–18, 2012.

[64] S. Raum and M. Weber. The full classification of orthogonal easy quantum groups. *Comm. Math. Phys.*, 341(3):751–79, 2016.

[65] S. Schmidt. Quantum automorphism groups of folded cube graphs. *Ann. Inst. Fourier*, 70(3):949–70, 2020.

[66] K. Tanabe. On the centralizer algebra of the unitary reflection group $G(m, p, n)$. *Nagoya Math. J.*, 148:113–26, 1997.

[67] P. Tarrago and J. Wahl. Free wreath product quantum groups and standard invariants of subfactors. *Adv. Math.*, 331:1–57, 2018.

[68] P. Tarrago and M. Weber. Unitary easy quantum groups: The free case and the group case. *Int. Math. Res. Not.*, 18:5710–50, 2017.

[69] T. Timmermann. *An invitation to quantum groups and duality: From Hopf algebras to multiplicative unitaries and beyond*. EMS Textbooks in Mathematics. European Mathematical Society, 2008.

[70] S. Wang. Free products of compact quantum groups. *Comm. Math. Phys.*, 167(3):671–92, 1995.

[71] S. Wang. Tensor products and crossed-products of compact quantum groups. *Proc. London Math. Soc.*, 71(3):695–720, 1995.

[72] S. Wang. Quantum symmetry groups of finite spaces. *Comm. Math. Phys.*, 195(1):195–211, 1998.

[73] M. Weber. On the classification of easy quantum groups: The nonhyperoctahedral and the half-liberated case. *Adv. Math.*, 245(1):500–33, 2013.

[74] M. Wilde. *Quantum information theory*. Cambridge University Press, 2013.

[75] S. L. Woronowicz. Compact matrix pseudogroups. *Comm. Math. Phys.*, 111(4):613–65, 1987.

[76] S. L. Woronowicz. Tannaka–Krein duality for compact matrix pseudogroups. Twisted $SU(N)$ groups. *Invent. Math.*, 93(1):35–76, 1988.

[77] S. L. Woronowicz. Compact quantum groups. In A. Connes, K. Gawędzki and J. Zinn-Justin, eds., *Symétries quantiques (Les Houches, 1995)*, pages 845–84, North-Holland, 1998.

Index

Printed in the United States
by Baker & Taylor Publisher Services